Clean Energy

RSC Clean Technology Monographs

Series Editor: J.H. Clark, *University of York, UK*

Advisory Panel: N.M. Edinberry (*Sandwich, UK*), J. Emsley (*London, UK*), S.M. Hassur (*Washington DC, USA*), D.R. Kelly (*Cardiff, UK*), T. Laird (*Mayfield, UK*), T. Papenfuhs (*Frankfurt, Germany*), B. Pearson (*Wigan, UK*), J. Winfield (*Glasgow, UK*)

The chemical process industries are under increasing pressure to develop environmentally friendly products and processes, with the key being a reduction in waste. This series introduces different clean technology concepts to academics and industrialists, presenting current research and addressing problem-solving issues.

Other titles in this series:

How to obtain future titles on publication

A standing order plan is available for this series. A standing order will bring delivery of each new volume upon publication. For further information please contact:

Sales and Customer Care
Royal Society of Chemistry
Thomas Graham House
Science Park
Milton Road
Cambridge
CB4 0WF
Telephone: +44(0)1223 420066, Fax: +44(0)1223 426017, Email: sales@rsc.org

RSC
CLEAN TECHNOLOGY
MONOGRAPHS

Clean Energy

Ronald M. Dell
Formerly Head of Applied Electrochemistry,
Atomic Energy Research Establishment, Harwell, UK

David A. J. Rand
CSIRO Energy Technology, Clayton South, Victoria, Australia

RS•C
advancing the chemical sciences

The cover illustration is of an experimental solar–thermal dish developed by CSIRO, Australia for the steam reforming of natural gas to hydrogen for use in fuel cells.

ISBN 0-85404-546-5

A catalogue record for this book is available from the British Library

Published by The Royal Society of Chemistry,
Thomas Graham House, Science Park, Milton Road,
Cambridge CB4 0WF, UK

Registered Charity Number 207890

For further information see our web site at www.rsc.org

Typeset by Alden Bookset, Northampton, UK
Printed by Athenaeum Press Ltd, Gateshead, Tyne & Wear, UK

Preface

Society and Energy in Transition

The second half of the 20th century was a time of unprecedented change in world affairs. Aside from all the political and social developments that took place, many of the key changes in society stemmed from advances in science and technology. Foremost among these changes may be cited the progress in materials science, especially solid-state physics, that has given rise to the electronics revolution. The advent of sophisticated and reliable solid-state electronic devices, such as integrated circuits, liquid crystal displays and lasers, has brought about a corresponding revolution in both computing and communications. Within the space of ten years, the power and speed of personal computers have improved beyond belief, while their cost has fallen dramatically. This has led to the 'information and communications technology (ICT) society' in which data is stored electronically on computers and much communication is conducted via e-mail – the concept of the 'paperless office' is nearing fruition. In parallel, the internet and the world-wide-web have also evolved at a phenomenal pace. Personal computers linked to the web are now commonplace in offices, homes, and schools. Thus, information that is in the public domain is accessible instantly.

The electronics revolution has also led to major advances in telecommunications and television. Telephone conversations across the world are as clear as those held locally, thanks to improvements in telecommunications satellites, while television pictures that show action as it happens are flashed around the globe. In the field of private communications, video-conference facilities exist, and soon third-generation mobile phones with video facilities will become more widely available. Refinements in optical fibres and in opto-electronics have introduced greater bandwidth in digital communications, which allows many more messages to be conveyed along a single cable. Broadband technology is providing much faster access to the internet. Within the space of twenty years, entertainment in the form of recorded music has moved from vinyl records, through audio-tapes, to compact discs. Similarly, video-tapes are giving way to Digital Versatile Discs (DVDs). Movie films can be downloaded from the web and viewed at leisure. In industry, micro-processors and robots have revolutionized the way in which operations are conducted and controlled automatically.

There have been equally dramatic advances in the bio- and medical sciences. Traditional plant breeding has led to strains of crops that are high yielding, while chemists have discovered agrochemicals (pesticides) that are effective in controlling pests at ever-lower doses. The result has been dramatic improvements in agricultural yield per hectare. Genetic manipulation holds out the prospect of improving still further the yields of crops so that nobody need go hungry. In the fields of human and veterinary health and medicine, chemists and biochemists, working together in the pharmaceutical industry, have formulated a host of new drugs for treating many illnesses. There are also good prospects that, in the not too distant future, there will be major breakthroughs in the treatment of certain forms of cancer. The human genome has been sequenced, and scientists are rapidly distinguishing the genes that control different traits and learning how to identify defective genes that cause hereditary disease. At the heart of all this work lie advances in biochemistry, organic chemistry, and analytical chemistry.

There have been parallel developments in transportation. The private motor-car (automobile) has become almost universal, while, as a consequence of greatly improved road systems, much freight traffic has transferred from the railways to the roads. Air transport has grown correspondingly fast; in the developed world, many people expect to fly to their holiday destinations, even several times a year. Freight transport by air has expanded as clients are no longer prepared to wait weeks for delivery by ship. Supermarkets import perishable foods and flowers from distant parts by air-freight. And all of these activities and products have become increasingly affordable with technical progress in road vehicles and aircraft, as well as in the infrastructures that support them. Truly, the 'global village' has become a reality.

It is interesting to observe that many, if not most, of these advances have depended upon multiple inventions – discovery feeds upon discovery. In the late 19th century, the advent of electric lighting was not possible until both the dynamo and the incandescent electric light bulb had been invented independently. In the mid 20th century, the rapid growth in civil aviation would not have been possible without the arrival of both the jet engine and radar. More recently, the remarkable progress made with computer technology has resulted from many separate improvements in information storage systems, and compact discs would not have been possible without the introduction of both lasers and microchips. Meanwhile, modern laptops and cellular phones rely on new types of battery, as well as on more sophisticated integrated circuits. The scientist makes the initial breakthrough in understanding and the engineer puts together the different knowledge branches to create useful technology.

Not surprisingly, there have also been major changes in the use of energy. A hundred years ago, most energy came in the form of two solid fuels – coal and wood – both of which were inconvenient to use. For mechanical power, it was necessary first to convert water to steam in a boiler and then to utilize the steam in a steam engine. Thus, in the first half of the 20th century most railway locomotives were steam-powered, as was much industrial machinery. Progressively, steam engines were replaced by internal-combustion engines fuelled by liquid petroleum and diesel. The jet engine was designed for use in aircraft, and the gas turbine for electricity generation. Such innovations in engineering led to improved efficiency of energy use and correspondingly reduced costs. At the same time, further resources of petroleum and natural gas were discovered and exploited, and the off-shore energy industry was established.

These various technological and societal developments are highly interactive and have consequences that may be either desirable or undesirable from the viewpoint of energy consumption and environmental degradation. For example, fast and easy transportation has led to a proliferation of business travel, which includes long-distance daily commuting. While such activity may be good for business, it is bad for energy consumption, for road traffic congestion, and for atmospheric pollution. More recently, however, improvements in electronic communications have resulted in more people working from home and only occasionally travelling to the office. Obviously, this saves energy and reduces both congestion and pollution. Increasing use of video-conferencing will further lessen the need for travel, as will internet publications and conferences.

Certain of the 'advances' in modern society have brought with them negative consequences that have yet to be faced squarely. As well as air pollution and global warming, which are thought to arise from the ever-expanding use of energy, there are other problems of a global nature. Inequalities in the geographic distribution of natural resources, especially petroleum, make some countries wealthy and others relatively poor. This may lead to political and social unrest, especially in a global society where communications are so extensive and immediate. Such various environmental, geo-political and socio-economic issues affect the entire world. In this book, we examine the extent to which clean fuels and alternative, sustainable forms of energy may contribute to the solution of these problems. We do not discuss in depth complementary energy conservation initiatives as these constitute a broad subject that merits a separate book.

Global atmospheric warming is now a generally accepted fact, although there is still some debate on the true nature of its origins. The global average temperature has risen by about 0.5 °C over the past 100 years and this phenomenon may be accelerating. In many parts of the world, records have been broken for the warmest ever temperatures. If this trend continues, it will have the most serious consequences for world society. In 1995, the Intergovernmental Panel on Climate Change (IPCC) predicted that by 2100 the average global temperature will probably have risen by a further 1.0 to 3.5 °C, with 2 °C as the best estimate. Storms, major floods (and droughts elsewhere) will become commonplace. Progressive melting of the polar ice-caps would lead to a rise in sea levels and the inundation of low-lying areas such as the huge delta of the River Ganges. Large areas of land suitable for cultivation would be lost and populations displaced. It is even possible that some low-lying islands could disappear altogether. The IPCC has also predicted a rise in average sea level by 2100 of 15 to 95 cm, with the best estimate being 50 cm. The IPCC warns, however, that these are all plausible estimates of what *might* occur, not necessarily of what *will* occur. Indeed, if the Arctic or Antarctic ice sheets were to melt the consequences would be inordinately worse.

Although there have been major climatic changes over the millennia, notably the periodic ice ages separated by relatively warm interludes, most informed scientific opinion considers that the current global warming has been of relatively short duration (~200 years) and is attributable, at least in part, to the anthropogenic combustion of fossil fuels. These fuels are laid down over geological periods of time, but it is only during the past two centuries that they have been extracted and burnt on a substantial scale. Carbon dioxide, the product of combustion, is a so-called 'greenhouse gas' that traps heat in the atmosphere. The effect is analogous to that of glass in a greenhouse. Carbon dioxide, like glass, allows sunlight to penetrate but absorbs the infra-red radiation (heat) that is re-radiated from the earth's surface. Thus, enhanced levels of carbon dioxide in the atmosphere result in more heat being trapped and a concomitant rise in ambient temperature. While it is true that there are other greenhouse gases, notably water vapour and methane (marsh or natural gas), most authorities believe that carbon dioxide from burning fossil fuels is the major culprit. Others take a contrary view, namely, that global warming is a natural phenomenon that leads to greater release of carbon dioxide into the atmosphere, both from the enhanced decay of vegetable and animal matter and from the oceans – the so-called, 'CO_2 thermometer'. For the present this remains a subject of debate. Nevertheless, the global warming situation is being exacerbated by the deforestation of huge areas of land that removes one of the principal natural sinks for carbon dioxide from the atmosphere. In principle, a contribution to reversing this trend towards global warming would be to reduce the consumption of fossil fuels, coupled with the introduction of a programme of re-forestation and afforestation of land that is not required for agriculture. In practice, as we discuss in Chapter 1, it is unlikely that within the next few decades there will be an overall reduction in the usage of fossil fuels, while any realistic amount of forestation would make only a small contribution towards solving the global problem.

Another adverse consequence of modern lifestyles is air pollution, particularly in cities. Before 1950, when the chief fossil fuel was coal and there were many 'smoke-stack industries', air pollution was a serious issue and gave rise to frequent 'smogs' (smoke-filled fogs) in urban areas. The smogs caused high death rates from chest infections, particularly among the elderly in winter. The advent of clean-air legislation and the manufacture of solid smokeless fuels went a long way towards resolving this problem in the UK. Now that coal has been largely replaced by natural gas for industrial and domestic heating, this form of pollution has vanished, only to be replaced by that emanating from car exhausts. The emissions from cars consist of un-burnt hydrocarbons, carbon monoxide, and nitrogen oxides (NO_x). These give rise to a different form of photochemical smog that, in urban areas, is often pronounced under conditions of sunlight and an atmospheric inversion layer. This layer traps the smog in natural basins and prevents it escaping. The phenomenon was first observed in the Los Angeles basin, but became widespread in cities around the world. The subsequent development of lean-burn engines and car exhaust catalysts has helped to ameliorate this problem. Clean-air legislation in the USA has stimulated world-wide interest in hybrid electric vehicles and in pure electric vehicles powered by fuel cells. It may be that such vehicles will play an increasingly important role in combating the global problem of urban air pollution.

A different form of air pollution arises from the burning of coal in power stations. Here, the principal pollutant is sulfur dioxide that derives from sulfur contained in the coal. This gas is normally dispersed from a high stack (about 150 to 200 m tall) and has comparatively little effect upon the immediate surrounding area. Rather, the pollutant gas may travel for many hundreds of kilometres before dissolving in atmospheric moisture and being precipitated as 'acid rain'. Forests in many countries are thought to have been damaged by this rain, while the acidification of many lakes and rivers has resulted in the depletion of fish stocks and other ecological damage. Some coal-fired power stations have fitted scrubbers to remove sulfur dioxide from the exhaust gases. This is a costly solution that may become mandatory. In the UK, power stations are licensed in terms of the quantity of sulfur dioxide that they can emit annually.

Another concern associated with the present, near-exponential growth in energy consumption is the depletion of stocks of fossil fuels. Over the past 30 years, many pessimistic forecasts have been made as to the reserves of economically recoverable fuels, particularly petroleum. By now, according to early projections, production of oil should have peaked and be in decline. In fact, this has not happened because of advances in geo-prospecting and, particularly, in the technology of drilling and tapping off-shore oil wells in deep water. New wells are constantly being found and exploited to compensate for those that have been worked out and closed. Moreover, methods of secondary treatment have been devised that allow a much greater percentage of the oil to be extracted before the well is closed. Thus, at the present time, there are plenty of oil products available in the world and there is no danger of resource depletion in the near-term. Nevertheless, it is self-evident that this situation cannot continue indefinitely and sooner or later petroleum will be in short supply.

Huge quantities of natural gas, a desirably clean fuel, have been discovered around the world and no shortage is expected for several decades to come. Even so, ultimately, on a time-scale of perhaps a century or so, this fuel too will become restricted in supply and will need to be replaced. Methane can be manufactured from coal and other low-grade fossil fuels, and gasification may well be a long-term solution to the cleaner utilization of vast reserves of coal.

Although the world today is largely dependent on fossil fuels for its energy needs, there are two other significant sources of energy, both of which are manifest as electricity. These are hydroelectric power and nuclear power; each contributes significantly to the global electricity supply. The scope for increasing hydroelectric power is limited by the proximity of mountain terrain to urban centres of population and by the availability of land suitable for reservoirs, while that for nuclear power is affected by political considerations in addressing the perceived issues of safety and security.

In the long term, then, the use of fossil fuels will be restricted, first, by concerns over atmospheric pollution (urban air quality, acid rain, and global warming), and later by resource depletion. The relative importance of these two factors is still not clear, nor are the time-scales. The latter will depend upon the rate of increase in energy demand as the world population grows and as the developing countries become industrialized and increase their energy usage per capita. Another significant factor is the extent to which unconventional sources of fossil fuel (such as asphalts and tar sands) are utilized. In essence, the challenge facing world society today is to move progressively from the fossil-fuel age to renewable forms of energy that are non-polluting and secure in supply. In 1987, The World Commission on Environment and Development defined sustainable development as a process that 'meets the needs of the present without compromising the ability of future generations to meet their own needs'. In energy and environmental terms, this reduces to 'devising a set of energy technologies that meets human needs on an indefinite basis without producing irreversible environmental effects' or, more simply, 'to substitute renewable forms of energy for fossil fuels'.

There are many alternative forms of energy that fall into the renewable category and that may be tapped, albeit with new technology and at possibly greater cost. The object of this book is to explore in outline these renewable forms of energy, to set them in the context of world production and use of energy – both present and predicted (Chapter 1) – and to discuss the technological barriers to their introduction. Since the transition from fossil fuels to renewable energy will inevitably be a slow process, it is first necessary to consider the possibilities for reducing pollution from fossil fuels by the use of different technology to prepare 'clean fuels'. This is the subject matter of Chapter 2. The overall title for this book – *Clean Energy* – is chosen to encompass both clean-burning fuels and renewable, sustainable forms of energy. Electricity, although often derived from fossil fuels, is itself a clean and convenient form of energy that will become increasingly important in the years ahead. Aspects of electricity generation are considered in Chapter 3. The various sources of renewable energy are discussed in Chapters 4 and 5, and are sub-divided into types that generate heat (Chapter 4) and those that generate electricity directly (Chapter 5).

Fossil fuels are remarkable in that they are not just fuels, but are also energy stores. Fossil fuels, when once mined and processed, may be transported and stored to where and when they are required for use. It is this vital storage capability that is lacking in most renewable forms of energy. Without such a facility, there is often a mismatch between where and when renewable energy is available, and where and when it is required. For instance, aero-generators ('windmills') work well when the wind is blowing, but electricity is also needed on windless days. Solar power is available during daylight hours, but much electricity use is at night. Finding an economic means for storing energy, particularly electricity, lies at the heart of the renewables problem.

The growing interest of electric utilities in energy storage and the future role for such technology in the electricity supply system is discussed in Chapter 6. Other major uses for stored electrical energy are also outlined. In Chapter 7, we describe the storage of energy in its various physical forms, namely: thermal, potential, kinetic, electromagnetic, and electrostatic. The alternative to storing energy in physical form is to store it as chemicals that can be re-converted to heat or electricity. One such chemical store is hydrogen, which may be used to produce electricity in a fuel cell. Another is methanol, closely related to hydrogen. In recent decades, hydrogen has come to be seen as the 'ultimate' fuel for the future Accordingly, Chapter 8 is devoted to a discussion of the so-called 'Hydrogen Economy'.

The other prime form of chemical storage is as electrochemical materials in rechargeable batteries. There are many different battery chemistries and many different types of battery, either available commercially or under development. In Chapter 9, an evaluation is given of the five broad categories of battery that we regard as holding most promise for large-scale energy-storage applications. The choice of battery depends upon the application and its technical specification, as well as upon cost considerations. It is not an easy matter to find a battery that matches the required specification in full.

Some of the applications for stored energy in the field of electric propulsion are reviewed in Chapter 10. This leads to Chapter 11 in which we present our vision of how the energy sector might progress over the next 20 years or so, together with an examination of the broader socio-political problems that need to be addressed if renewable energy is to fulfill its potential on this time-scale.

Because energy supply and consumption are ubiquitous in modern society, and because the various technologies employed and the applications are so numerous, it has proved difficult to sub-divide *Clean Energy* into self-contained Chapters and Sections. To aid the reader, we have therefore made extensive use of cross-referencing between Chapters.

Acknowledgements

The authors have consulted many sources of information that have included books, reports, reviews, and web sites. As the present volume is simply an overview of the subject of *Clean Energy*, it is not seen fit to quote detailed references to all these sources. Nevertheless, our indebtedness to the work of others must be acknowledged. Similarly, we are grateful to holders of copyright who have kindly consented to the use of their illustrations. Should any omissions have inadvertently occurred, sincere apologies are offered.

The authors also wish to express their special appreciation of the dedication and expert skills of Ms Rita Spiteri (CSIRO) for producing the complete text and redrafting most of the illustrations. Final thanks must go to authors' respective wives, Sylvia Dell and Gwen Rand, who have provided sustained encouragement during the researching and writing of this book.

R.M. Dell
D.A.J. Rand

Biographical Notes

The authors are both senior research chemists who have spent their entire professional careers working in the energy field.

Ronald Dell Ph.D. D.Sc. CChem. FRSC graduated from the University of Bristol. He lived for several years in the USA when he worked as a research chemist in the petroleum industry. Upon returning to Britain, he joined the Atomic Energy Research Establishment at Harwell. During a tenure of 35 years, Ron investigated the fundamental chemistry of materials used in nuclear power and managed projects in the field of applied electrochemistry, especially electrochemical power sources. Since retiring in the mid-1990s, he has interested himself in the developing world energy scene, particularly with regard to environmental factors, and in technical writing.

David Rand Ph.D. Sc.D. was educated at the University of Cambridge. Shortly after graduating, he emigrated to Australia and has spent his research career working at the government's CSIRO laboratories in Melbourne. During the 1970s, he investigated the electrochemistry of noble metals and their alloys as electro-catalysts for fuel cell systems. In the late 1970s, David also set up the CSIRO battery research group and he is now an acknowledged world expert on lead–acid technology. Since its inception in 1987, he has served as the Chief Battery Technical Officer of the World Solar Challenge, the trans-Australia event for cars powered exclusively by solar photovoltaic electricity. David's present work is directed towards the use of hydrogen as a global energy vector.

Both authors have published their research findings extensively in the scientific literature and have also co-authored two previous books:
Batteries for Electric Vehicles, (Research Studies Press, 1998) ISBN 0 86380 205 2.
Understanding Batteries, (RSC Paperbacks 2001), ISBN 0 85404 605 4.

Contents

Abbreviations, Symbols and Units Used in Text

Abbreviations

a.c.	alternating current
AFC	alkaline fuel cell
AFV	alternate fuelled vehicles
AGM	absorptive glass-mat
AGR	advanced gas-cooled reactor
ALABC	Advanced Lead – Acid Battery Consortium
BESS	battery energy storage systems
BEV	battery electric vehicle
BP	British Petroleum Company
BWR	boiling water reactor
CAES	compressed air energy storage
CBM	coal bed methane
CCGT	combined-cycle gas turbine
CHP	combined heat and power (co-generation)
CNG	compressed natural gas
CRT	continuously regenerating trap
d.c.	direct current
DG	distributed generation
DHW	domestic hot water
DMFC	direct methanol fuel cell
DoD	depth- of- discharge (for batteries)
DVDs	digital versatile discs
EG	embedded generation
ESES	electrostatic energy storage
EV	electric vehicle

FBR	fast breeder reactor
FC	fuel cell
FCV	fuel cell vehicle
FGD	flue gas desulfurization

GTL	gas-to-liquid (conversion)

HDR	hot dry rocks
HEV	hybrid electric vehicle
HHV	higher heating value
HRSG	heat recovery steam generator
HT	high temperature
HTGCR	high temperature gas-cooled reactor

ICEV	internal-combustion engine vehicle
ICT	information and communications technology
IEA	International Energy Agency
IGCC	integrated gasification combined cycle
IPCC	Inter-governmental Panel on Climate Change
ITER	International thermonuclear experimental reactor

JET	Joint European Torus

LED	light emitting diode
LEV	low emission vehicle
LH_2	liquid hydrogen
LHV	lower heating value
LNG	liquefied natural gas
LPG	liquefied petroleum gas
LRP	Lead replacement petroleum
LT	low temperature

MCFC	molten carbonate fuel cell
MOX	mixed oxide fuel
Mm	misch metal
MRI	magnetic resonance imaging
MSW	municipal solid waste
MT	medium temperature

NASA	National Aeronautics and Space Administration
NGV	natural gas vehicle
NO_x	nitrogen oxides

OECD	Organisation for Economic Co-operation and Development
OPEC	Organization of Petroleum Exporting Countries
OPVC	organic photovoltaic device

| OTEC | ocean thermal energy conversion |
| OWC | oscillating water column (wave energy device) |

PAFC	phosphoric acid fuel cell
PBMR	pebble bed modular reactor
PCs	personal computers
PEMFC	proton exchange membrane fuel cell
PNGV	Partnership for a New Generation of Vehicles
PSoC	Partial state-of-charge
PV	photovoltaic
PWR	pressurized water reactor

| RAPS | remote-area power supply |
| RDF | refuse derived fuel |

SDG&E	San Diego Gas & Electric
SMES	superconducting magnetic energy storage
SNG	synthetic natural gas
SoC	state- of- charge (for batteries)
SOFC	solid oxide fuel cell
SPEFC	solid polymer electrolyte fuel cell (same as PEMFC)
SRC	short rotation coppicing

UCC	ultra-clean coal
UN	United Nations
UNFCCC	United Nations Framework Convention on Climate Change
UPS	uninterruptible power supplies

| VRLA | valve-regulated lead-acid (batteries) |

| WEC | World Energy Council (also: Wave Energy Converters) |

| YBCO | superconducting material of formula $YBa_2Cu_3O_7$ |

| ZEV | zero emission vehicle |

Symbols and Units

Sub-units

d	deci	10^{-1}
c	centi	10^{-2}
m	milli	10^{-3}
μ	micro	10^{-6}
n	nano	10^{-9}

Multiple units

k	kilo	10^{3}
M	mega	10^{6}
G	giga	10^{9}
T	tera	10^{12}
P	peta	10^{15}

| A, A_2B, AB, AB_2, AB_5 | families of metal hydrides |

A	ampere
A	area
A h	ampere-hour
atm	atmosphere ($=101.325$ kPa)
bar	unit of pressure ($=0.1$ MPa)
bhp	brake horsepower ($=745.7$ W)
°C	degree celsius
C	coulomb ($=1$ A s)
C	capacitance
cal	calorie ($=4.184$ J)
cm	centimetre
d	distance between plates in a capacitor
e^-	electron
ε	dielectric constant (or relative permittivity)
eV	electron volt
F	farad ($=1$ C V^{-1})
F	faraday ($=96\,485$ C mol^{-1})
g	gram
G	Gibbs free energy (J mol^{-1})
ΔG	change in Gibbs free energy (J mol^{-1})
H	enthalpy (J mol^{-1})
ΔH	change in enthalpy (J mol^{-1})
ΔH_f^0	standard molar heat (enthalpy) of formation (J mol^{-1})
hp	horsepower ($=745.7$ W)
I	current (A, mA, μA)
I_{MP}	current at maximum power point of a photovoltaic cell
I_{SC}	short-circuit current
J	joule ($=1$ W s)
K	degree kelvin
L	litre
m	metre
min	minute
mol	mass of 6.02×10^{23} elementary units (atoms, molecules, etc.) of a substance
Mtoe	million tonne (of) oil equivalent

N	newton (SI unit of force $= 1$ kg m s^{-2})
n	number of electrons involved in electrode process
Ω	ohm
P	pressure
Pa	pascal ($=1$ N m^{-2}; $=9.8692 \times 10^{-6}$ atm)
q	charge on plates of a capacitor (measured in coulombs)
Q	quantity of heat
R	gas constant ($=8.3145$ J K^{-1} mol^{-1})
s	second
S	entropy (J K^{-1} mol^{-1})
ΔS	change in entropy (J K^{-1} mol^{-1})
t	tonne
T	temperature ($^\circ$C, K)
T_c	critical temperature
V	volt
V_c	potential difference between plates of a capacitor
V_{MP}	voltage at maximum power point of a photovoltaic cell
V_{OC}	open-circuit voltage of a cell
V_r	reversible voltage of a cell
V_r°	standard reversible voltage of a cell
W	watt
W_e	watt, electrical power
W_p	peak power output, *e.g.* from a wind generator or solar module
Wh	watt-hour
wt%	percentage by weight
W_{th}	watt, thermal power
x	variable in stoichiometry

Glossary of Terms

Active material	The material in the electrodes of batteries that takes part in the electrochemical reactions to store and deliver the electrical energy.
Adiabatic process	A process that takes place without heat entering or leaving the system.
Alternating current	Electric current that flows for an interval of time (half-period) in one direction and then for the same time in the opposite direction. The normal waveform is sinusoidal.
Angular velocity	Quantitative expression of the amount of rotation that a spinning object undergoes per unit time.
Anion	Ion in an electrolyte that carries a negative charge and that migrates towards the anode under the influence of a potential gradient. See **Anode, Ion**.
Anode	An electrode at which an oxidation process, *i.e.* loss of electrons, is occurring. During electrolysis, the anode is the positive electrode. In a rechargeable battery, the anode is the positive electrode on charge and the negative electrode on discharge.
Aquifer	Underground water-bearing, porous rock strata that yields economic supplies of water, sometimes heated, to wells or springs.
Automotive battery	A battery designed to provide electrical power for an internal-combustion-engined vehicle.
Band-gap energy	The energy gap, generally measured in electron-volts (eV), between the top of the valence band and the bottom of the conduction band in a

	crystalline solid. See **Conduction band, Energy band, Valence band**.
Barrel	A measure of crude oil (petroleum), approximately 159 L.
Battery	A multiple of electrochemical cells of the same chemistry housed in a single container. In this book, the term 'battery' refers specifically to rechargeable batteries.
Battery cycling	Repeated charging and discharging of a secondary battery.
Battery electric vehicle (BEV)	A vehicle driven by an electric motor that is powered by rechargeable batteries.
Battery energy-storage system (BESS)	An assembly of large batteries that is used for the bulk storage of electricity, especially in grid systems.
Battery pack	A number of batteries connected together to provide the required power and energy for a given application.
Beta particle	A particle emitted from an atom during radioactive decay. Beta particles are either electrons (negative charge) or positrons (positive charge).
Bio-energy	Energy derived from combustible waste materials or crops.
Biofuel	A gaseous, liquid or solid fuel that is derived from a biological source.
Biogas	A mixture of methane and carbon dioxide that results from the anaerobic decomposition of waste matter.
Biomass	The term used to describe all biologically produced matter at the end of its life. This includes both waste matter and crops that are specially grown as a source of energy.
Bipolar	A design of battery or fuel cell in which the component cells are connected through plates that each, in turn, act as the current-collector for the positive electrode in one cell and for the negative in the adjacent cell.
Breakdown voltage	The voltage at which a discharge occurs across the dielectric in a capacitor.
Capacity	The amount of charge, usually expressed in ampere–hours, that can be withdrawn from a fully charged battery under specified conditions.

	Alternatively, the charge on a capacitor (measured in farads).
Carnot cycle	The most efficient cycle of operation for a reversible heat engine. It consists of four operations, as in the four-stroke internal-combustion engine, namely: isothermal expansion, adiabatic expansion, isothermal compression, and adiabatic compression to the initial state.
Catalyst	A substance that increases the rate of a chemical reaction, but that is not itself permanently changed.
Cathode	An electrode at which a reduction process, *i.e.* gain of electrons, is occurring. During electrolysis, the cathode is the negative electrode. In a rechargeable battery, the cathode is the negative electrode on charge and the positive electrode on discharge.
Cation	Ion in an electrolyte that carries a positive charge and that migrates towards the cathode under the influence of a potential gradient. See **Cathode, Ion**.
Chain reaction	A process in which one nuclear transformation (such as a fission) sets up conditions for a similar nuclear transformation in another atom.
Charge-acceptance	The ability of a battery to convert active material during charge into a form that can be subsequently discharged; it is quantified as the ratio, expressed as a percentage, of the charge usefully accepted during a small increment of time to the total charge supplied during that time.
Chemical reactor	Engineering equipment in which a chemical reaction takes place.
Coal gas	A fuel gas, which contains around 50% hydrogen and varying amounts of methane, carbon monoxide and nitrogen, produced when coal is heated in the absence of air (so-called: destructive distillation) or pyrolysis. It is a by-product in the preparation of coke and coal tar. Coal gas was a major source of energy in the late 19th and early 20th centuries and was also known as town gas. The use of this gas declined with the increasing availability of natural gas. See **Pyrolysis**.

Co-generation See **Combined heat and power (CHP) system**

Combined-cycle gas turbine (CCGT) A technology employed in a natural-gas-fired power station. The gas is first burnt in a gas turbine and the waste heat contained in the exhaust gases is then recovered and used to raise steam to drive a steam turbine.

Combined heat and power (CHP) system An installation where there is simultaneous generation of usable heat and power (usually electricity) in a single process. The term is synonymous with co-generation.

Conduction band Partially filled or empty energy levels in a crystalline solid where electrons are free to move and thus allow the solid to conduct an electrical current.

Critical temperature In the science of superconductors, the temperature below which a metal or alloy loses all of its electrical resistance.

Current density The current flowing per unit electrode area.

Cycle A single charge–discharge of a battery.

Cycle-life The number of cycles that can be obtained from a battery before it fails to meet selected performance criteria.

Depth-of-discharge The ratio, usually expressed as a percentage, of the ampere–hours discharged from a battery at a given rate to the available capacity under the same specified conditions.

Dielectric A substance that is a non-conductor of electricity (*i.e.* an insulator). An electric field in a dielectric substance gives rise to no net flow of electricity. Rather, an applied field causes electrons within the substance to be displaced and, thereby, creates an electric charge on the surface of the substance. See **Dielectric constant**.

Dielectric constant Quantitative expression for the reduction in field strength within a dielectric substance when subjected to an electric field. The dielectric constant of a vacuum is unity.

Direct current Electric current that flows in one direction only, although it may have appreciable pulsations in its magnitude.

Drive-train
The elements of the propulsion system that produce and transmit mechanical energy to drive the wheels of an electric or a hybrid electric vehicle.

Duty cycle
The operating regime of a battery in terms of rate and time of both charge and discharge, and time in standby mode.

Electric vehicle (EV)
See **Battery electric vehicle (BEV)**.

Electrochemical capacitor
A capacitor that stores charge in the form of ions rather than electrons, adsorbed on materials of high surface area. The ions undergo *redox reactions* during charge and discharge. Also known as super-capacitors and ultra-capacitors.

Electrode
An electronic conductor that acts as a source or sink of electrons that are involved in electro-chemical reactions.

Electrode potential
The voltage developed by a single electrode, either positive or negative. The algebraic difference in voltage of any pair of electrodes of opposite polarity equals the cell voltage.

Electrolysis
The production of a chemical reaction by passing a direct electric current through an electrolyte.

Electrolysis cell
An electrolytic cell in which the electrochemical reactions are caused by supplying electrical energy.

Electrolyte
A chemical that conducts electricity by means of positive or negative ions. Electrolytes are solids, molten ionic compounds, or solutions containing ions, *i.e.* solutions of ionic salts or of compounds that ionise in solution.

Electrolytic capacitor
A capacitor in which a very thin film of oxide is formed electrolytically on a metal such as aluminium or tantalum and serves as the dielectric.

Electrolytic cell
An electrochemical cell that consists of a positive and a negative electrode and an electrolyte.

Energy efficiency
For a battery, the fraction (usually expressed as a percentage) of the energy used in charging a battery that is delivered on discharge.

Endothermic reaction   A chemical reaction in which heat is absorbed.

Energy   The ability to do work or produce heat (measured in joules).

Energy band   The range of energies that electrons can have in a solid. In a single atom, electrons exist in discrete energy levels. In a crystal, where large numbers of atoms are held closely together in a lattice, electrons are influenced by a number of adjacent nuclei and the sharply defined energy levels of the atoms become bands of allowed energy that are separated by bands of forbidden values. In a metal, there is a continuous energy band.

Energy crops   Trees and grasses grown specially for use as a fuel or for extracting plant oils or alcohols that may be used as fuels in internal-combustion engines.

Energy density   Stored energy per unit volume, usually expressed in MJ m^{-3} or kWh m^{-3}.

Enthalpy   A thermodynamic quantity (H) equal to the total energy of a system when it is at constant pressure. The gain or loss of energy of a system when it reacts at constant pressure is expressed by the change in enthalpy, symbolized by ΔH. When all the energy change appears as heat (Q), the change in enthalpy is equal to the heat of reaction at constant pressure, *i.e.* $\Delta H = Q$. The values of ΔH and Q are negative for exothermic reactions (heat evolved from system) and positive for endothermic reactions (heat absorbed by system).

Entropy   A thermodynamic quantity representing the amount of energy in a system that is no longer available to do useful work. When a closed system undergoes a reversible change, the entropy change (ΔS) equals the energy lost from, or transferred to, the system by heat (Q) divided by the absolute temperature (T) at which this occurs, *i.e.* $\Delta S = Q/T$. At constant pressure, the amount of heat (Q) is equal to the change in enthalpy (ΔH). For a more detailed explanation see Box 8.1.

Equilibrium potential   See **Reversible potential**.

Equilibrium voltage   See **Reversible voltage**.

Exothermic reaction   A chemical reaction in which heat is evolved.

Fast-breeder reactor   A nuclear reactor that has no neutron moderator, uses fast (energetic) neutrons to cause fission in plutonium, and has the capability of 'breeding' more plutonium than it consumes.

Fissile material   Any material capable of undergoing fission by thermal (or slow) neutrons. For example, uranium (U^{235}) and plutonium (Pu^{239}).

Fission   Usually, the division of a heavy nucleus into two unequal masses and the emission of neutrons, gamma radiation, and a great deal of energy.

Fission products   The unequal masses that result from fission and the subsequent decay products. When formed initially, most fission products are very unstable, highly radioactive, have short half-lives and emit copious quantities of beta rays and gamma rays over a range of energies.

Flow battery   A battery system in which the active materials of one or both electrode polarities are stored externally and pumped to the battery during operation.

Fuel cell   An electrochemical device for generating low voltage d.c. electricity from a fuel (often hydrogen) and air or oxygen.

Fusion   The formation of a heavier nucleus from two lighter ones (such as hydrogen isotopes) with an attendant release of energy (as in a fusion reactor).

Galvanic cell   An electrolytic cell in which chemical energy is converted into electrical energy on demand.

Gasification   A special type of pyrolysis where thermal decomposition takes place in the presence of a small amount of air or oxygen. See **Coal gas, Pyrolysis**.

Gasoline   Term used for petrol in the USA.

Gibbs free energy   The energy liberated or absorbed in a reversible process at constant pressure and constant temperature. The change in free energy, ΔG, in a chemical reaction is given by

$\Delta G = \Delta H - T\Delta S$, where ΔH is the change in enthalpy and ΔS is the change in entropy. This is known as the Gibbs equation. See **Enthalpy**, **Entropy**.

Half-life   The characteristic time in which a radionuclide decays to half its initial radioactivity.

Higher heating value of a fuel (HHV)   The maximum heat of combustion ($MJ\ kg^{-1}$) of a fuel, based on complete combustion to carbon dioxide and water at 25 °C.

Hybrid electric vehicle (HEV)   A vehicle that has two power sources, one of which is used for electric propulsion through an electric motor.

Hydrothermal reservoir   An aquifer containing water or brine at more than 100 °C and therefore pressurized.

Inductance   The magnitude of the capability of a component (inductor), such as a wire loop or coil, in an electrical circuit to store energy in the form of a magnetic field. An inductance of one henry is produced when one volt is induced by a change in current of one ampere per second.

Integrated gasifier combined-cycle (IGCC)   A technology employed in some coal-fired power stations. Instead of feeding powdered coal directly to the boilers to raise steam, it is first converted to gas, which is a mixture of carbon monoxide, hydrogen and nitrogen. The gas is then combusted in a gas turbine.

Internal resistance   The opposition to current flow that results from the various electronic and ionic resistances within the battery.

Internal short-circuit   Same as **Short-circuit.**

Ion   An atom that has lost or gained one or more orbiting electrons, and thus becomes electrically charged.

Ionization   Any process by which an atom, molecule or ion gains or loses electrons.

Isotopes   Nuclides that have the same atomic number but different mass numbers. Different isotopes of the same element have the same chemical properties, but somewhat different physical properties.

Latent heat   The heat absorbed or released by a substance when it changes state (*e.g.* from solid to liquid, or vice versa) at constant temperature and pressure. The term **Specific latent heat** denotes the heat absorbed or released per unit mass of a substance in the course of its change of state.

Load-levelling Storage of surplus electricity generated (usually overnight) to accommodate increased demand (usually during the day).

Lower heating value of a fuel (LHV) The heat of combustion ($MJ\,kg^{-1}$) of a fuel, based on complete combustion to carbon dioxide and steam at $100\,°C$.

Magnetohydrodynamics The study of the motions of electrically conducting fluids and their interactions with magnetic fields. Examples of such fluids include plasmas and liquid metals.

Maximum power point The point on the current–voltage curve for a photovoltaic cell at which the cell generates maximum power.

Megawatt (MW) Unit of energy equal to one million watts. MW_e denotes electrical output and MW_{th} denotes thermal heat output.

Molar Terminology to denote that an extensive physical property is being expressed per mole of a substance. (An extensive variable is proportional to the size of the system, *e.g.* volume, mass, energy.) Alternatively, a concentration of one mole per litre.

Mole The amount of a substance (expressed in grams) that contains as many elementary units as there are atoms (6.02×10^{23}) in $0.012\,kg$ of the carbon isotope ^{12}C. The elementary units may be atoms, molecules, ions, electrons, *etc.*

Moderator A material used in a nuclear reactor to slow down high-velocity neutrons and thus increase the likelihood of further fission. Examples of moderators include ordinary water, heavy water, and graphite.

Monopolar The conventional method of battery construction in which the component cells are discrete and are externally connected to each other.

Municipal solid waste Solid domestic/household waste.

Negative electrode   The electrode in an electrolytic cell that has the lower potential.

Neutron   An uncharged elementary particle with a mass slightly greater than that of the proton and found in the nucleus of every atom except ordinary hydrogen. Neutrons are the links in a chain reaction in a nuclear reactor.

Nuclear reactor   A structure in which a fission chain reaction can be maintained and controlled. It usually contains fuel, coolant, moderator, control absorbers, and safety devices. It is most often surrounded by a concrete biological shield to absorb neutron and gamma ray emission.

Nucleus   The positively charged core of an atom. It is about 1/10 000 the diameter of the atom but it contains nearly all the mass of the atom. All nuclei contain protons and neutrons, except the nucleus of ordinary hydrogen, which consists of a single proton.

Nuclide   An atom as defined by its atomic number and atomic mass.

Oil   In this book, oil and petroleum are used synonymously for crude (unrefined) oil.

Open-circuit voltage   The voltage of a power source, such as a battery, fuel cell or photovoltaic cell, when there is no net current flow.

Organic Rankine Cycle   Replacement of water by an organic compound in the Rankine cycle. Superheating is not necessary and results in higher efficiency. See **Rankine cycle**.

Overcharge   The supply of charge to a battery in excess of that required to return all the active materials to the fully charged state.

Overdischarge   The discharge of a battery beyond the level specified for correct operation.

Overpotential   The shift in the potential of an electrode from its equilibrium value as a result of current flow.

Peak power   The sustained pulsed power that is obtainable from a battery under specified conditions, usually measured in watts over a period of 30 s.

Peak-shaving Electricity storage capacity in the network to accommodate short peaks in the daily demand curve. Also used for electricity storage on the customer side of the meter.

Petrol Term used in the UK for a light hydrocarbon liquid fuel for spark-ignition internal-combustion engines. Other terms for such fuel are gas, gasoline, and motor spirit.

Petroleum See **Oil**.

Photosynthesis The chemical process by which green plants synthesize organic compounds from carbon dioxide and water in the presence of sunlight.

Photovoltaic (PV) cell A semi-conductor device for converting light energy into low-voltage d.c. electricity.

Platinum-group metals The three members of the second and third transition series immediately proceeding silver and gold, *i.e.* ruthenium, rhodium, palladium, and osmium, iridium, platinum.

Polarity Denotes the electrode that is positive and the electrode that is negative.

Positive electrode The electrode in an electrolytic cell that has the higher potential.

Power density The power output of a battery per unit volume, usually expressed in $W\ L^{-1}$ or $W\ dm^{-3}$ and quoted at 80% depth-of-discharge.

Primary battery A battery designed to deliver a single discharge.

Producer gas A mixture of carbon monoxide and nitrogen made by passing air over very hot carbon. Usually some steam is added to the air and the mixture contains hydrogen. The gas is used as a fuel in some industrial processes.

Proton An elementary particle with a single positive electrical charge and a mass approximately 1837 times that of the electron. Also, the nucleus of an ordinary or light hydrogen atom. Protons are constituents of all nuclei.

Pyrolysis Thermal decomposition of a substance at elevated temperatures in the absence of air or oxygen.

Radioactivity The ability of certain nuclides to emit particles, gamma rays or X-rays during their spontaneous transformation into other nuclei. The final outcome of radioactive decay is a stable nuclide.

Radio-nuclide   An isotope that is radioactive.

Rapid reserve   Stored electrical energy that is available for transmission virtually instantaneously.

Rankine cycle   A thermodynamic cycle used to generate electricity in many power stations. Superheated steam is generated in a boiler, and then expanded in a steam turbine. The turbine drives a generator to produce electricity. The remaining steam is then condensed and recycled as feedwater to the boiler.

Recombination   For lead–acid batteries, this term refers to the reaction of oxygen evolved on charge with the active material of the negative electrode. Also known as the **internal oxygen cycle**.

Redox battery   A battery in which the chemical energy is stored as dissolved redox reagents. See **Redox reaction**.

Redox reaction   A chemical reaction that involves the transfer of an electron from one species (which is thereby oxidised) to another (which is thereby reduced). The species are known as **redox reagents**.

Regenerative braking   The recovery of some fraction of the energy normally dissipated during braking of a vehicle and its return to a battery or some other energy-storage device.

Reversible potential   The potential of an electrode when there is no net current flowing through the cell.

Reversible voltage   The difference in the reversible potentials of the two electrodes that make up the cell.

Secondary battery   A battery that is capable of repeated charging and discharging.

Self-discharge   The loss of capacity of a battery under open-circuit conditions as a result of internal chemical reactions and/or short-circuits.

Sensible heat   The heat absorbed by a substance that gives rise to an increase in temperature of the substance. See **Latent heat**.

Separator
An electronically non-conductive, but ion-permeable, material that prevents electrodes of opposite polarity from making contact.

Shedding
The loss of active material from battery grids.

Short rotation coppicing (SRC)
The planting and harvesting of trees especially for use as an energy crop. Trees that have rapid growth (*e.g.* willow, poplar) are planted and cut down (coppiced) on a rotation of 2–4 years.

Specific energy
Stored energy per unit mass, expressed in MJ kg^{-1}, Wh kg^{-1}, or kWh kg^{-1}.

Specific heat
The quantity of heat that unit mass of a substance requires to raise its temperature by one degree. It is expressed in units of J kg^{-1} K^{-1}.

Specific latent heat
See **Latent heat**.

Specific power
The power output of a battery per unit weight, usually expressed in Wkg^{-1}.

Spinning reserve
Lightly loaded electricity generators that are available at short notice to meet sudden demands for electricity, or to take over in the event of a breakdown.

Standard cell voltage
The reversible voltage of an electrochemical cell with all active materials in their standard states. See **Reversible voltage**.

Standard electrode potential
The reversible potential of an electrode with all the active materials in their standard states. Note, usual standard states specify unit activity for elements, solids, 1 M-solutions, and gases at a pressure of 101.325 kPa. See **Reversible potential**.

State-of-charge
The fraction, usually expressed as a percentage, of the full capacity of a battery that is still available for further discharge, *i.e.* state-of-charge = [100−(% depth-of-discharge)]%.

Steam reforming
The reaction of fossil fuels with steam at high temperature to generate a mixture of hydrogen and carbon monoxide.

Synthesis gas
A mixture of carbon monoxide and hydrogen made by reacting natural gas with steam and air or oxygen.

Synthetic natural gas	Methane produced by the catalytic reaction of carbon monoxide with hydrogen or from coal by reaction with hydrogen.
Thermal reactor	A nuclear reactor in which the fission chain reaction is sustained primarily by slow or thermal neutrons.
Thermolysis	The dissociation or decomposition of a molecule by heat.
Town gas	See **Coal gas**.
Traction battery	A battery designed to provide motive power.
Valence band	The range of energy levels of electrons that bind atoms of a crystal together. When electrons are excited from the valence band to the conduction band, the resulting electron hole is mobile and gives rise to p-type conduction in the valence band. See **Conduction band**, **Energy band**.
Voltaic efficiency	The ratio, usually expressed as a percentage, of the average voltage during discharge to the average voltage during charge.
Water gas	A mixture of carbon monoxide and hydrogen produced by passing steam over hot carbon (coke). The reaction is strongly endothermic but may be combined with the exothermic reaction for producer gas. See **Producer gas**, **Water gas shift reaction**.
Water gas shift reaction	The reaction of water gas with steam to yield hydrogen and carbon dioxide.
Wind farm	A collection of wind turbines, grouped together to form a single generating unit.

CHAPTER 1

Energy Production and Use

It is generally held that the present production and use of energy pose a serious threat to the global environment, particularly in relation to the emission of greenhouse gases (principally, carbon dioxide, CO_2) and the perceived influence of these gases on the Earth's climate. Accordingly, industrialized countries are examining a whole range of new policies and technology issues to make their energy futures 'sustainable'. That is, to maintain economic growth and cultural traditions whilst providing energy security and environmental protection. Clearly, the world is set to make major changes so as to maintain adequate supplies of *Clean Energy*.

1.1 A Brief History of Energy Technology

The mastery of energy has always been the key to a better world. Ironically, though, the concept of energy is difficult to grasp; it is an abstract quantity that manifests itself in many forms, *e.g.* chemical, electrical, mechanical, radiant, nuclear, and thermal energy. In an electrical power station, for example, fossil fuel (chemical energy) is converted *via* steam to mechanical energy and then, *via* an alternator, to electrical energy. In an electric vehicle, a battery is used to convert chemical energy into electrical energy, which is then converted to mechanical energy by a motor. The scientific use of the term 'energy' was introduced by Thomas Young (1773–1829), an English physicist, physician, and Egyptologist who also provided the most astute definition to date, namely: 'energy is the ability to do work.' It is commonly understood that 'work' means the application of effort to accomplish a task, and the rate at which work is performed is called 'power'. Thus, machines consume energy, perform work, and provide power. A simple example of the relationship between these quantities is shown in Figure 1.1. The 'efficiency' of a machine is a measure of its performance obtained from the ratio of energy output to energy input. The efficiency must always be less than 100% (which would imply perpetual motion).

Until the advent of the Industrial Revolution in the 18th century, humankind derived its power mainly from its own exertions, from animal muscle (horses, oxen, camels, *etc.*), from the wind (windmills and sailing ships), and from water (watermills). Even with these limited resources, however, some of humankind's

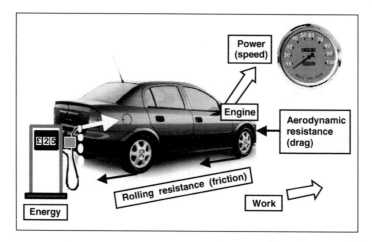

Figure 1.1 *An example of the relationship between energy, work, and power. Petroleum provides energy to propel a car against resistive forces. The greater the rate of work performed by the engine (i.e. the greater its power), the faster the speed of the car. It should be noted that energy is also lost as heat during combustion of the fuel and this decreases engine efficiency.*

achievements were remarkable. Consider, for instance, the ancient pyramids of Egypt, prehistoric Stonehenge, or the great cathedrals of Europe built in the Middle Ages. As late as the 18th and early 19th centuries, an extensive system of canals was constructed across England to permit the conveyance of freight by horse-drawn barge. Such canals, which are still in use today for recreational boating, were all hand-dug by labourers with spades and wheelbarrows. These feats are truly awe-inspiring when viewed from the comfort of the present mechanized age.

Sources of power began to change with the development of the 'atmospheric' engine in the early 18th century by Thomas Newcomen (1663–1729), who was inspired by the earlier work of Denis Papin (1647–1712) and Thomas Savery (1650–1716). Newcomen was assisted in his experiments by John Calley and, in 1705, they devised the first reliable engine. This invention was brought to practical realization in 1712 when the first working engine was installed at a colliery near Dudley Castle, in Tipton, Staffordshire, UK (Figure 1.2(a)). In Newcomen's engine, atmospheric pressure drives a piston down into a vacuum, which is created by condensing steam introduced into the cylinder space below; hence, the description 'atmospheric' engine. The piston is raised again by the over-balancing weight of the pump rods.

James Watt (1736–1819) subsequently recognized that the 'atmospheric' engine is very inefficient – energy is wasted by having to reheat the cylinder after each stroke of the piston. Watt solved this problem by using a separate condenser and driving the engine by the pressure of steam itself. Thus, in 1769, he patented the first real 'steam' engine, which offered superior performance in terms of both energy efficiency and economy. Further developments followed, but Watt's most decisive invention was to make the steam engine rotative, by using a rigid rod to

(a) **Thomas Newcomen (1663 –1729)**

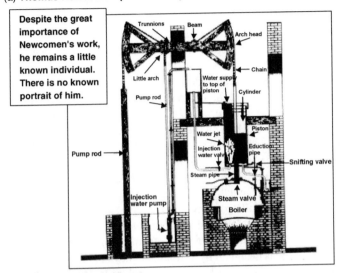

Despite the great importance of Newcomen's work, he remains a little known individual. There is no known portrait of him.

(b) **James Watt (1736 –1819)**

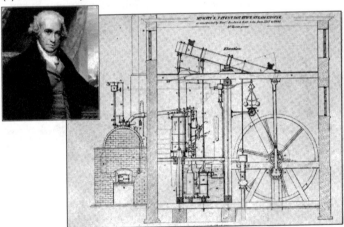

Figure 1.2 (a) *Principal features of the Newcomen atmospheric engine in section.* (b) *Drawing of the 10 hp rotative steam engine produced at the Boulton & Watt works between 1787 and 1800.*

connect the outer end of the beam to a crank below (Figure 1.2(b)). Watt also added a flywheel and devised a 'governor' by which the speed of the engine could be kept constant. In contrast to the up and down movement of the reciprocating engines of his day, the rotative motion made possible a much smoother action and a far greater range of industrial applications. The engines were used to pump water from mines, to drive machinery in factories, to dig tunnels, and to thresh corn. By 1800, there were about 2500 rotative steam engines operating in the UK. One-third of these were made by Watt and his partner Matthew Boulton (1728–1809) at the Soho

Works in Birmingham, UK. The oldest surviving engine was built in 1785 for the London Brewery of Samuel Whitbread. After providing service for 102 years, the engine was presented in 1887 to the Power House Museum in Sydney, Australia, where it has been restored to a working condition.

It was only a short while before the utility of the steam engine was extended to the propulsion of ships, railway locomotives, and road tractors (Figure 1.3). Engines working on the principle developed by Watt were used to propel boats from as early as the 1780s in both France and the USA. Britain's first practical steamboat was a tug, the *Charlotte Dundas,* designed by William Symington (1763–1831) and in use in 1802 on the Forth–Clyde canal. The world's first steam-powered factory, the Blockmills, was opened in 1802 in Portsmouth Dockyard, UK, to mass-produce pulley blocks for sailing ships. Meanwhile, in early 1804, Richard Trevithick (1771–1833) produced the first steam engine to run successfully on rails; on 21st February 1804, it hauled 10 tonnnes of iron, 70 passengers and 5 wagons at an 'easy' 4 mph on a 9-mile journey from the ironworks in Pen-y-Darren, South Wales to the Merthyr–Cardiff Canal. Unfortunately, the engine proved too heavy for the rails, and the era of the practical steam locomotive began in 1813 with *Puffing Billy,* which was designed by William Hedley (1779–1843) and was the first to run on smooth rails, instead of the previous rack rails. By the early 19th century, steam was also replacing water to power cotton mills (see Chapter 5) and *The Times* newspaper was printed in London on a steam press as early as 1814. Thus, such

The *Fighting Téméraire's* tugged to her last berth to be broken up 6 September 1838

Puffing Billy, 1813

Isambard Kingdom Brunel's SS *Great Britain,* 1843

Steam traction engines

Figure 1.3   *J.M.W. Turner's poignant painting of the Fighting Téméraire's inglorious last journey is the defining image of the triumph of the steam-driven industrial age that saw the introduction of railway locomotives, iron ships, and road tractors.*

engines turned steam into a universal source of power and heralded the beginning of the fossil-fuel age. By the end of the 19th century, there were steam-driven cars in London, Paris, and New York as well as lorries and trams. These competed with electric vehicles and petrol-driven vehicles. Eventually, the internal-combustion engine proved superior (see below) and steam vehicles were mostly phased out, although their use in agriculture continued for at least another 20 years.

Coal

Coal was the first fossil fuel to be exploited to produce power. Previously, wood and coal had been used principally for space heating and cooking and, in the form of charcoal and coke, for metallurgical purposes (*e.g.* in the smelting and casting of iron). Later, in the 19th century, the pyrolysis of coal yielded coal gas ('town gas'), which was distributed in cities for lighting lamps and cooking, and coal tar, which was the early raw material for the organic chemicals industry (explosives, dyes, drugs, *etc.*). As recently as 1937, coal accounted for three-quarters of energy consumption world wide through its use as a fuel for space heating, cooking, industrial processes, and transportation (steam trains and ships).

Petroleum

In 1855, Benjamin Silliman (1816–1885), chemistry professor at Yale University in the USA, investigated the properties of crude oil and predicted that 90% of its contents could be distilled into saleable products (Figure 1.4(a)). Following this epochal discovery, the Seneca Oil Company was founded and started to search for oil by the novel method of drilling. Edwin L. Drake (1819–1880), a retired railway conductor, was put in charge of operations (Figure 1.4(b)). With an assistant, he erected an engine-house and derrick on a farm near Titusville, Pennsylvania, and on 27 August 1859 oil began coming to the surface. The extraction of crude oil from the ground was found to be remarkably simple and inexpensive, and large-scale distillation methods were soon developed to yield light and heavy fractions. Hence was born the petroleum ('oil') industry, based on the second-major fossil fuel.

The petroleum industry was transformed from a thriving local activity into one of global influence when, in 1876, the four-stroke internal-combustion engine was built by the German engineer, Nikolaus Otto (1832–1891) (Figure 1.5).[*] Otto's engines ran at slow speeds and it was not until 1885 that a suitable power unit for motor cars (automobiles) became available – the high-speed engine invented by Gottlieb Daimler (1834–1900). Shortly afterwards, in 1892, Rudolf Diesel (1858–1913) introduced the diesel engine. These engines quickly led to the widespread development of the motor car, just over 100 years ago. Moreover, the success of internal-combustion engines operating on petroleum ('gasoline') was such that they rapidly replaced the steam engine (an external combustion engine)

---

[*] Subsequently, it was established that the French engineer Alphonse Beau de Rochas was the true inventor of the four-stroke cycle in 1862 and Otto's patent was revoked in 1886. In fact, Beau de Rochas never built a working model and it can be said that it was Otto's later design of the engine that largely enabled the creation of automobiles, powerboats, motorcycles, and even aeroplanes.

(a)

**Benjamin Silliman, Jnr.
(1816–1885)**

*'From the rock-oil might be
made as good an illuminant
as any the world knew. It also
yielded gas, paraffin, lubricating
oil. In short, your company have
in their possession a raw material
from which, by simple and not
expensive process, they may
manufacture very valuable
products. It is worthy of note
that my experiments prove
that nearly the whole of the
raw product may be manufactured
without waste, and this solely by a
well-directed process which is in
practice in one of the most simple
of all chemical processes.'*
(Extract from Silliman's report, 1855)

(b)

**Edwin L. Drake
(1819–1880)**

**Top-hatted Edwin Drake stands
in front of his historic well
with his friend Peter Wilson at
Titusville, Pennsylvania, 1861**

(c)

Modern oil production in the North Sea

Figure 1.4   *Oil production from early days to modern times. Edwin Drake's well produced
less than a barrel a day from a depth of 21.2 m (69.5 ft). The design capacity of
today's off-shore platforms can allow for a daily production of ~300 000
barrels from wells drilled to depths of over 2 km. The Troll platform, which
stands in the North Sea, became one of the two man-made objects visible with
the naked eye from the surface of the moon.*

for almost all other applications although, of course, steam turbines are still used in
electricity generation.

It is interesting to note that, at first, there was considerable hesitation and doubt
about the wisdom of developing the internal-combustion engine for road vehicles,

Figure 1.5 *Pioneers of the automobile industry.*

as is made clear by the following quotation, said to be from the Congressional Record of the USA for 1875.*

"A new source of power, which burns a distillate of kerosene called gasoline, has been produced by a Boston engineer. Instead of burning the fuel under a boiler, it is exploded inside the cylinder of an engine. This so-called internal combustion engine may be used under certain conditions to supplement steam engines. Experiments are under way to use an engine to propel a vehicle.

This discovery begins a new era in the history of civilisation. It may some day prove to be more revolutionary in the development of human society than the invention of the wheel, the use of metals, or the steam engine. Never in history has society been confronted with a power so full of

---

* We have been unable to verify this quotation, but feel that it is an interesting anecdote worthy of inclusion, especially as its announcement of the production of an internal-combustion engine precedes that reported by Nikolaus Otto in 1876.

potential danger and at the same time so full of promise for the future of man and for the peace of the world.

The dangers are obvious. Stores of gasoline in the hands of the people interested primarily in profit, would constitute a fire and explosive hazard of the first rank. Horseless carriages propelled by gasoline engines might attain speeds of 14 or even 20 miles per hour. The menace to our people of vehicles of this type hurtling through our streets and along our roads and poisoning the atmosphere would call for prompt legislative action even if the military and economic implications were not so overwhelming. The Secretary of War has testified before us and has pointed out the destructive effects of such vehicles in battle. Furthermore, our supplies of petroleum, from which gasoline can be extracted only in limited quantities, make it imperative that the defence forces should have first call on the limited supply. Furthermore, the cost of producing it is far beyond the financial capacity of private industry, yet the safety of the nation demands that an adequate supply should be produced. In addition, the development of this new power may displace the use of horses, which would wreck our agriculture.

– the discovery with which we are dealing involves forces of a nature too dangerous to fit into any of our usual concepts."

Clearly, such fears did not quench a natural human desire to outpace and outdistance the horse-drawn carriage. The issue was finally put to rest in 1908 when Henry Ford (1863–1947) introduced the world's first mass-produced, low-cost car – the Model T – which became even more affordable when assembly-line production was implemented in 1913 (Figure 1.5). Through such streamlining, the car could be put together in 98 min and sold for only US$ 400.

Just how far the motor car has come in a little more than 100 years is quite remarkable, although on the pollution front we have merely replaced one environmental problem (horse manure) with another (tailpipe emissions). The 20th century saw the widespread adoption of the internal-combustion engine for transport and for power applications. The use of this engine has become so extensive that grave concern has arisen over the pollution it causes in cities, and also over its contribution to global warming through enhancement of the 'greenhouse effect', which is attributable in part to the carbon dioxide (CO_2) produced by combustion. The greenhouse effect is essentially the trapping of heat in the lower levels of the Earth's atmosphere. Most of the short-wave and visible radiation from the Sun is transmitted through the atmosphere to the Earth's surface where it is largely absorbed, while part is re-radiated as long-wave infra-red radiation. Because of the anthropogenic build-up of gases such as carbon dioxide, methane, nitrous oxide, and ozone, however, a substantial amount of the long-wave rays are absorbed by these gases and re-radiated back to the Earth's surface. Therefore, the lower levels of the atmosphere are heated to higher temperatures. In satisfying energy and agricultural needs, humankind has probably increased the amount of carbon dioxide alone by some 30% since the Industrial Revolution. It is widely believed that long-term climatic changes may result from such global warming.

Electricity

Another major change that took place towards the end of the 19th century was the growth of the electrical industry. The basic components required to establish this industry – the electric motor, the transformer, the dynamo – were provided by the earlier research of the English scientist Michael Faraday (1791–1867) on electromagnetic induction, see Figure 1.6. Following the development of the incandescent electric light bulb in the 1870s by Henry Woodward and Mathew Evans in Canada, Joseph Swan (1828–1914) in England, and Thomas Edison (1847–1931) in the USA electricity was increasingly used for lighting. This

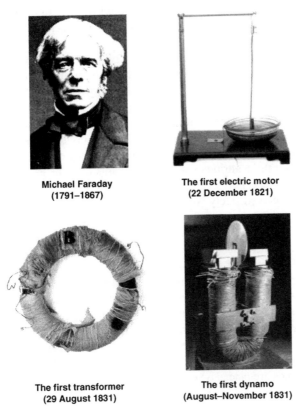

Michael Faraday
(1791–1867)

The first electric motor
(22 December 1821)

The first transformer
(29 August 1831)

The first dynamo
(August–November 1831)

Figure 1.6 *By making the first electric motor, transformer, and dynamo, Michael Faraday provided the key components required to create an electrical industry – a remarkable achievement from a man who was born into poverty, malnourished in childhood, and rudimentarily educated, and who failed to acquire a knowledge of mathematics beyond simple arithmetic.*

demand for electricity led to the construction of large, coal-fired, power stations and the establishment of transmission networks.

Natural Gas

Natural gas, formerly known as 'marsh gas', is formed in marshes by the anaerobic decay of natural vegetation. In the late 19th and early 20th centuries, when oil reservoirs began to be exploited, it was found that natural gas was generally associated with the oil. This was often flared off as there was little use for it in the neighbourhood of oil fields such as those in the Middle East. Later in the 20th century, it was realized that there are many reservoirs of natural gas in porous, sedimentary rock structures that are not associated with oil. This gas, like the oil, was formed by the decay over geological time of countless small marine organisms. The reservoirs that contain oil or gas are located beneath a layer of impermeable rock, which forms a cap (or dome) and prevents escape of the gas. When the cap is drilled through, the pressurized gas forces the oil (if any) to the surface.

During the second half of the 20th century, there has been wholesale exploitation of natural gas – the third major fossil fuel. The convenience of this fuel is such that it is now preferred for the space heating of both commercial and domestic buildings. Huge reserves of natural gas have been discovered and extensive networks of underground pipes have been constructed to convey the gas to consumers. Over the past decade or so, there has been a marked swing towards the combustion of gas for the centralized generation of electricity. This may be explained by the abundance, accessibility and relatively low cost of natural gas, its convenience of transport and use, and the high efficiency of the combined-cycle gas turbine. Contrast these attributes with the concerns over the environmental aspects of burning coal (especially coals with high-sulfur contents) to generate electricity, and also the perceived safety issues of nuclear power and radioactive wastes.

Energy from the three great fossil fuels – coal, petroleum and natural gas – has provided the means by which our industry and our civilization have steadily progressed since the Industrial Revolution. Coal now constitutes a much smaller proportion of the energy supply. Today, the world is largely dependent upon petroleum for its transportation, and upon both petroleum and natural gas for its agriculture, its chemicals and materials, its space heating and, not inconsiderably, for its electricity supply. In the long-term, this is obviously an unsustainable situation as fossil fuels are being extracted at a rate that grossly exceeds the rate at which they are laid down. Moreover, the atmosphere will no longer accept unlimited combustion of fossil fuels. These problems will worsen as the world's population expands and ever-growing demands are placed on energy and power supplies.

Clearly, properly-used energy technologies serve as instruments for improving our material well-being, but a continuation of current trends could lead to a degraded environment. How can we solve this problem? The obvious answer is to move away from fossil fuels, towards sustainable energy supplies of a non-polluting nature, while maintaining the standard of living of the developed world and dramatically improving that of developing nations. Quite a challenge!

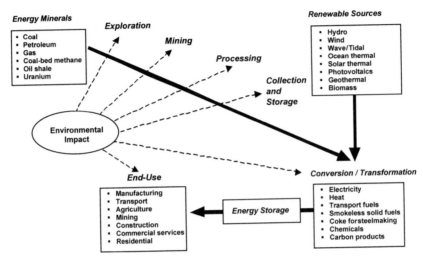

Figure 1.7   *Interaction between energy consumption and environmental impact.*

Fortunately, nature has bestowed upon us a bountiful supply of such benevolent energy in the forms of hydro energy, wind energy, wave energy, tidal energy, ocean thermal energy, solar energy, geothermal energy, and biomass. The task, therefore, is to capture and utilize greater amounts of these renewable energies in an efficient and economical way so as to minimize the environmental impact of energy consumption by the industry, transportation, agriculture, commerce and domestic sectors (Figure 1.7).

We may summarize this brief historical introduction by drawing attention to the interplay between energy, science and technology that has produced three major changes in civilization. During the early 18th century, the dominant technologies were coal mining, the smelting and casting of iron, and steam-driven rail and marine transport. Through the creation of a transportation infrastructure, the provision of materials and the invention of machines to run factories, rapid industrialization was possible. Then, towards the end of the 19th century and early in the 20th century, came electric power, internal-combustion engines, motor vehicles, aeroplanes, and the chemical and metallurgical industries. During this phase, crude oil emerged both as a fuel and as a feedstock for the petrochemicals industry. Now, as we enter the 21st century, the world has embarked on a third era of advancement – one that is characterized by a shift to computers, advanced materials, optical electronics, biotechnology, and nanotechnology. The first two of these changes in civilization were each accompanied by a massive increase in the consumption of energy. The impact that the present and third change will have on future global patterns of energy supply and demand has still to be determined.

1.2 World Energy Supply – The Present Scene

In industrialized countries, energy demand comes from three major sectors: (i) industry and agriculture; (ii) transport; (iii) domestic and commerce. In many countries, these each account for about one-third of the total energy flow, although the size of domestic demand depends very much on climate. In Australia, for example, domestic demand is relatively small, whereas in Canada it is extremely large because of the more extensive and severe periods of cold weather. It is also possible to identify the demands for particular purposes within these three sectors, namely: (i) low-temperature heat ($<110\,^\circ$C) for water and space heating in houses and industry; (ii) high-temperature heat ($>110\,^\circ$C) for industrial processes; (iii) lighting; (iv) power for factories and appliances; (v) mobile transport for public and private use. Primary energy sources are not always of a form that is suitable for these end-uses. Instead, the sources have to be converted once or twice before they become appropriate for their intended use. Unfortunately, each step in a conversion process involves losses, or more correctly, the production of waste energy. The most important, versatile and useful of these 'converted' or 'secondary' energy sources is electricity.

Overwhelmingly, fossil fuels continue to meet the global demand for energy – only about 10% of energy needs are satisfied by other supplies. This situation has persisted despite the fact that the contribution from coal peaked around 1920, when it accounted for more than 70% of all fuel consumed, and that oil usage peaked in the early 1970s at around 45%. By contrast, the share provided by natural gas is expected to rise. Overall, fossil fuels are being depleted at a rate that is five orders of magnitude faster than the rate at which they are being formed.

The extent of the challenge in moving towards global energy sustainability and the reduction of CO_2 emissions can be assessed by consideration of the trends in primary energy supplies. Such information for 1973 and 2001 is provided in Table 1.1 for both the world and the OECD countries (Organisation for Economic Co-operation and Development – a consortium of 29 developed countries). The data are published by the International Energy Agency (IEA)[1] and is the latest information available to the authors. The totals are expressed in million tonnes of oil equivalent (Mtoe), which is a convenient unit for comparing very large quantities of energy in different forms.

What may we deduce from these data?

- The total energy supply of the world has increased by 66% in 28 years, while that of OECD nations has increased by 42%. The difference represents the faster growth of many developing nations that start from a lower energy base.
- While the production of oil has increased everywhere, the expansion in activity has been fairly modest compared with natural gas, with the result that oil now provides a significantly smaller percentage share of the total energy supply.
- Coal is used mainly to generate electricity and, despite the fact that the industry is using more gas, the overall increase in electricity consumption has resulted in only a small decline in the percentage contribution made by coal to total world energy.
- The production of natural gas has risen appreciably following the discovery and opening up of new fields. Nevertheless, again because of the overall growth in energy demand, the percentage contribution of natural gas has increased only modestly. (Note: since the late nineties, there has been a 'dash for gas' in electricity production, using combined-cycle gas turbine

Table 1.1 *Total primary energy supply by fuel in 1973 and 2001*[1]
(Supply based on Mtoe for each fuel and given in % terms.)

Energy supply	The world[a]		OECD[b]	
	1973	2001	1973	2001
Oil	45.1	35.0	53.1	40.8
Coal	24.8	23.3	22.4	20.8
Gas	16.2	21.2	18.8	21.3
Combustible renewables (biomass) and waste	11.1	10.9	2.1	3.3
Nuclear	0.9	6.9	1.3	11.2
Hydro	1.8	2.2	2.1	2.0
Other[c]	0.1	0.5	0.2	0.7
Total (%)[d]	100.0	100.0	100.0	100.1
Total (Mtoe)[e–h]	6034	10 029	3757	5332

[a] Excludes international marine bunkers and electricity trade.
[b] Excludes international electricity trade.
[c] Includes geothermal, solar, wind, and heat.
[d] IEA data have rounding errors of 0.1%.
[e] The statistics for world energy *consumption* in Mtoe are only about 70% of these for total primary energy supply. This is because of losses in conversion and transmission, especially in the conversion of fossil fuels to electricity.
[f] To convert Mtoe to TJ, multiply by 4.18×10^4.
[g] Other sources quote oil production in standard barrels. There are 7.3–7.4 barrels per tonne, as dictated by the oil density.
[h] Other energy and volume conversion factors are summarized in the Appendix.

technology. In the UK, for instance, gas production rose by 15% between 1997 and 1999).

- Although the use of combustibles and waste to generate heat has increased slightly in OECD countries, it is still far lower than in the rest of the world. The latter reflects the shortage and cost of fossil fuels in many non-OECD countries. Traditional forms of biomass account for 90% or more of the total energy consumed in some low-income, developing countries.
- Over the period under consideration, the nuclear generation of electricity has increased by more than 7-fold. For the present, this growth has largely stopped, and may well be in decline.
- Hydroelectric power ('hydro power') makes only a small contribution to world energy supply, but its significance for electricity production is considerable. This source of power is limited to regions with mountainous terrain. In some countries, however, it may be the dominant means of generating electricity (in Norway and Brazil, for example).
- The contribution of all the non-combustible forms of renewable energy ('Other', Table 1.1) is exceedingly modest, but is growing rapidly.

The ultimate goal of global energy sustainability implies the replacement of all fossil fuels (oil, coal, natural gas) by renewable energy sources. This is indeed a monumental undertaking! For example, total non-fossil energy in 2001 was only 20.5% (including nuclear) throughout the world, and was even less (17.2%) in OCED countries. Furthermore, there are limited hydro sites left to be exploited, and if the nuclear programme contracts, as seems likely in many countries, and the total world demand for energy rises, as is virtually certain, the proportion of energy provided by non-fossil sources could fall below its present low level.

1.3 World Energy Supply – Projections to 2020

The IEA's forecast of the world demand for primary energy in 2010[1] and 2020[2] is shown in Table 1.2. Compared with the situation in 2001, the IEA predicts a 21% increase in 2010 (12 100 Mtoe) and a 48% increase in 2020 (14 800 Mtoe).[3] Fossil fuels (oil, coal, natural gas) will continue to provide about 80–85% of the world's primary energy right through to 2020. Oil, in the form of petroleum, will be the dominant fuel and will meet 35–37% of global energy needs. This will require an increase in production from around 70 million barrels per day (Mb d^{-1}) now to around 110 Mb d^{-1} in 2020.* This reflects a substantial increase in the demand for transportation fuels. Shell, for example, has predicted that oil consumption by road vehicles in 2020 will be 40% higher than today. Whether or not such increased production will take place is in part a political question. Nuclear power output will decline as a proportion of the total energy supply as older reactors in Europe and North America are retired. The apparent increases in 'renewables', categorized under 'Other' in Table 1.2, is misleading. If the estimated contribution from wood and dung in developing nations is deducted, the expected contribution from 'new' renewable sources (geothermal, solar, wind, tidal, *etc.*) will increase from 0.5% to between 2 and 3% of the total primary energy supply.

Of course, all such predictions over extended periods are subject to large uncertainties and depend upon the growth assumptions used in the model. In the case of energy supply, these uncertainties may be less than in other fields. This is because reasonable guesses may be made for both the demand side (population growth, effects of globalisation, and the aspirations of developing countries) and the supply side (known and likely reserves of fossil fuels). With respect to oil reserves in the ground, it is worth noting that the present-day recovery from reservoirs is, on average, only about 30–40% of the oil they contain. Given improvements in technology and expected rises in the price of oil as supplies dwindle, it will become practicable and economic to increase this recovery. The Shell Company has set a recovery target of 70–80% by 2010. At present, however, it is not commercially attractive to rework old wells once they have been closed.

Even major swings in the prices of fuels do not seem to impact their consumption greatly. This is because people are prepared to pay for 'essentials' such as comfort (heating, air-conditioning) and leisure pursuits (transportation) at the expense of

---

* tonne of crude oil approximates to 7.4 barrels.

Table 1.2 *Total primary energy supply by fuel for world in 2001[1] and forecast for 2010[1] and 2020[2]*
(Supply based on Mtoe for each fuel and given in % terms.)

	Year		
Energy supply	2001	2010	2020
Oil	35.0	35.3	37.0
Coal	23.3	22.3[a]	22.6[a]
Gas	21.2	23.1	23.9
Combustible renewables (biomass) and waste[a]	10.9	–	–
Nuclear	6.9	6.2	4.2
Hydro	2.2	2.3	2.3
Other[b]	0.5	10.9	10.0
Total (%)[c]	100.0	100.1	100.0
Total (Mtoe)	10 029	12 100	14 800

[a] Includes combustible renewables and waste for OECD countries only.
[b] The 2001 data separate out combustible renewables (mostly wood and dung) and waste from 'coal'. The IEA predictions for 2010 and 2020 report this item for OECD countries under 'coal', and for non-OECD countries under 'other'. This somewhat curious procedure explains the apparent increase attributed to renewables, which is mostly illusionary.
[c] IEA data have a rounding error of 0.1%.

other purchases. Meanwhile, commercial and industrial use of fuel continues unabated to provide people with employment and, therefore, income. To attempt a radical reduction in consumption and a change to renewable forms of energy, as is implied by energy sustainability, seems barely feasible.

1.4 Fossil Fuel Reserves

Oil and Gas

Oil and gas reserves are classified as 'proven', 'probable', or 'possible'. In the past, prospecting has tended to be limited by having an adequate bank of proven reserves in hand for the reasonable future. For this reason, over the past 30 years or more, world reserves have been repeatedly quoted as being sufficient to meet the demand for the next 10–25 years. This is illustrated in Table 1.3, which provides data for the UK in terms of oil and gas reserves from the North Sea.[4] It is seen that the UK production of oil is at, or near, its peak and known reserves are declining. From 1980 to 1998, the proven reserves of natural gas remained fairly constant as new discoveries offset production, but since 1998 there has been a sharp decline in reserves. It may be noted that in 2002, the UK produced 3.2% of the world's crude oil and 4.1% of its natural gas.[1]

Table 1.3 *Oil and gas production and reserves in the UK*[4]

Fuel	1980	1990	2000	2002
Oil ($\times 10^6$ t)				
Cumulative production	263	1374	2570	2799
Estimated remaining reserves	2300	1815	1490	1345
Total	2563	3189	4060	4144
Gas ($\times 10^9$ m^3)				
Cumulative production	382	752	1518	1726
Estimated remaining reserves	1560	1785	1630	1330
Total	1942	2537	3148	3056

Note: 1×10^9 m^3 of gas $\cong 0.9 \times 10^6$ t of oil equivalent (0.9 Mtoe).

Reliable estimates of the world reserves of oil and gas are difficult to obtain for a number of reasons, namely: (i) new oil and gas fields are constantly being discovered due to continuous improvements in geo-prospecting techniques as well as the exploration of fresh sedimentary rock basins; (ii) the technology for exploiting offshore fields in deep water is evolving; (iii) there are promising developments for increasing the recovery rates from fields by secondary injection techniques. Despite these factors, BP made an estimate of the reserves of oil in various parts of the world at the end of 2002.[5] The distribution is shown in Figure 1.8; the numbers represent thousand million barrels of oil. It is evident that the greatest reserves are in the Middle East. This domination of reserves by one geographic region causes much concern about the security of oil supply in the future.

In 1971, M. King Hubbert predicted that world production of oil would peak in the period 1990–2000, that the total ultimate recoverable reserves would lie in the range 1.4–2.1×10^{12} barrels, and that 80% of these reserves would be processed in about 60 years, *i.e.* from 1970 to 2030.[6] While it now appears that this prediction was unduly pessimistic, it is only a matter of degree. At current production rates, recent estimates suggest that world oil will last about another 40 years, and natural gas about 60 years.[5,7] Replotting King Hubbert's data on a scale of four centuries shows that fossil oil is but a transitory phenomenon (Figure 1.9). Even aside from environmental factors, it seems likely that native petroleum will become a scarce commodity within the lifespan of children alive today.

What is known with greater certainty is the amount of crude oil that is being produced annually by different countries around the world. The major producers are listed in Table 1.4, along with their estimated reserves and the number of years these reserves would last at the respective production rates in 2002. Six countries – Saudi Arabia, Iraq, Kuwait, United Arab Emirates, Iran, Venezuela – dominate the reserves, although there may be further substantial reserves in the Russian Federation and the People's Republic of China still to be determined.

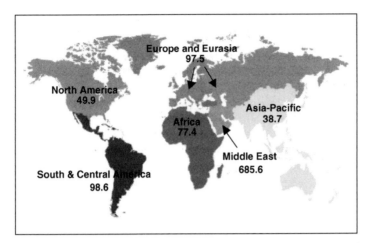

Figure 1.8 *BP estimate of world reserves of oil at the end of 2002 (1000 million barrels).*[5]

Natural gas is distributed widely around the world, both in association with petroleum and also in 'gas wells' that contain little, or no, higher hydrocarbons. Frequently, the gas is almost pure methane. The three principal producers of natural gas are the Russian Federation (22% of world total), the USA (21.7%), and Canada (7.3%) – together these provide half of the world's production (see Table 1.5). The North Sea contributes around 7%, mostly from the UK and Norway.

By far the bulk of the proven reserves of natural gas is located in just three countries, namely, the Russian Federation, Iran, and Qatar. Russian gas is exported to many northern European countries, while southern Europe is supplied with Algerian gas. There are no large markets immediately at hand for Middle Eastern

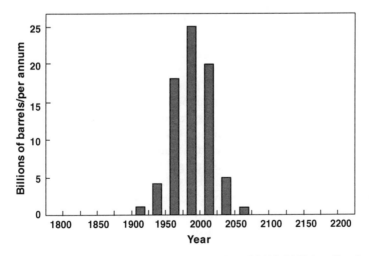

Figure 1.9 *World oil production 1800–2000 (actual) and 2000–2200 (predicted).*

Table 1.4 *Production and reserves of crude oil at end of 2002*[5]

Countries: major producers[a]	Annual production (Mt)	% World total	Countries: greatest reserves	Reserves (Gt)	Reserves/ production (years)
Saudi Arabia	418.1	11.8	Saudi Arabia	36.0	86
Russian Federation	379.6	10.7	Iraq	15.2	>100
USA	350.4	9.9	Kuwait	13.3	>100
Mexico	178.4	5.0	United Arab Emirates	13.0	>100
PR China	168.9	4.8	Iran	12.3	74
Iran	166.8	4.7	Venezuela	11.2	74
Norway	157.4	4.4	Russian Federation	8.2	22
Venezuela	151.4	4.3	Libya	3.8	59
Canada	135.6	3.8	USA	3.8	11
UK	115.9	3.3	Nigeria	3.2	33
Rest of the world	1334.3	37.5	Rest of the world	22.7	17
Total	3556.8	100.2[b]		142.7	

[a] Iraq is omitted because its production in 2002 was low for political reasons. Its reserves, however, at 15.2 Gt are the second largest in the world after Saudi Arabia.
[b] Data have a rounding error of 0.1%.

and Nigerian gas and, for this reason, much is liquefied and shipped to market by tanker (see Section 2.1, Chapter 2). The only major producers (and consumers) of natural gas whose resources are getting short (*i.e.* reserves/production <10 years) are Canada, the UK, and the USA. No doubt, more gas will be discovered in these countries, but it is clear that indigenous supplies will become strictly limited by 2020.

There are at least two other large, prospective sources of natural gas that are quite distinct from that found with petroleum. The first is associated with coal, namely coal-bed methane (CBM). Despite being a huge potential resource, CBM has not, to date, been greatly exploited, although the coal fields in Queensland, Australia are being actively explored and mapped.

The second potential source of natural gas is more speculative and futuristic. A number of gases of small molecular size, *e.g.* methane, chlorine, carbon dioxide and hydrogen sulfide, have the ability to be trapped or occluded in ice crystals to form so-called 'clathrate' compounds.[*] In these crystalline solids, the ice takes up a characteristic cubic structure, rather than its usual hexagonal structure. The 'guest

[*] The term 'clathrate' is derived from the Latin 'clathratus', which means 'enclosed by bars'.

Table 1.5 *Production and reserves of natural gas at end of 2002*[5]

Countries: major producers[a]	Annual production (10^9 m^3)	(Mtoe)	% World total	Countries: greatest reserves	Reserves (10^{12} m^3)	Reserves/ production (years)
Russian Federation	554.9	499.4	22.0	Russian Federation	47.6	81
USA	547.7	492.9	21.7	Iran	23.0	>100
Canada	183.5	165.2	7.3	Qatar	14.4	>100
UK	103.1	92.8	4.1	Saudi Arabia	6.4	>100
Algeria	80.4	72.3	3.2	United Arab Emirates	6.0	>100
Indonesia	70.6	63.5	2.8	USA	5.2	10
Norway	65.4	58.9	2.6	Algeria	4.5	56
Iran	64.5	58.1	2.6	Venezuela	4.2	>100
Netherlands	59.9	53.9	2.4	Nigeria	3.5	>100
Saudi Arabia	56.4	50.7	2.2	Iraq	3.1	>100
Rest of world	741.2	667.0	29.3	Rest of world	37.9	44
Total	2527.6	2274.7	100.2[b]		155.8	

[a] Iraq is omitted because its production in 2002 was low for political reasons.
[b] Data have a rounding error of 0.1%.

molecule' is trapped at the centre of this structure and held there by van der Waals forces.

In recent years, growing evidence has emerged that large quantities of methane hydrates have been formed over geological time by anaerobic bacterial degradation of organic matter. These clathrate compounds are found in two geographic locations, in the permafrost of continental polar regions and beneath the beds of oceans. The conditions for stability of the methane hydrates are low temperature (broadly below 0 °C) and high pressure (>5 MPa or 50 atm.). The temperature and pressure range of stability is interrelated. Estimates of the amount of methane available as methane hydrate to the USA alone range from 3×10^{15} to 19×10^{15} m^3. This is sufficient to fuel the USA for more than 300 years. By comparison, the corresponding reserves of methane gas lie in the range from 10^{12} to 10^{13} m^3. It is claimed that world-wide reserves of methane hydrate may be as high as 11×10^{19} m^3, compared with 15×10^{13} m^3 of methane gas. Evidently, all such estimates are preliminary and quantitatively suspect, but it does appear to be established that very large amounts of methane hydrate do exist. Nevertheless, how to exploit this resource – both technically and economically – is still far from being understood.

Tar Sands, Asphalts, Oil Shales, and Bitumen

Looking beyond the era of crude oil, there are many other fuels of lower grade remaining to be exploited. Foremost among these are the tar sands, huge deposits of which exist in Alberta, Canada. It is estimated that the reserves are equivalent to 180–300 billion barrels of oil (30–40 Gt). From such material, it is possible to extract bitumen that may then be upgraded to liquid fuels. Some development work was carried out on this process in the 1970s and the 1980s, but it was shown not to be economic at the prevailing oil prices and, therefore, was not pursued. Recently, the technology has been re-activated, principally through a desire by countries to utilize indigenous energy sources and so to be less dependent upon imports. A plant has been built that is capable of extracting 155 000 barrels of bitumen a day, equivalent to 10% of Canada's need for crude oil. Extraction involves agitating the sand with warm water (30–40 °C) followed by processing of the bitumen released. The warm water is supplied by a combined heat and power plant that is fired by natural gas; the electricity is exported to the grid. The bitumen is separated from the water, treated with hydrogen at high temperature for upgrading to lighter hydrocarbons, and then refined conventionally.

Unfortunately, there are many problems associated with exploiting such huge reserves of fossil fuel. The first is that the tar sands lie under 60–100 m of overburden that has to be removed before the operation can begin. This is both highly energy-intensive and costly, as is the restoration of the land after the bitumen has been extracted. Next, the large quantities of hydrogen that are required have to be produced from natural gas, which is also the source of heat for the warm water. It is said to take almost as much energy to mine, process, refine and upgrade the bitumen as the energy contained in the light oil that is produced. Thus, from a purely energetic viewpoint, the value of the operation is questionable, although on economic grounds it can be profitable since, effectively, low-value natural gas (of which Canada has plenty) is converted to high-value liquid fuels (see gas-to-liquids, Section 2.1, Chapter 2).

A further problem is associated with the amount of greenhouse gases that are produced during the processing of tar sands. It is estimated that these emissions are five-to-ten times more than those released when processing conventional oil. If carbon taxes or tradable emission permits are introduced under the Kyoto Protocol (see Section 1.5 and Section 11.1, Chapter 11), then tar sands may well become unattractive financially. This disadvantage is compounded by the above-mentioned environmental and energetic reservations.

In summary, there are major problems associated with the exploitation of the tar sands:

- a probable misuse of resources of energy and money;
- environmental destruction – forests, wildlife habitat and water sources could well be ruined;
- water pollution to lakes and rivers;
- possible air pollution with the spreading of toxic and carcinogenic compounds over a wide area;
- climate change implications (greenhouse gas emissions);
- lack of the vast quantities of water required for the extraction process.

Many of these difficulties would be circumvented if an economic technique could be developed to treat the tar sands *in situ*, without mining, by injecting high-pressure steam and recovering and separating the heavy oil. Altogether, despite their enormous potential, the true value of tar sands remains questionable.

Other sources of natural hydrocarbons are native asphalts and oil shales. At one time, some petroleum was derived commercially from West Indian asphalt. Extraction from solid oil shales, which are widely distributed in Western USA, poses formidable technical and cost problems. There are three essential steps in the extraction of oil from shales, namely: (i) mining; (ii) purification by distillation in the absence of air ('retorting'); (iii) disposal of solid residue. The materials handling that is involved in mining and conveyance to a central processing point, followed by return of the spoil is very energy intensive as also is the retorting process because shales need to be pyrolyzed at high temperature. A further difficulty with oil shales is that loss of volatiles is accompanied by exfoliation of the rock and a substantial increase in its bulk volume. This poses a major environmental problem of landfill disposal. For this and other reasons, it appears that oil shale are even less likely than tar sands to be exploited in the near future.

In Venezuela, there are huge reserves of natural bitumen that are said to have an energy content equivalent to 64 Gt of coal. This would be sufficient to sustain the production of 200 Mt of coal equivalent for over 300 years. The bitumen may be extracted by steam injection. Using this technique, BP developed a stable emulsion of bitumen (70 vol.%) in water (30 vol.%), known as 'Orimulsion', which could be burnt in existing power stations. Unfortunately, however, Orimulsion is high in sulfur content (2.7 wt.%) and is a 'dirty' fuel. It is no longer used in the UK electricity industry.

Coal

There are huge reserves of coal and these are widely distributed around the world. This fossil fuel is of plant origin and was laid down in geological time, especially during the Carboniferous Period between 345 and 280 million years ago. The first product of decay and consolidation is peat, which has a relatively low carbon content and a high moisture content. Under forces of heat and pressure, peat gradually converts to coal. Coal is not a unique commodity like natural gas, but exists in many different grades and qualities. The grades include in terms of increasing carbon content – lignite (or brown coal), soft bituminous coal, anthracite. Each type of coal is suitable for different applications. The quality of coal is determined by its calorific value, its moisture and volatile hydrocarbon contents, its ash content and, given present concerns over the effect of emissions on the environment, its sulfur content. Coal is by far the most abundant of the conventional fossil fuels, with proven economically recoverable reserves close to 10^{12} t world wide; about a quarter of this is in the USA, with large deposits also in the Former Soviet Union, the People's Republic of China, India, and Australia.[5] This is equivalent to the energy in 3.5×10^{12} barrels of oil, and represents about 250 years of coal production at today's rates.[7] There is also the natural gas associated with some of this coal (see above). At present, the world mines some 3800 Mt of coal a year, of which rather more than one-half is employed in electricity generation.[1,7] There are many ways in which coal may be converted to cleaner and more versatile fuels, as discussed in Chapter 2.

1.5 The Overall Energy–Environment Problem

The continued availability of fossil fuels and the ability to use them wisely, together with renewable forms of energy, are critical to the long-term future of the human race. This is not, however, the only challenge in moving towards true global energy sustainability; the two key issues of population growth and local/global environment impacts must also be addressed (Figure 1.10).

At the start of the Industrial Revolution, the world population was only a few hundred million. Today, it stands at almost 6.4 billion people who collectively occupy around a billion dwellings, drive 700 million motor vehicles, and expend much effort to produce a wide variety of industrial products to further their well being. Many of these people (mainly in the developing countries) still do not have access to commercial energy and, in particular, to a reliable source of electricity. In fact, of the two billion people living without electric power, about one billion have no supplies of commercial energy in *any* form – not even petroleum or diesel oil. These people operate entirely on wood fuels and other biomass resources. Moreover, the disparity will intensify. It is generally expected that the world population will grow to an estimated eight billion in 2020, and 90% of this growth will take place in the developing nations. Thus, over the next few decades, it is clear that global energy-supply issues will move from the industrialized to

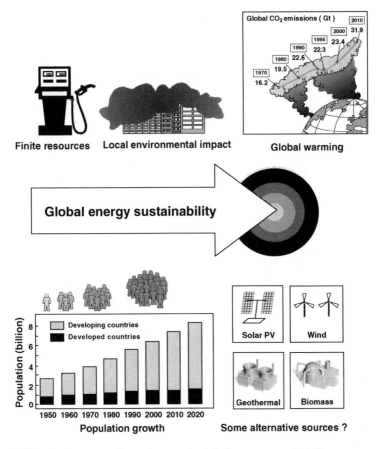

Figure 1.10 *Key issues on the path towards global energy sustainability.*

the developing countries, many of which have serious social, economic, and environmental problems. Those that have coal will wish to burn it, despite the environmental consequences.

Of equal concern are the destructive effects of the growing levels of energy conversion and usage on the earth's biosphere. In global terms, the energy sector is the single largest source of anthropogenic greenhouse gases, with emissions of: (i) carbon dioxide from the combustion of coal, oil, and natural gas in electricity generation, in transport, and in the direct use of such fuels by industry, commerce and households; (ii) methane from the production, transport and end-use of natural gas, and from coal mining; (iii) nitrous oxide, which is one of the nitrogen oxides (NO_x) formed during the combustion of fossil fuels. At the local level, energy use in motor vehicles is a major contributor to the degradation of urban-air quality. The exhaust pollutants include carbon monoxide (which displaces oxygen in the blood), NO_x (which combine with water to form corrosive nitric acid) and hydrocarbons (which react with NO_x in the presence of sunlight to form ozone, *i.e.* photochemical smog, a lung irritant). Energy use in large stationary applications such as power stations is also a serious source of both sulfur oxides (SO_x) and NO_x, which are the major precursors to 'acid rain'. Other environmental concerns include: (i) local impacts associated with land use, mining and processing (including dust from coal stockpiles, oil spills, disposal of solid wastes, emissions of volatile organic compounds and trace elements); (ii) electromagnetic fields; (iii) nuclear waste.

A first step towards reducing CO_2 emissions to the levels that existed pre-industrialization was agreed at the United Nations Framework Convention on climate change that was held in Kyoto in December 1997. The resulting 'Kyoto Protocol' called for the industralized nations to reduce the average of their individual emissions by at least 5% below baseline 1990 levels by 2008–2012. The global baseline level for 1990 was about 22 Gt per year. (Global emissions of CO_2 rose from 16 Gt per year in 1970 to over 23 Gt per year in 2001. More than 50% of the emissions still emanate from the industrialized nations, see Figure 1.11.[1]) The specific target reductions proposed for the European Union, Japan, and the USA were 8, 6, and 7%, respectively. Compared with projected emission levels in the absence of any mitigating action, these targets represent reductions of 20–30%. Targets were not set for developing countries.

Not only do the emission restrictions proposed at Kyoto for developed countries imply a marked reduction in their use of fossil fuels, particularly coal, but also such restrictions are not immediately achievable by burning biomass and waste, two of the most promising renewables in the short term. It is true that the substitution of coal by natural gas for electricity generation, *via* the use of efficient combined-cycle gas turbine technology, plays a significant role in reducing CO_2 emissions. The extent to which this is possible, however, is dictated by considerations of resource availability, politics, and economics. In the field of politics, for example, the Bush administration in the USA has backed away from the Kyoto Protocol as a result of short-term power deficits in California and worries about the effect of reducing emissions on US industry and its competitiveness. Rather, the USA

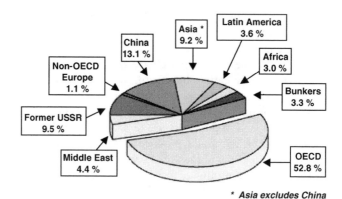

World total CO_2 emissions = 23.7 Gt

Figure 1.11 *Regional shares of carbon dioxide emissions from fuel consumption only in 2001.*[1]

proposed that its target for CO_2 reduction should be linked to the nation's economic output and should fluctuate as this output ebbs and flows. Uniquely, the UK government has set a target to reduce its CO_2 emissions by 20% (compared with 1990 levels) by 2010, which is well in excess of its Kyoto obligation. This is to be achieved through the recent construction of gas-fired power stations to replace coal, through an ambitious programme of off-shore wind turbines, and by the issuing of carbon-emission permits to major fuel-consuming industries.

Finally, it is remarkable that many developing nations foresee substantial *increases* in CO_2 emissions as their economies evolve. For instance, without major changes in its present policies, the People's Republic of China expects to be emitting three times as much CO_2 in 2020 as in 1995, *i.e.* more than the USA. In January 2002, however, China announced a national programme to reduce pollution and especially CO_2 emissions.

The implementation of revolutionary energy policies will not be easy in any country, and the IEA estimates that global CO_2 emissions will increase by 60% over the period 1997–2020 in line with increased fuel usage.[8] Emissions from developing countries will continue to grow as they become industrialized, and there is a danger that present industrialized countries beyond the USA will also become disillusioned about adopting costly solutions to reduce their own emissions. There is potential for political conflict between those who take a long-term global view (the 'Greens') and those who take a short-term national view based on economics and national prosperity. There is also the sheer practicality and long time-scale involved in effecting major changes to the world pattern of energy consumption and fuel use. Meanwhile, although there is a growing consensus that anthropogenic greenhouse gases are contributing to the observed rise in atmospheric temperature and the resultant change in climate, there is as yet no unequivocal scientific proof of the extent of these effects since they are superimposed on long-term natural cycles in the world climate that are only imperfectly understood.

The conclusion to be drawn from this brief review of the overall energy scene is re-assuring in terms of the potential supply of fossil fuels for the next few decades, possibly 50 years or more, but is less so in terms of the prospects for renewable energy. Unless there are major changes in the relative economics of the different energy sources – which is possible – it is difficult to see how substantial movement towards global energy sustainability is likely in the next 20 years. Even though the use of renewable energies may well grow rapidly, their contribution to the overall world energy scene and to a reduction in CO_2 emissions will still be minimal. The IEA takes a somewhat less gloomy outlook and points to potential reductions in CO_2 emissions that may result from the trading of carbon permits (if they are introduced), improved energy efficiency, the use of new fuels in the transportation sector, and the switching of electricity generation from coal to gas and nuclear. The IEA admits, however, that economic and political obstacles will not allow such changes to be put in place rapidly and they mostly relate to the post-2020 era.

In order to meet the longer term goals for reduced emissions of CO_2, it will be necessary to address several issues. These are summarized in Figure 1.12. The first is to improve the efficiency with which fossil fuels are extracted, refined, and converted to end products. This applies especially to utilizing the waste gas that is flared from oil

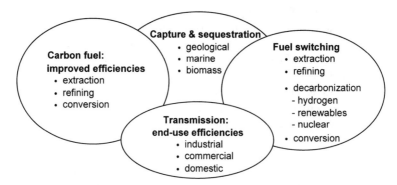

Figure 1.12   *Strategies for the reduction of carbon dioxide emissions.*

wells and to improving the recovery rate of petroleum. The second issue is fuel switching. This includes moving progressively from high-carbon to lower-carbon fuels, as well as introducing hydrogen fuel, renewable forms of energy and, probably, further nuclear-generated electricity. A third long-term goal is to develop procedures for the capture and sequestration of CO_2. Ideas to be investigated as regards their practicality include disposal in disused oil or gas wells, or in other geological formations (particularly, aquifers). Another option under active consideration is direct injection into the deep ocean, although this poses many environmental uncertainties. Once CO_2 has escaped into the atmosphere, the only practical means of recovery is through photosynthesis by plants and this suggests reforestation and afforestation of suitable land not required for agriculture. Finally, improvements may be made in the efficiency of energy transmission and utilization. This includes the many measures for energy conservation that are outside the scope of this book.

Energy storage, a principal topic of this book, is relevant to electricity generation and use, in particular to electricity from renewable sources. In later chapters, we shall focus our attention on this sector of the overall energy scene, but it is important to bear in mind that the renewable electricity sector is very small at present and will probably remain so for some considerable time. Thus, in Chapter 2, we address the near-term prospects for 'clean' fuels and for improved efficiency in their use as interim measures while the world moves slowly from an economy based on fossil fuels to a sustainable economy based on renewable sources of energy.

1.6 Prospects for Renewable Energy

Fossil fuels were laid down over prehistoric millennia and may be seen as the world's 'energy capital account'. By contrast, renewable energy is produced today and represents 'energy income'. As in matters financial, it is easier to raid the capital account than work hard to earn income. But that is what ultimately the world will have to do and it is time to start preparing.

Renewable energy sources can, in principle, meet many times the present world energy demand; it is a matter of harvesting and exploiting them economically.

In broad terms, the various forms of renewable energy have a number of attractive features. They:

- enhance diversity in energy supply;
- secure long-term sustainable sources of energy;
- reduce local and global atmospheric emissions;
- open up energy supplies to rural areas and developing nations;
- create new employment opportunities;
- offer possibilities for the local manufacture of equipment;
- conserve fossil fuels for use by future generations.

This is an impressive list of potential benefits, but for such benefits to be realized – both practically and on a large scale – it is essential that renewable forms of energy should become cost-competitive. This should come about gradually, as a consequence of technology development, economies-of-scale in manufacture, and the imposition of taxes on conventional fuels in order to penalize the user for polluting the atmosphere. Since the damage to the environment of discharging gaseous wastes is not normally included in cost calculations, this is generally referred to as a 'externality'. The imposition of a discrete emission tax (carbon tax) would go some way to rectifying this omission and would thus encourage the uptake of renewable forms of energy.

There are many different types of renewable energy that in the longer term should be capable of being harvested to provide a more sustainable energy future. The principal renewable forms of energy are listed in Table 1.6, together with their category and likely time-scale of early commercial use.

Combustible materials (biomass) may be used either for direct heating applications or for electricity generation in conventional power stations. At present, the extent of biomass combustion is limited largely by the quantity of feedstock that is available and by economic considerations. There is, however, considerable scope for increasing the supply through restricting the disposal of waste as landfill and through growing energy crops. Nevertheless, any exploitation of biomass will be principally a matter of economics. Similarly, if the economic conditions were favourable, the solar heating of water and of buildings could be increased substantially and, thereby, could achieve savings in fossil fuels. In particular, hot water in lagged containers may be stored for a considerable time, as may the passive heat in a well-insulated building. The latter is an area where architectural design and building methods have key roles to play.

The global scope for renewable forms of energy in the years ahead has recently been the subject of major studies undertaken by the IEA, the World Energy Council (WEC), the United Nations (UN), and Shell International. Inevitably, these studies are based on various predictive models and energy scenarios. By way of illustration, Figure 1.13 shows the WEC projections for the contribution of renewables to electricity generation (measured in PWh; note the different scales) for the following two scenarios:[9]

Table 1.6 *Renewable energy sources and means of utilization*

Energy source[a]	Energy form	Availability
Agriculture and forestry waste	Combustion process	Now
Energy crops	Combustion process	Now
Landfill and sewage gas	Combustion process	Now
Municipal solid waste	Combustion process	Now
Direct solar (active and passive)	Heating	Now
Geothermal	Heating/electricity	Now/limited scope
Hydro power	Electricity	Now
Wind power	Electricity	Now and developing
Hydrogen/fuel cells[b]	Electricity	Now and developing
Solar photovoltaic	Electricity	Now and developing
Tidal power	Electricity	Now/limited scope
Wave power	Electricity	Medium-/long-term
Solar – thermal	Electricity	Medium-/long-term

[a] Although the distinction between 'energy' and 'power' is scientifically rigorous, in general discussion of renewable energy sources there is a tendency to use the terms interchangeable. We prefer to use 'energy' where stored energy is implied (*e.g.* geothermal, biomass, hydrogen, batteries), and 'power' where a machine or device is rated in power output (*e.g.* hydro power, wind power, fuel cells).

[b] Note, hydrogen is essentially a secondary form of energy but should be included as it is widely considered to be the ultimate conduit (the so-called 'Hydrogen Economy') between the primary renewable source and its conversion to electricity, ideally *via* a fuel cell (see Chapters 8 and 10).

- The 'current policy' scenario (Figure 1.13 (a)) assumes continuation of existing trends, *i.e.* low increases in the price of fossil fuels, steady increases in energy efficiency, modest penetration of renewables, *etc.* On this basis, the renewables shown in the scenario are predicted to contribute 2.1 PWh per year to electricity production in 2020. When traditional large-scale hydro is added (predicted to be 4.4 PWh), the contribution rises to 6.5 PWh. Note that hydro production in 2001 was 2.65 PWh.[1]
- The 'ecologically driven' scenario (Figure 1.13(b)) assumes faster and more extensive penetration of renewables that may arise from a range of measures, *i.e.* greater cost reduction, enhanced environmental concerns, higher costs of fossil fuels, *etc.* Under these assumptions, the 'new' renewables (*i.e.* excluding conventional hydro) are expected to grow at an increasing annual rate and lead to a global output of 5.2 PWh in 2020. Adding in large-scale hydro, which will not grow so rapidly in this scenario, it is expected that electricity production from renewables will be 9.2 PWh in 2020. The 'renewables intensive scenario' proposed by the UN gives 11.0 PWh by 2020.[9] For comparative purposes, it may be noted that the total world production of electricity in 2001 was 15.5 PWh, with the following percentage breakdown: coal 38.7%; hydro 16.6%; nuclear 17.1%; gas 18.3%; oil 7.5%; other renewables 1.8%. Electricity

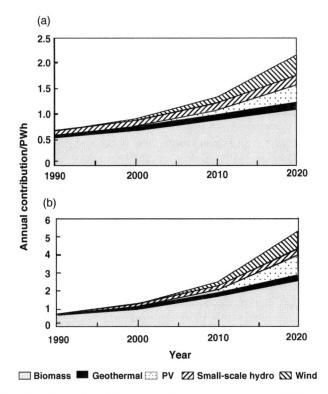

Figure 1.13 *Growth in electricity output from renewables:* (a) *WEC current policies scenario;* (b) *WEC ecologically driven scenario.*[9] 1 PWh = 10^6 GWh = 10^{15} Wh.

represented 15.6% of the world's energy consumption.[1] The use of gas to generate electricity is growing at the expense of oil and nuclear energy.

The predicted world demand for electricity in 2020 is around 20 PWh.[8] According to which scenario is adopted, total renewables (including large-scale hydro) will contribute between one-quarter and one-half of all electricity generated in 2020. If we consider only the 'new' renewables (*i.e.* excluding traditional large-scale hydro), the data given in Figure 1.13 show that roughly one-half of the electricity of this type will come from burning biomass in power stations, either directly or *via* intermediate gasification. This helps to compensate for the declining nuclear power programme, but does not reduce direct emissions of CO_2 and places far too much emphasis on biomass at the expense of non-polluting renewables. Moreover, traditional uses of biomass for domestic heating and cooking are also expected to grow as the world population expands.

The portfolio of renewables is well diversified geographically to an extent that depends on the resources available, the state of development of the various technologies, local preferences and politics, and the level and structure of energy demand. It is not anticipated that the various renewables will diffuse uniformly throughout the world. To cite an obvious example, tropical countries will tend to favour solar energy, while countries at high latitudes will prefer wind energy. Countries with excess hydroelectric potential will be inclined to use this for non-traditional uses, *e.g.* electric road vehicles and space heating. Some of the technologies, for instance, photovoltaics, are likely to be introduced *via* a series of niche applications, and will grow and diffuse gradually as experience develops and costs fall. Even intervention at government level to encourage new technologies may have comparatively little impact on the rates of their implementation and deployment. This is because good operating practice generally develops gradually and it is also necessary to establish a strong manufacturing base. Other factors that may delay the uptake of renewables are the failure of governments to have an overall policy for dealing with siting decisions, considerations of local environmental impact, delays in granting planning permission, local public opposition, non-availability of risk capital, and public reluctance to provide financial support in the early years until the technology is fully commercialised.

To summarize this discussion on energy production and use, the global demand for energy will continue to increase and the prospect of climate change will impose a major long-term constraint on fuels that emit greenhouse gases. A genuinely sustainable energy system that promotes sustainable economic growth with an improved standard of living for all does however present a major challenge. Consideration must be given to new methods of energy conservation, to the more efficient use of energy, to the introduction of cleaner fossil fuels and renewable forms of energy, and to the time-scales over which these changes may be effected economically. Furthermore, these initiatives need to take account of national and local government legislation, the operational procedures of industry, and the predilections of consumers. Clearly, sustainability is a complex problem in an increasingly complex world. To guide us on the road to a more secure future, the following must become the main ingredients of the energy mix: (i) energy conservation, especially in the heating, lighting, and cooling of buildings; (ii) reduction in petroleum consumption by road transportation; (iii) aggressive efficiency of energy use in all other situations; (iv) innovation in the way electrical power is presently generated and supplied, with an emphasis on distributed generation and storage; (v) continued integration and greater penetration of renewables; (vi) development of technologies to enable the continued use of fossil fuels with greater efficiency until the transition to sustainability is completed.

These, then, are the challenges in energy production and use that the world faces. The further and faster we can travel along this road, the nearer we shall be to ensuring a sustainable future for the generations yet to be born.

1.7 References

1 *Key World Energy Statistics from the IEA*, 2003 edn, International Energy Agency, Paris, France, 2003.
2 *Key World Energy Statistics from the IEA*, 2002 edn, International Energy Agency, Paris, France, 2002.

3 *World Energy Outlook 2001 Insights*, International Energy Agency, Paris, France, 2001.

4 *UK Energy in Brief*, UK Department of Trade and Industry, London, UK, November 2000 and July 2003.

5 *BP Statistical Review of World Energy 2003*, June 2003; see: www.bp.com.

6 M. King Hubbert, *Scientifc American*, 1971, **225**(4), 61–70.

7 Shell Company website: www.Shell.com (November 2001).

8 *World Energy Outlook 2000*, International Energy Agency, Paris, France, 2000.

9 *New and Renewable Energy: Prospects in the UK for the 21st Century*, UK Department of Trade and Industry, London, UK, March 1999.

CHAPTER 2

Clean Fuels

Since it is clear that the replacement of all fossil fuels by renewable sources of energy is going to take a very long time, interim measures must be devised to utilize the remaining reserves in as efficient a manner as possible and to minimize the pollution that will arise from their combustion. In this Chapter, we discuss the actions that are already under way and the future prospects for developing cleaner forms of fossil fuels.

By 'clean fuels', we mean clean-burning fuels that emit less pollution to the atmosphere. Among common pollutants are sulfur dioxide formed from sulfur in the fuel, nitrogen oxides (NO_x), produced during combustion of hydrocarbons in air, and the 'greenhouse gas' carbon dioxide that is an inevitable consequence of burning any hydrocarbon fuel. Other possible pollutants are carbon monoxide from incomplete combustion and unburnt hydrocarbon vapours. With correct control of the ignition conditions, the amounts of nitrogen oxides, carbon monoxide and unburnt hydrocarbons should be kept to a minimum. Finally, it should be noted that the combustion of waste materials and plastics can give rise to dioxins, which are highly toxic, and to chlorocarbons, which are damaging to the stratospheric ozone layer that protects the Earth from the sun's harmful ultraviolet rays.

Sulfur may be removed from liquid fuels by catalytic processes at the refinery. For solid fuels, the only practical process may be scrubbing of sulfur dioxide from the combustion gases. In the case of internal-combustion engines, the residual nitrogen oxides, along with unburnt hydrocarbon and carbon monoxide, are effectively removed from the exhaust by catalytic converters, which are universal on modern petrol-engined cars.[*]

At present, there is no practical way of collecting carbon dioxide from combustion processes and so it is desirable to restrict its formation. One way to accomplish this is to burn fuels with as high a hydrogen-to-carbon ratio as possible. From this standpoint, the ideal fuel is pure hydrogen. The next best, and most immediately useful, is natural gas (methane, CH_4) that has four hydrogen atoms

---

[*] 'Petrol' is the UK term for a light hydrocarbon liquid fuel for spark-ignition engines and is a blend of different components obtained by refining crude petroleum ('oil'). Other terms for such fuel are 'gas', 'gasoline', and 'motor spirit'. Diesel engines, which achieve ignition by the heat of compression of the air charge use heavier and less volatile hydrocarbons as fuel.

to every carbon. After methane comes ethane (C_2H_6) – and then propane (C_3H_8) and butane (C_4H_{10}) that, either separately or together, form liquefied petroleum gas (LPG). Next are the paraffin hydrocarbons of composition C_5H_{12}, C_6H_{14}, C_7H_{16} and C_8H_{18} that along with certain performance-enhancing additives, constitute 'petrol'. When crude petroleum is separated into its component parts by distillation, it is the higher boiling fractions that constitute the 'middle distillates' and 'fuel oils' and these have progressively lower hydrogen-to-carbon ratios, see Figure 2.1.

The world has large reserves of lower-grade fossil fuels (tar sands, asphalts, oil shales, bitumens and coals), as discussed in Section 1.4, Chapter 1. These are composed of more complex organic compounds with still lower hydrogen-to-carbon ratios. Most are intrinsically 'dirty' fuels with relatively high levels of included elements, especially sulfur and nitrogen that give rise to toxic gases on combustion of the fuels. The challenge now facing the world is to develop economic processes for the conversion and utilization of these raw materials as clean fuels.

2.1 Clean-burning Fuels

Natural Gas

From the mid 20th century onwards, natural gas has increasingly been identified as a valuable fuel, not only for heating purposes but also for chemical manufacture,

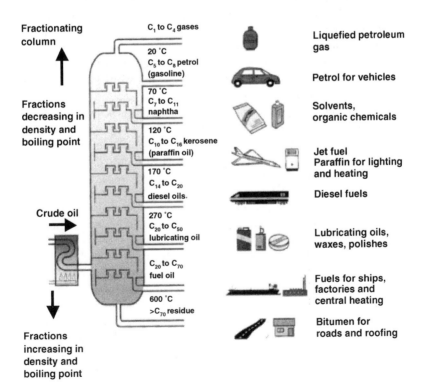

Figure 2.1 *Distillation products of crude petroleum.*

and gas fields have been exploited in many countries. The fields may be inland or off-shore on the Continental Shelf. The consumption of natural gas varies greatly from country to country. As a percentage of total energy consumed, Russia is first among the major countries, with over 50% of its energy needs being met by natural gas. In the UK, the figure is about 40% and in the USA about 25%.

Much natural gas is exceptionally pure, with very low concentrations of sulfur-containing compounds. It consists predominantly of methane (CH_4) with lesser amounts of ethane (C_2H_6). Small quantities of helium, formed by the radioactive decay of uranium or thorium, are found in some gas wells. This constitutes a commercial source of helium, but this gas is inert and therefore has no significance for natural gas as a fuel.

In recent years, major pipelines have been constructed to convey natural gas from the oil/gas fields to centres of population where it is needed. This method is being used to export natural gas from the Russian Federation (Siberia) to Western Europe, and from Algeria to Southern Europe. In the UK sector of the North Sea, the oil fields are situated predominantly off the coast of Scotland. The gas fields lie to the south and pipes have been laid to a central collection point in Norfolk, from where the gas is distributed across the country.

Natural gas is not only a clean form of energy, but also its distribution by pipeline has only a minor impact on the landscape. Once the pipes have been laid in trenches and backfilled, the distribution system is invisible and the land may be restored to agricultural or other use. Contrast this with electricity distribution where there is the visual intrusion of pylons and transmission lines across the countryside.

Often, there is natural gas associated with petroleum in oil fields. Where these fields are remote from centres of population, as in much of the Middle East, it has been traditional to flare this gas to waste for lack of a suitable market. The flaring of natural gas is an exceedingly wasteful practice, with huge quantities of energy being lost and carbon dioxide formed. In recent years, plants have been constructed to harness this remote gas by liquefaction (liquefied natural gas, LNG; the boiling point of methane is $-164\,°C$) for export to centres of population in ships equipped with insulated, cryogenic tanks. This has necessitated the construction of port facilities close to the liquefaction plant, pipelines to convey the gas from the fields to the port, and a fleet of cryogenic tankers (Figure 2.2). New liquefaction plants and port terminals in Nigeria and Oman have come on-stream. A further major facility is being built on Sakhalin Island, off the Russian Siberian coast, and will be capable of producing 4.8 Mt per year to supply the growing markets in Asia and the west coast of the USA. These plants will considerably increase world trade in LNG and end the routine flaring of gas from the associated oil wells.

Where the gas fields are off-shore, on the Continental Shelf, or where port facilities are not available, it is possible to build floating liquefaction plants (Figure 2.3). These will load the LNG directly onto ships that convey it to market. Shell has plans to construct such a floating LNG platform in the remote ocean some 450 km off the north coast of Australia and 150 km from East Timor. This facility would be able to process and store more than 200 000 cubic metres of LNG[*] that

---

[*] 200,000 m^3 LNG $\simeq$ 90,000 t LNG $\simeq$ 0.11 Mtoe

Figure 2.2 *LNG tanker.*

would be periodically off-loaded onto LNG bulk carriers moored alongside. Such an operation combines three well-established technologies, namely: floating production platforms for petroleum, LNG plant design, and LNG shipping. Together, these constitute a complex and capital-intensive chain from gas reservoir to customer in a far-off land.

In 2002, the total world trade in LNG was 150 billion cubic metres of gas, which amounts to 135 Mtoe.[1] More than half of the total world exports go to Japan, with Korea and France also being significant importing nations. The principal exporting countries are Indonesia, Algeria, Malaysia, Qatar, and Australia. The development of the world trade in LNG between 1978 and 1998 is shown in Figure 2.4. In 1998, LNG accounted for only 4% of world gas consumption, but was one-quarter of internationally traded gas.

A 60% growth in the Asia–Pacific market for LNG is predicted over the period 2000–2010 and new import facilities are being constructed throughout the region.[2] There is also scope for increased imports into Europe and the USA and huge

Figure 2.3 *Schematic of off-shore liquefaction of LNG.*
(Courtesy of Shell)

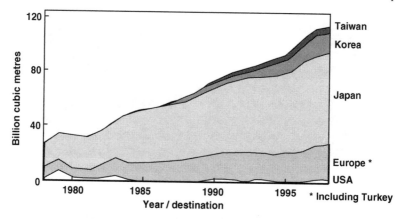

Figure 2.4 *World import trade in LNG 1978–1998*[2].
(Courtesy of Shell)

potential markets in the People's Republic of China and India have hardly been tapped. All this activity stems from the increased availability of natural gas, its clean-burning properties as a fuel, and the world's insatiable demand for more energy.

An alternative option to deal with natural gas from fields that are remote from the coast (and from liquefaction plants), and that are uneconomic or difficult to access *via* pipelines, is to process the 'stranded' gas into clean liquid fuels or other chemicals that can easily be transported by tanker. The gas-to-liquids (GTL) option is the subject of much present interest, particularly by the petroleum industry. The first stage in a GTL process is the production of a mixture of carbon monoxide and hydrogen by reacting methane with steam and air (or oxygen) over a nickel-based catalyst, *i.e.*

$$CH_4 + H_2O \xrightarrow[\text{Ni catalyst}]{900\,°C} CO + 3H_2 \qquad (2.1)$$

$$2CH_4 + O_2 \longrightarrow 2CO + 4H_2 \qquad (2.2)$$

The mixture is known as 'synthesis gas' (or 'syngas') because it may be used for the preparation of a range of commercial products that include hydrogen, ammonia, methanol, and various organic chemicals. The reaction of methane with steam, Equation 2.1, is known as 'steam reforming' and is discussed in more detail in Section 2.5 below.

The GTL process is completed by Fischer–Tropsch synthesis (see Equations 2.7 and 2.8, later) in which the hydrogen and the carbon monoxide in the synthesis gas are recombined to form a mixture of hydrocarbons and alcohols. Issues that must be resolved in the development of an off-shore GTL process include: (i) scaling the size of the operation to meet the dimensions of a floating facility; (ii) maintaining economic credibility in terms of the capital cost per barrel of liquid fuel produced per day, particularly given that an off-shore GTL plant will probably have a lower production throughout; (iii) determining whether it is more advantageous to make the simplest chemical (probably methanol) on board the platform and transport it for further processing on land; (iv) restricting environmental damage through, for example, underground sequestration of carbon dioxide emissions.

The momentum behind the world-wide expansion in the use of natural gas, including LNG, is enormous in political, financial, industrial and environmental terms, and it seems inevitable that this will be a major source of Clean Energy throughout much of the 21st century. To replace natural gas by renewable forms of energy on this time-scale will be a monumental challenge. It should be noted, however, that Western Europe will become increasingly dependent upon imported natural gas as the reserves from the Continental Shelf are depleted. Imports will come from the Russian Federation and possibly Algeria by pipeline, and from elsewhere in the form of LNG. By about 2020–2030, it is likely that only Norway, among the Western European nations, will have substantial remaining supplies of natural gas. It is widely believed, as discussed in Section 1.4, Chapter 1 that there are huge reserves of natural gas in the ocean beds. These are in the form of a crystalline hydrate, which is produced under intense pressure. Whether or not the gas is ultimately exploitable remains to be seen.

One possible new application for natural gas is as a fuel for motor vehicles. For this purpose, the gas is compressed to high pressure in a steel cylinder that is carried in the boot (trunk) of the car – or elsewhere on-board a commercial vehicle. Although extensive trials have been conducted, it has been found less convenient to use compressed natural gas (CNG) as opposed to liquefied petroleum gas (LPG, see below). The gas cylinders are heavier and occupy much greater volume. There may, however, be locations where natural gas is more readily available than LPG and is possibly cheaper. Natural gas also liberates less carbon dioxide per km travelled. In 2001, there were over 2-million natural gas vehicles (NGVs) in the world, with approaching 5000 refuelling stations. The most recent development is that of vehicles fuelled with petroleum synthesized from natural gas (GTL technology). GTL fuel produced by Shell is being evaluated in a London bus and there are other trials in Berlin, Tokyo, and California. If these prove successful and cost-effective, it seems likely that in the years ahead, road transport will prove to be a major new market for natural gas.

Liquefied Petroleum Gas

LPG is a relatively new motor fuel that is starting to gain prominence. It consists of propane (C_3H_8, boiling point $-42\,°C$), butane (C_4H_{10}, boiling point $-0.5\,°C$), or a mixture of the two. At normal atmospheric pressure and temperature, these two hydrocarbons are gases, but they may be liquefied under pressure and stored as liquids in pressurized containers. Because of the comparatively low vapour pressure of LPG, light-gauge steel containers can be employed and also more energy may be accommodated within a given volume. Moreover, no compressor is needed at the point where fuel is dispensed. Thus, in terms of energy content and convenience, LPG is a more acceptable fuel for road vehicles than CNG. The lower and upper flammability limits of both gases in air are around 2 and 10 vol.%, respectively.* The calorific value of LPG per unit volume of gas is about 2.5 times that of natural gas.

In liquid form, both propane and butane have long been used as fuels for cooking and heating in situations where natural gas is unavailable. They may also be used for refrigeration, air-conditioning, or as a propellant in spray cans. There are still many commercial buildings and private residences that are remote from the gas grid. These often have large LPG tanks, which are refilled from road tankers. The choice of propane or butane is determined in part by the ambient temperature. In cold climates, propane is more suitable, since its vapour pressure is higher, whereas in warm climates butane is preferred to maintain the operating pressure at a more acceptable level. These gases are also employed where a portable source of heat is required, *e.g.* for cooking and space heating in caravans and boats. LPG stored in pressurized containers is more expensive than piped natural gas and the latter is generally preferred where it is available.

From an environmental viewpoint, LPG is a relatively clean-burning fuel that causes less pollution than either petrol or diesel. This is the prime reason for advocating its use in internal-combustion engines. It is said to produce 90% less NO_x and 90% fewer particulates (soot) than diesel engines and the exhaust gases are odourless. The reduction in emission of ultra-fine particles, which are so damaging to health, can be as high as 99%.[3] LPG engines also have advantages in terms of reduced formation of carbon monoxide, compared with petrol engines, and much lower noise levels compared with diesel engines. With respect to carbon dioxide formation, per unit of energy, LPG is intermediate between natural gas and petrol or diesel.

The attributes of low exhaust emissions are especially important in cities and this is why many public-service passenger vehicles (buses and taxis) have switched from diesel to LPG. This is particularly so in parts of continental Europe (Italy, Austria, France, the Netherlands). South Korea and Australia also have shown enthusiasm for LPG fuel. Altogether, there are more than 8 million LPG vehicles world-wide. Vienna has operated buses on LPG successfully for over 30 years and now its entire bus fleet of over 400 buses runs on LPG. The UK has been relatively slow to switch, but at present has some 100 000 LPG vehicles that include small numbers of buses operating in several cities.[3] It is predicted that this number will rise to 250 000 by the end of 2005. The conversion to LPG does of course depend upon installing storage tanks and pumps at garages and service stations. These dispensers are essentially similar to those used for conventional petrol, except that the storage tanks are pressurized and there is a sealed connection between the hose and the vehicle's tank when refuelling. To date, the safety record world-wide has been very high. In part, this may be attributed to the robust fuel storage tank, which is far stronger than the normal petrol tank and is much less likely to rupture in a traffic accident.

---

* Below the lower flammability limit, there is an insufficient percentage of fuel in air to maintain combustion. Above the upper limit, there is too much fuel and insufficient air.

The rate of conversion of private cars to LPG has been slower than that of commercial vehicles and buses. There are two reasons for this. The first is a classic 'chicken-and-egg' situation. Nobody will buy an LPG vehicle unless the fuel is widely available, but there is no incentive for the petroleum companies to set up the required facilities until there are customers for the fuel. LPG pumps are gradually being installed at some of the major service stations, especially for use by commercial vehicles that have large storage tanks. For private cars, the interim solution is to have 'dual-fuel' vehicles that can operate on either petrol or LPG. This extends the range between refuelling and makes allowance for the fact that LPG is not yet universally available. A car is converted to dual-fuel operation by installing a LPG tank in its boot, with a switch on the dashboard to change from one fuel to another. This permits the use of LPG where available, and petrol in more remote areas.

The second major impediment to the use of LPG by the private motorist lies in the relatively higher cost of LPG vehicles and, similarly, in the cost of converting existing vehicles to dual-fuel use. If governments wish to encourage the wider use of LPG, then they must provide incentives for the car owner to offset this extra cost. This may be in the form of lower duties or taxes on LPG compared with petrol or diesel fuels, and/or a grant towards the cost of a LPG vehicle or the conversion of an existing vehicle to dual-fuel operation. Vehicle manufacturers and petroleum companies also have a role to play. With increased production of LPG vehicles, the unit price should fall through economies of scale. Moreover, the widespread installation of LPG pumps at service stations will encourage motorists to convert to this fuel. A precedent was set by diesel engines. Originally confined mostly to commercial vehicles, diesel engines became more popular as the vehicle price fell through the mass production of small diesel engines, and as the fuel became more widely available. Further stimulus was provided by those governments that imposed a lower rate of tax on diesel fuel than on petrol. Some countries also tax LPG at a lower rate than petrol.

Another option that is open to governments to encourage greater uptake of LPG vehicles is through the enactment of clean air legislation. A notable example of such action has taken place in California, where government authorities have been instrumental in the introduction of vehicle exhaust catalysts and have encouraged the development of electric and hybrid electric cars. If regulations are stringent enough, vehicle owners will be forced to abandon petrol and diesel engines in favour of LPG or electric traction. This legislative option could be either an alternative to fiscal incentives or a complementary measure. Under the stimulus of the Californian initiative, there has already developed a world-wide interest in electric vehicles and hybrid electric vehicles (see Sections 10.2 and 10.4, Chapter 10, respectively).

2.2 Developments in Petroleum Technology

Petrol consists predominantly of saturated hydrocarbons (paraffins) from the C_5H_{12} to C_8H_{18} range. The composition may vary somewhat depending upon the climate. In cold weather, *i.e.* below $0\,°C$, C_4H_{10} hydrocarbons (butane and isobutane) may be added to give greater volatility. Conversely, in hot climates it may be desirable to eliminate the C_5H_{12} content since too high a volatility may lead to vapour locks. In parts of the world that experience extremes of climate, such as Canada and Northern USA, the composition of petrol may vary with the season.

Internal-combustion engines are subject to the phenomenon of 'engine knock'. This occurs when the last portion of the fuel–air charge, *i.e.* that farthest from the spark plug, self-ignites with explosive violence and sets up a high-frequency wave that causes vibration of the combustion chamber. Knocking can be avoided by an engine design that increases turbulence in the combustion chamber and, thereby, increases flame velocity. Alternatively, a fuel with high performance may be used to ameliorate the problem. The 'anti-knock performance' of petrol is measured by its octane number, where pure iso-octane (iso-C_8H_{18}) has a value of 100. With the development of high-compression engines, it was necessary to increase the octane number in order to prevent engine knock. Some petroleum companies improved the rating by adding aromatic compounds such as benzene and toluene. More commonly, this was achieved by the addition of lead tetra-ethyl, an anti-knock agent. A sizeable chemical industry developed to manufacture this compound which when added to petroleum distillate raised the octane rating to around 100 and made it suitable for use in high-compression engines.

The practice of using lead tetra-ethyl continued for 30–40 years until concern began to be voiced over the emission of lead compounds from vehicle exhausts as lead can damage the human brain, particularly in children. In the 1990s, in response to this concern, the manufacture and use of lead compounds in petrol was gradually phased out and other, cleaner anti-knock agents (such as methyl tertiary butyl ether) were employed as octane improvers. High-performance engines were also developed that operate satisfactorily with petrol of octane rating 95–96, *i.e.* premium grade fuel. Regular grade petrol has an octane rating of 90–91. For older vehicles, which need fuel with a higher octane rating to operate well, special 'Lead Replacement Petroleum' (LRP) is on sale in the UK.

Much progress has also been made with suppressing other pollutants from vehicle exhausts. This work was stimulated by the increasing levels of atmospheric pollution that developed in many cities of the world. Degradation of air quality is particularly prevalent when there is a combination of: (i) a city situated in a basin surrounded by hills, and (ii) a natural temperature inversion, which leads to entrapment of pollutants in the basin. As the fleet of internal-combustion-engined vehicles rises, the pollution worsens and people can develop symptoms of running eyes, coughing, chest pains, *etc.* The classic example has been Los Angeles that as early as the 1940s, began to experience significant atmospheric pollution ('smog'). Other cities to be likewise affected have included Athens, Rome and Tokyo.

Automotive exhaust consists, typically, of carbon dioxide and water, together with unburnt hydrocarbons, partially combusted products (carbon monoxide and aldehydes), and oxides of nitrogen (designated collectively as 'NO_x'). Under conditions of strong sunlight, a photochemical reaction occurs between hydrocarbons and NO_x to give rise to ozone (O_3) and other undesirable products such as peroxyacetyl nitrate, a strong lachrymator irritant. The overall phenomenon is known as 'photochemical smog'; it is a different type of smog to that formed in earlier years from a combination of smoke and fog, when there were fewer cars and hence less emissions of NO_x. Engine modifications were made to provide some reduction of emission levels, in particular the development of 'lean burn' engines in which the air-to-fuel ratio was above stoichiometric and also the recycling of exhaust gases from the crankcase (note, in older engines, there was a crankcase 'breather' to liberate to the atmosphere any exhaust gas that passed the piston rings and finished up in the crankcase). The development of oxygen sensors in the 1970s facilitated control of the air to fuel ratio. These were placed in the gas, close to the exhaust manifold, to provide feedback electronic control of the carburettor. It was soon discovered, however, that these engineering improvements were insufficient to meet the US Clean Air Legislation that was introduced in 1970 and mandated a 90% reduction in exhaust emissions. There was no alternative but to fit a catalytic device to treat the gas to remove the pollutants.

Exhaust Catalyst Technology

In the late 1960s, the Johnson Matthey Company in the UK was developing platinum-group catalysts to remove gaseous pollutants from industrial effluents, such as NO_x from the tail gas of nitric acid plants. It was found that the NO_x could be chemically reduced by methane (CH_4) over such a catalyst and that rhodium was particularly effective for this reaction. The Company, therefore, directed its research effort to address the new problem of developing catalysts for the treatment of vehicle exhausts. Initially, two catalysts were employed. The engine was run slightly rich to suit the first catalyst, platinum–rhodium supported on a ceramic matrix, which reduced NO_x to nitrogen. The chemical reactions may be represented as:

$$4C_nH_m + (8n + 2m)NO \longrightarrow (4n + m)N_2 + 4nCO_2 + 2mH_2O \qquad (2.3)$$

$$2NO + 2CO \longrightarrow N_2 + 2CO_2 \qquad (2.4)$$

Air was introduced between the first and the second catalyst. The latter (platinum–palladium) then effected the oxidation of the remaining C_nH_m and CO to CO_2 and H_2O. It was important that the first catalyst did not reduce further the nitrogen to ammonia (NH_3) as this would be oxidized back to NO_x over the second catalyst. Catalytic converters were introduced around 1975 and were fitted to cars in the USA and Japan.

Subsequent research at Johnson Matthey led to the development of the so-called 'Three-way Catalyst' to control simultaneously all three pollutants, *viz.*, C_nH_m, CO and NO_x. This was a major advance that was facilitated by increasingly sophisticated fuel-management systems such as direct fuel injection. The problems to be overcome included not only pollution control under steady-state operation, but also the development of a catalyst that could achieve rapid heat-up on starting and maintain good stability for the life of the vehicle. Moreover, exhaust catalysts are rapidly poisoned by lead and this, in addition to the above-mentioned health hazard, has been a further reason for the demise of lead additives in petrol. Today, more than 50 million cars per year are fitted with exhaust catalysts. Meanwhile, to meet the ever-increasing demands of environmental legislation, the research on catalysts is continuing.

The treatment of pollutants from diesel engines has lagged behind that for petrol engines. The most serious pollutant from diesel engines is particulates (soot) and traps have now been developed for the larger diesel engines used in commercial vehicles and buses. Johnson Matthey again pioneered this field with their 'continuously regenerating trap (CRT)$^{®}$, which collects and burns the soot in diesel exhaust and also removes carbon monoxide and unburnt hydrocarbons. Despite this progress, however, most light vehicles with diesel engines still do not have particle traps.

Diesel engines produce less nitrogen oxides than petrol engines, which is why their exhaust systems do not corrode nearly so rapidly. Nevertheless, new catalyst technology is being evolved to meet the challenge of future heavy-duty diesel emission legislation. To remove NO_x, it will be necessary to reduce it chemically to nitrogen (as with petrol vehicles), and catalytic converters combined with soot traps are being developed for this purpose. Light diesel engines, as used in cars and vans, have been greatly improved in terms of reduced emissions. To date, it has not been required to fit particulate traps or catalytic converters to these vehicles. Future legislation may change this position.

It should be noted that since diesel vehicles are more economical than petrol counterparts (typically, around 20% more km per litre of fuel, for engines of similar power), their emissions of carbon dioxide are correspondingly less. In the UK, vehicles are now taxed on the basis of such emissions, which are measured in $g\,km^{-1}$. This legislation favours small cars and, particularly, diesel cars. In France, diesel fuel is substantially cheaper than petrol.

Another undesirable component of both petrol and diesel fuel is small quantities of sulfur-containing compounds. These cause the production of minor amounts of sulfur dioxide (SO_2), another acid pollutant. This gas is also detrimental to the life of the catalytic converter. For both these reasons, the major petroleum companies have developed and brought to the market ultra-low sulfur ('green') petrol and diesel, which is the latest advance in petroleum technology. The action to reduce dramatically sulfur levels in motor fuel has also been made in response to approved regulations in the USA and Europe (Figure 2.5). Proprietary additives can also inhibit corrosion of the fuel system and, in the case of diesel, can improve combustion, which leads to greater fuel efficiency and a reduction in white smoke and noise.

At present in Europe, 5-seater saloon (sedan) cars equipped with direct-injection diesel engines are capable of returning around 20 km per litre. In 1993, the US government launched its Partnership for a New Generation of Vehicles (PNGV) with a target of delivering by 2004, production-ready vehicles with a fuel economy of 34 km per litre.* It was soon realized that this was only likely to be achieved with direct-injection diesel engines. Good progress has been made by the automotive companies towards achieving the technical target, but this has not

---

* 34 km per litre = 2.94 litres per 100 km = 96 miles per UK gallon = 80 miles per US gallon.

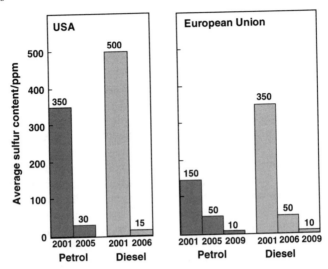

Figure 2.5 *USA and European Union regulations and schedule for introduction of maximum sulfur levels in fuels for road vehicles.*

been matched by any government-backed initiatives to wean US motorists from petrol-engined cars and sports utility vehicles that consume large quantities of cheap fuel.

2.3 Clean Coal Technologies

Coal is generally perceived as being a 'dirty' fuel. This is not only in terms of its handling, but also on account of the volatile matter liberated on burning. Among the products of combustion are aliphatic and aromatic hydrocarbons, which include carcinogenic poly-cyclic compounds, partially oxidized hydrocarbons, toxic gases such as SO_2 and NO_2, tar, soot, and smoke. After the dreadful urban smogs of the 1940s and early 1950s, clean air legislation was introduced into various countries. In Britain, the burning of bituminous (soft) coal on open fires was banned and householders were required to switch to 'smokeless fuel', such as anthracite (hard coal) of much lower volatile content. This legislation, together with the changeover to natural gas for space heating, completely transformed the urban environment. (There followed a period when urban air quality deteriorated again as a result of exhaust pollution from the growing number of vehicles, before the introduction of catalytic converters for the clean-up of vehicle exhaust.)

Around half of all coal mined today is used in electricity generation; see Section 1.4, Chapter 1. This includes both lignite and bituminous coals. The principal environmental problem here is the release of sulfur dioxide, as discussed further in Section 3.2, Chapter 3. As higher-grade fuels, such as natural gas and petroleum become depleted, it is expected that the world will turn again to its huge reserves of coal as a source of energy. Fortunately, coal need not be a dirty fuel and numerous clean technologies exist, or are being developed, to utilize coal on an industrial scale. These technologies fall broadly into two categories: coal gasification and coal liquefaction.

Coal Gasification

The gasification of coal has long been practised. When heated in a restricted supply of air, coal or coke is converted to carbon monoxide that is heavily diluted by nitrogen. This is a low-grade fuel known as 'producer gas' and has been employed in industry as a reducing atmosphere. Because of the low calorific value of producer gas, transportation costs are an important factor and thus it is mainly produced close to where it is needed. During World War II, when petrol was in short supply, some buses in the UK were converted to operate on producer gas. The bus towed a trailer equipped with a small anthracite or coke oven and the gas was stored in an inflatable bag carried on top of the bus. Conventional petrol engines were employed, although their performance was heavily degraded.

When heated coal or coke is reacted with steam, the 'water-gas reaction' occurs:

$$C + H_2O \longrightarrow CO + H_2 \tag{2.5}$$

Water-gas found widespread use before World War II in producing hydrogen for the manufacture of ammonia *via* the Haber process. Today, most hydrogen for this purpose is obtained from synthesis gas, which is made from natural gas by steam reforming (Equation 2.1); it is a cleaner and cheaper option. The water-gas reaction is highly endothermic (heat absorbing) and thus soon ceases unless heat is supplied. Conversely, the combustion of coal or coke in air is highly exothermic (heat evolving). It is, therefore, usual to pair off the two reactions so as to balance the heat evolved with that absorbed. The two reactions may either be conducted consecutively, in short bursts or, more usually, simultaneously by feeding a mixture of air and steam to the heated bed. The resulting gas is a mixture of CO, H_2, CO_2, and N_2. A gas of higher calorific value can be obtained by using oxygen rather than air, but for many applications this is not economic. From time to time, proposals have been advanced for the underground gasification of coal as a more efficient and safer alternative to deep mining, but such an approach runs into similar problems, namely, air gives a gas that is too dilute to be useful, while oxygen is too expensive.

The gas produced by the water-gas reaction may be upgraded in terms of hydrogen content by the 'water-gas shift reaction' (note, synthesis gas, Equations 2.1 and 2.2 can be similarly upgraded). The gas is reacted with steam over a catalyst that converts carbon monoxide to carbon dioxide and increases the amount of hydrogen, *i.e.*

$$CO + H_2O \longrightarrow CO_2 + H_2 \tag{2.6}$$

The carbon dioxide is then removed from the gas by scrubbing to leave a mixture of predominantly hydrogen and nitrogen. This enhances the usefulness of water gas in the synthesis of ammonia.

By adjusting the fuel, the gases used and the operating conditions, it is possible to tailor-make a gas of desired composition. Two examples are: (i) 'chemical synthesis gas', a mixture of H_2 and CO used as chemical feedstock for the production of ammonia (as discussed above), methanol and other organic compounds; (ii) 'synthetic natural gas' (SNG), which is largely methane and is produced by the reverse of Equation 2.1. The latter gas may also be made directly by the reaction of hydrogen with coal.

The process engineering of coal gasification is quite complex and several large-scale processes have been developed. One of the best known is the Lurgi gasifier. In this equipment, a bed of crushed coal moves downwards through the vertical reactor to meet an ascending flow of gas, a mixture of oxygen and steam, pressurized to about 3 MPa (30 atm.). As the coal descends, the temperature rises progressively. Pyrolysis takes place in the top zone, then gasification and, finally, combustion of the char in the bottom zone where the temperature approaches 1200 °C. The calorific value of the resulting gas depends upon whether pure oxygen is used, or a mixture of oxygen and air.

The integrated gasification combined-cycle (IGCC) process for electricity generation from coal has the potential to increase thermal efficiencies to over 50% with greatly reduced emissions of greenhouse gases (see Figure 3.6, Chapter 3). Pulverized coal is reacted with oxygen and steam at temperatures up to 1850 °C to produce an impure mixture of hydrogen and carbon monoxide. Hydrogen sulfide (formed from sulfur impurity in the coal) is removed by adsorption in a polyethylene-based solvent to leave a mixture of hydrogen and carbon monoxide (synthesis gas) (Figure 2.6). The mineral matter in the coal forms a slag at these high temperatures and is removed from the gasification reactor. The gas is further

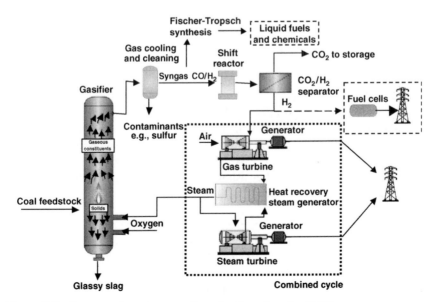

Figure 2.6 *Concept of a poly-generation plant based on the IGCC process.*

reacted with additional steam (shift reaction, Equation 2.6) to convert the gas to a mixture of hydrogen and carbon dioxide. The carbon dioxide may then be separated in a form suitable for sequestration (*e.g.* in geological structures), while the hydrogen is used for power generation in gas and steam turbines, and/or in fuel cells. The ultimate aim is 'zero-emissions power from coal'. To this end, in February 2003, the USA government announced FutureGen – a 10-year research project to build the world's first coal-fired station to generate electricity and hydrogen with zero emissions. This will employ coal gasification technology to produce an equivalent gross electricity output of 275 MW with the capture of carbon dioxide and its sequestration in deep underground geologic formation(s). Such facility could also be used for the production of synthetic liquid fuels and chemicals, a so-called 'poly-generation plant'.

Another promising process is fluidized-bed combustion in which finely powdered coal, mixed with limestone, is fluidized by a stream of upward moving air. This allows control of the temperature of combustion and thereby reduces the formation of nitrogen oxides. The sulfur in the coal is oxidized to sulfur dioxide and this reacts with the limestone to form calcium sulfate. Between 90 and 95% of the sulfur dioxide is removed from the exhaust gases.

Ultra-clean Coal

For coal to be acceptable as a direct feed into gas turbines, it must maintain strict levels of purity, that is, it must contain very little mineral matter. Otherwise, damage to the turbine blades will occur through the following processes.

- Erosion: caused by the abrasive action of ash particles produced either by direct combustion or gasification of the coal. Studies have shown that the rate of erosion decreases with decrease in particle size, and approaches zero for particles smaller than 5 μm.
- Corrosion: chemical attack of the protective oxide film on the blade surface by elements ('impurities') in the coal such as sodium and potassium (as sulfates) and vanadium (as V_2O_5).
- Fouling: build-up of deposits of resolidified molten ash. With the performance of gas turbines being raised to higher and higher efficiencies (Figure 3.6, Chapter 3), the temperature at which the gas enters is increasing (*i.e.* from 1150 °C for 44% efficiency to 1500 °C for 52% efficiency), even to above the melting point of the ash. Molten ash will adhere to the slightly cooler surfaces of the blades and hence interfere with operation of the turbine.

In order to minimize these three problems, research is being conducted at CSIRO Energy Technology, Australia, to develop 'ultra-clean coal'. The process involves purifying raw coal by chemical and hydro–thermal leaching to produce a fuel that has much reduced levels of elemental impurities – particularly sodium, potassium, vanadium – as well as a very low ash content (0.1 wt.% compared with 8 wt.% for untreated coal) with a particle size ($<5\,\mu$m and a fusion temperature of 1500 °C.

These exacting targets have been met in the laboratory and the project has moved to the pilot-plant stage.

Coal Liquefaction

Commercial plants for the manufacture of petrol from coal were operated in Germany during World War II and, later, in South Africa (the Sasol process). In the 1970s, there was a plant operating at 5000 t coal per day in South Africa for the domestic production of petrol. The conversion process of coal to so-called 'syncrude' requires the addition of hydrogen to the coal so as to raise the hydrogen-to-carbon ratio. There are three basic methods (Figure 2.7), each of which has been investigated on a significant scale to give, typically, three barrels of petroleum from one tonne of coal. These methods are as follows:

- Direct hydrogenation at high temperatures (450 °C) and pressures (13–27 MPa) in the presence of a catalyst.
- Solvent refining, which involves dissolution of the coal under pressure in a solvent (anthracene oil), followed by separation of the ash and catalytic hydrogenation and hydrodesulfurization of the solution.
- Conversion of coal to water gas ($CO + H_2$), followed by the production of liquid hydrocarbons and alcohols by catalytic synthesis in the presence of added hydrogen according to the general equations:

$$nCO + (2n + 1)H_2 \xrightarrow[\text{Ni, Co or Th catalyst}]{200\,°C} C_nH_{2n+2} + nH_2O \tag{2.7}$$

$$nCO + 2nH_2 \longrightarrow C_nH_{2n+1}OH + (n - 1)H_2O \tag{2.8}$$

This catalytic process for making fuels from carbon monoxide and hydrogen was invented in 1923 by Franz Fischer and Hans Tropsch. The resulting hydrocarbon

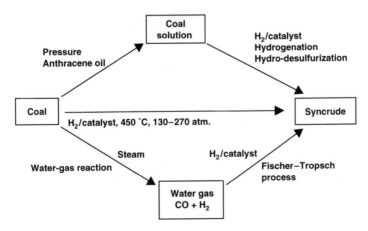

Figure 2.7 *Possible routes for the synthesis of petroleum from coal.*

mixture is separated into a higher boiling fraction for diesel engines and a lower boiling fraction for petrol engines. The latter fraction contains a high proportion of straight-chain hydrocarbons (alkanes) and has to be reformed ('cracked') by catalytic action into branched-chain alkanes for use in motor fuel. The process used at the Sasol plant in South Africa is based upon Fischer–Tropsch synthesis.

Both coal gasification and coal liquefaction have tended to go into abeyance with the discovery of large supplies of petroleum and natural gas (the production of synthesis gas from natural gas is economically more attractive). When it becomes necessary to call upon coal reserves for our energy supplies, these processes will be available to manufacture clean-burning gaseous and liquid fuels.

2.4 Alcohols and Bio-fuels

The alcohols methanol (CH_3OH) and ethanol (C_2H_5OH) are clean-burning fuels that potentially have a role to play in future energy scenarios.

Methanol

Methanol has been described, erroneously in chemical terms, but usefully in energy terms, as 'two molecules of hydrogen made liquid by one molecule of carbon monoxide'. It may be viewed as a portable form of hydrogen. Originally, methanol was made by distillation from waste wood products. Nowadays, it is invariably produced from natural gas *via* synthesis gas ($CO + H_2$, Equations 2.1 and 2.2), *i.e.*

$$CH_4 + O_2/H_2O \rightarrow CO + 2H_2 \xrightarrow[\text{Cu-based catalyst}]{200-300\,^\circ C,5\,MPa} CH_3OH \qquad (2.9)$$

In 2002, there were over 90 plants world-wide, with a capacity to produce over 11 billion gallons of methanol annually.

In principle, biomass may be allowed to ferment directly to methane at ambient temperature, as happens in marshes and in landfill sites under anaerobic conditions, and the gas could then be collected and catalytically oxidized to methanol. Thus, methanol may be produced from either fossil or non-fossil sources. It would appear, however, that practical processes for the manufacture of methanol from biological sources have yet to be developed.

Under present-day circumstances of energy resources and economics, there is no incentive to manufacture methanol for use as a fuel. Rather, it is a basic chemical commodity. The excess natural gas that is flared at oil fields could, in principle, be catalytically converted to methanol *via* a GTL process (see Section 2.1), rather than liquefied as LNG, for transport to market. Commercial units that can produce 5000 t per day are feasible. The demand for methanol as a chemical does not seem to be sufficiently large to justify this course of action (the total utilization rate of existing plants world-wide is only 80%), while as a fuel, natural gas or petroleum are preferable to methanol for most purposes. The use of methanol as a liquid fuel is, therefore, seen as part of a future scenario.

Methanol has been used experimentally as a motor fuel. One shortcoming is that its energy density is only half that of petrol. Other problems are its toxicity and its miscibility with water, which results in corrosion of fuel tanks and feed pipes. An alternative is to add methanol (or ethanol) to petrol at the 10–30 vol.% level to produce 'gasohol'. A positive feature for methanol is that it has a comparatively high octane number, which makes it suitable for engines with a high-compression ratio.

There is also considerable interest in the use of methanol in fuel cells on account of the great convenience of having a portable liquid fuel (see Section 8.3, Chapter 8). Fuel-cell vehicles are presently being developed by several automobile companies. In summary, as illustrated in Figure 2.8, methanol holds potential as a clean and versatile fuel for the future. It may be manufactured from almost any fossil fuel or, in principle, from biomass or biological waste. Being a liquid, it is conveniently transported and stored, and may be employed in many different ways for transportation applications, for heating purposes, for electricity generation, or as a commodity chemical.

Ethanol

Ethanol may be made by the enzymatic fermentation of cereal crops such as maize, potatoes and sugar cane, followed by distillation. Although internal-combustion engines will run on pure ethanol, this requires some engine modifications. There are certain technical disadvantages (sensitivity to water uptake, cold-start problems), but ethanol is less corrosive and less toxic than methanol. Rather, it is better to use 5–20 vol.% ethanol as a petrol extender, 'gasohol' (see Section 4.2, Chapter 4). This has been done extensively in Brazil where millions of vehicles are fuelled by a petrol–ethanol mix. In the USA, around three billion litres of ethanol are produced each year from corn.

From a purely energetics viewpoint, the case for absolute ethanol as a fuel is questionable. The problem lies not in the fermentation but in the heat requirement of the distillation process, which is considerable. This energy loss may be justifiable

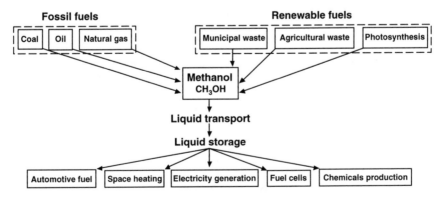

Figure 2.8 *Methanol as an energy vector and fuel.*

when ethanol is to be consumed as a spirit or used as a motor fuel, but is questionable when ethanol is to serve as a general fuel. Economic viability as a general fuel, therefore, depends on finding alternative ways of separating water and ethanol.

Bio-diesel

Certain crops have long been grown for non-food applications in industry. These include cotton, flax and hemp for use in textiles and rope making, and other crops such as linseed and oilseed rape (also known as canola) from which oils may be extracted. Linseed oil was traditionally used in paint. Vegetable oils are widely employed in making margarines, but may potentially be useful as feedstocks for manufacturing lubricants, resins, polymers, *etc.* They can also be converted to liquid fuels.

Vegetable oils, like animal fats, are glycerides of long-chain aliphatic acids (so-called 'fatty acids'). They are extracted from a variety of oil seeds, which include rape, sunflower, soybean, cottonseed, and coconut. In the raw state, these oils are unsuitable for use as a fuel in internal-combustion engines, but may be converted to a useable fuel by processing. The processing conditions depend upon the chosen feedstock. Bio-diesel is a diesel substitute that is made from these vegetable oils by esterification. This involves reacting the oil chemically with methanol or ethanol. Glyerol is produced as a by-product for use in the pharmaceutical industry. The resulting bio-diesel is mixed with conventional diesel, often in a 1:5 ratio, for commercial use. Trials have been conducted with buses or trucks in Australia, Austria, France, the USA, and elsewhere, although there is a question mark over the net energy ratio (energy output compared with energy input) when account is taken of the total energy consumed in the growing/harvesting, transport and conversion processes.

The attractions of bio-diesel may be summarized succinctly as:

- provides a transport fuel from a renewable source;
- reduces dependence on imported fuel;
- can be distributed *via* the existing petroleum-supply infrastructure;
- introduces a new rural industry and a new market for farmers;
- can be used in standard, unmodified diesel engines;
- contains no sulfur or aromatic compounds;
- is biodegradable.

There are, however, certain disadvantages associated with the large-scale development of liquid bio-fuels. At the present time, they are simply not cost-competitive with liquid fossil fuels. Bio-diesel is up to twice the price of conventional diesel, although it is open to governments to support bio-diesel by levying a lower rate of tax if they so choose. In the longer term, there are concerns over the sustainable yield of bio-fuel per hectare. Issues include maintenance of a good annual yield, year after year without crop rotation, and the amount of energy required in the form of artificial fertilizers. Furthermore, the land may be required for food growing as the world population expands. And there may be undesirable environmental effects such as lowering of the water table, increasing the demand for nutrients with consequent flow-off to water-courses, unanticipated effects on the habitat for wildlife, and visual impact on the landscape. Already, there are those who object to the introduction of rape fields in the UK, even at the level needed for use in the food industry. The serious production of bio-diesel for road transportation would require a vastly greater area of land. All these factors need careful evaluation before being too enthusiastic for bio-fuels.

2.5 Hydrogen – The Ultimate Fuel

Hydrogen is the ideal fuel from an ecological viewpoint. It may, in principle, be derived from water using a non-fossil energy source (*e.g.* solar, geothermal, nuclear) and combusted back to water in a closed chemical cycle involving no release of carbonaceous pollutants. Hydrogen also has the potential to provide a storage component for renewable forms of energy and to transport this energy, *via* underground pipelines, from where it is produced to where it is needed. These topics are explored in more detail in Chapter 8. Here, we are concerned with the combustion of hydrogen.

Today, most hydrogen is produced by the steam reforming of natural gas, coal or naphtha, and by the partial oxidation of heavy oils to produce synthesis gas. Steam reforming of natural gas (Equation 2.1) is the most efficient and widely used process and is combined with the water-gas shift reaction (Equation 2.6) to increase the amount of hydrogen produced. The final products are hydrogen and carbon dioxide. Steam reforming is very energy intensive since it operates at high temperatures (850–950 °C) and high pressure (3.5 MPa). The thermal efficiency can reach 60–70%. The catalyst is suspended in an array of tubes mounted in a hot box. Heat is provided by radiant transfer to the exterior of the tubes and is generated by the combustion of natural gas. The whole reactor–heater system is large and bulky. Efforts are being made to reduce the size of the reformer and improve the efficiency of heat transfer. Obviously, the process is not environmentally benign due to large emissions of carbon dioxide. Typically, a steam reformer plant has a capacity of 10^4–10^5 t of hydrogen per year (*i.e.* 10^8–10^9 m^3). In the USA alone, it is said that 90 billion cubic metres of hydrogen per year are produced by steam reforming of fossil fuels, predominantly methane, for use in the petrochemicals and related industries. This is 5% of the natural gas used in the USA.

In the alternative method for hydrogen production, *i.e.* the partial oxidation process, a fuel and oxygen (and sometimes steam) are combined in proportions such that the fuel is converted into a mixture of hydrogen and carbon monoxide. The amount of hydrogen is only about 75% of that produced by steam reforming. Depending on the composition of the feed and the type of the fossil fuel used, the partial oxidation process is carried out either catalytically or non-catalytically. The latter approach operates at high temperatures (1100–1500 °C) and can be applied to any possible feedstock, including heavy residual oils and coal. By contrast, the catalytic process is performed at a significant lower range of temperatures (600–900 °C) and, in general, uses light hydrocarbon fuels as feedstocks, *e.g.* natural gas and naphtha. The drawback to partial oxidation is that it requires the use of expensive oxygen (rather than air, which would dilute the product hydrogen with nitrogen).

Only a few percent of world hydrogen is produced by electrolysis and this is mostly as a by-product of the chlor-alkali process for the manufacture of chlorine and sodium hydroxide. More speculative methods of hydrogen generation include: the thermolysis of water and hydrogen sulfide; the photochemical and photoelectro-chemical decomposition of water; biological processes, *e.g.* solid biomass gasification, liquid biomass fermentation, algae photosynthesis, bacterial fermenta-tion (see Section 8.1, Chapter 8).

At current prices, hydrogen is used almost exclusively for the synthesis of ammonia, methanol, and other petrochemicals – generally in a plant situated in the same petrochemical complex as the reformer – and for petroleum refining (hydro-cracking, hydro-desulfurization, *etc.*). A small proportion of hydrogen is compressed into gas cylinders and sold for small-scale use in industrial plants and laboratories. Typical applications include the heat treatment of metals, the hydrogenation of oils to fats, the reduction of metallic oxides (as in the manufacture of refractory metals, nuclear fuels, *etc.*). With the growing importance of fuel cells, there will also be an increasing need for hydrogen as a fuel (see Section 8.3, Chapter 8). At present, however, the production of hydrogen as an all-purpose fuel is simply not economic, with so many cheap fossil fuels available.

The relevant technical properties of hydrogen as a fuel, in comparison with those of conventional and synthetic fuels are listed in Table 2.1. The most characteristic features of hydrogen are its low boiling point and its exceptionally low density in both the gas and the liquid state. By virtue of this latter feature, hydrogen has an extremely high heating value on a unit mass basis. On the other hand, the heating value of liquid hydrogen per unit volume is less than that of other liquid fuels.

The properties listed in the lower half of Table 2.1 are relevant to safety considerations. Hydrogen is notable for its very rapid diffusivity in air, its high value for the upper flammability limit in air, its exceptionally low ignition energy, and its remarkably high flame velocity. This combination of properties makes it a unique fuel from both utilization and safety aspects. Hydrogen has a poor safety reputation, largely on account of the Hindenburg airship disaster in 1937. It has been recently shown, however, that the Hindenburg fire was more to do with the flammability of the fabric chosen for the airship's envelope than with the fact that it was filled with hydrogen. Nevertheless, airships and balloons are now filled with helium and not hydrogen. There is, however, considerable industrial experience of the safe handling of hydrogen gas in bulk, both in refineries and in chemical plants. The general opinion of those involved in such operations is that hydrogen is safe provided that its properties are clearly understood and well-defined safety regulations are followed. A positive safety feature of hydrogen is that in the event of a fire, the low luminosity of the flame restricts the emission of thermal radiation to less than one-tenth of that from hydrocarbon flames. Thus, bystanders are much less likely to suffer radiation burns.

The low density and high diffusivity of hydrogen results in a very rapid dispersal of liquid hydrogen after a spillage, so that the risk of fire persists for a much shorter period than with other liquid fuels. All confined spaces in which hydrogen is

Table 2.1 *Technical comparison of hydrogen with other fuels*

	Hydrogen	Petroleum	Methanol	Methane	Propane	Ammonia
Boiling point, K	20.3	350–400	337	111.7	230.8	240
Liquid density, $kg\,m^{-3}$	71	702	797	425	507	771
Gas density, $kg\,m^{-3}$, s.t.p.[a]	0.08	4.68	–	0.66	1.87	0.69
Heat of vaporization, $kJ\,kg^{-1}$	444	302	1168	577	388	1377
Lower heating value[b] (mass), $MJ\,kg^{-1}$	120.0	44.38	20.1	50.0	46.4	18.6
Lower heating value (liquid) (volume), $MJ\,m^{-3}$	8960	31 170	16 020	21 250	23 520	14 350
Diffusivity in air, $cm^2\,s^{-1}$	0.63	0.08	0.16	0.20	0.10	0.20
Lower flammability limit vol.% (in air)	4	1	7	5	2	15
Upper flammability limit vol.% (in air)	75	6	36	15	10	28
Ignition temperature in air, °C	585	222	385	534	466	651
Ignition energy, MJ	0.02	0.25	–	0.30	0.25	–
Flame velocity, $cm\,s^{-1}$	270	30	–	34	38	–

[a] s.t.p.=standard temperature (273.15 K) and pressure (101.325 kPa).
[b] There are two ways to define the energy content of a fuel. The 'higher heating value' includes the full energy content by bringing all products of combustion to 25 °C. By contrast, the 'lower heating value' neglects the energy in the water vapour formed by the combustion of hydrogen in the fuel. This water vapour typically represents about 10% of the energy content. The higher heating value represents the true (thermodynamic) heat of combustion, but the lower heating value is more relevant because a steam condenser is not used in most practical applications. In this table, the lower heating value is the heat of combustion ($MJ\,kg^{-1}$) of a fuel, based on complete combustion to carbon dioxide and steam at 100 °C. For further details of higher and lower heating values, see Box 8.2 in Chapter 8.

handled must be well ventilated. The low ignition energy makes it necessary to exclude all sources of sparks, such as electric motors, synthetic garments, and steel tools. When these and similar precautions are observed, experience with handling hydrogen in bulk has been favourable.

The combustion characteristics of hydrogen are quite different from those of natural gas or LPG and this requires a modified burner design. The principles involved are well understood and the design of a burner for pure hydrogen presents no serious difficulties. The low ignition energy of hydrogen favours the use of catalytic burners that are of higher efficiency and lower flame temperature than conventional burners.

The use of hydrogen as a fuel for internal-combustion engines was first demonstrated in the 1930s. From the 1970s to the present day, there has been a continuing and strong interest in 'hydrogen energy' and numerous research institutes have investigated hydrogen-fuelled engines. In general, hydrogen engines tend to exhibit pre-ignition, back-fire, and knock. These phenomena are caused by the low ignition energy and high flame speed of hydrogen. These problems have now been largely overcome. By using lean mixtures of hydrogen in air, it is possible to reduce NO_x emissions to well below that of a conventional petrol engine. In principle, therefore, the practical realization of pollution-free, hydrogen-fuelled engines appears to present no major difficulties.

Several car companies are experimenting with models fitted with hydrogen-fuelled engines. Ford has demonstrated a mid-size saloon ('Mondeo') with a modified 2-litre petrol engine. The hydrogen fuel is stored under pressure (about 30 MPa) in cylinders that are located in the boot (trunk). Meanwhile, BMW has equipped several of its large (7-series) cars with dual-fuel capability, *viz.*, hydrogen and petrol. In this case, liquid hydrogen is used and is stored in cryogenic tanks of 120-litre capacity, again located in the boot.

The problem of how to store hydrogen for use with internal-combustion engines is formidable. Pressurized gas cylinders are heavy and bulky to accommodate, but recently lightweight gas cylinders based on carbon-fibre composites and capable of storing hydrogen at over 30 MPa pressure have been developed. Liquid hydrogen is expensive to produce, both in monetary terms and in terms of the energy lost in the liquefaction process (see Section 8.2, Chapter 8). The electricity consumed in the liquefaction of hydrogen depends on the size of the plant and ranges from 10 to 20 kW h per kg of hydrogen. Also, the cryogenic storage tanks and associated equipment are costly and take up considerable space. A third option, that of storage as a metal hydride, is discussed in Section 8.2, Chapter 8. Finally, as far as road vehicles are concerned, there is the problem of providing a network of refuelling stations at which either high-pressure hydrogen or liquid hydrogen can be supplied. This will be a much more difficult task than the current exercise of providing a distribution infrastructure for LPG.

Some interest has been expressed in the possibility of producing a new passenger aircraft, possibly supersonic, fuelled by hydrogen. Liquid hydrogen fuel is attractive to aircraft designers on account of its low mass, which would permit either increased range or payload. On the other hand, there is a space problem. In rounded terms, the data given in Table 2.2 show that liquid hydrogen occupies four

Table 2.2 *Mass and volume comparisons of kerosene and liquid hydrogen (lower heating values)*

	Mass		Volume	
	(kg per GJ)	Index[a]	(m^3 per GJ)	Index[a]
Liquid hydrogen	8.33	0.38	0.112	4.00
Kerosene	22.0	1.0	0.28	1.0

[a] Taking mass and volume for kerosene as unity.

times the volume of hydrocarbon fuels (*i.e.* kerosene, the conventional aircraft fuel), but has little more than one-third of the mass. A major design problem, therefore, is where to accommodate the liquid hydrogen on the aircraft. Possibilities include the use of tanks ('pods') suspended below the wings (taking advantage of the low mass), or storage above or even within the fuselage. There are also operational problems, mainly the logistics, costs and safety of supplying liquid hydrogen in the required quantities at major airports. These ideas were first discussed in the early 1970s. Later, Lockheed Corporation proposed to the US government that two of the Company's 'Tristar' freighters should be converted to liquid hydrogen (LH$_2$) and operated across the Atlantic as a demonstration project.

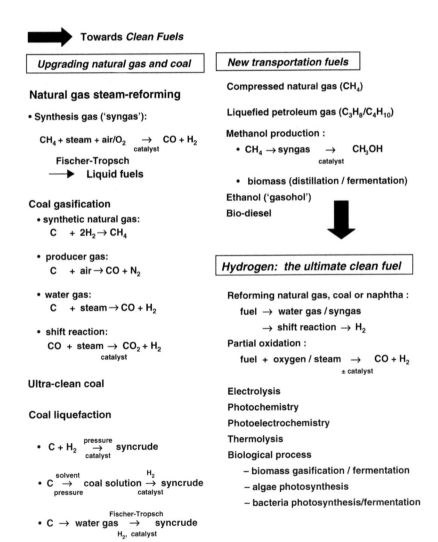

Figure 2.9 *Summary of synthetic fuels.*

Nothing came of this and the idea was abandoned, as was another proposal in 1980 for 'An International Research and Development Programme on LH_2-fuelled Aircraft'. At present, there is little prospect of this activity being revived, thanks to the availability and low cost of conventional (fossil) fuels, although it is pertinent to point out that there is some interest in a new generation of supersonic aircraft, and also that there is general concern over the release of greenhouse gases in the stratosphere by supersonic aircraft.

To assist the reader, a brief summary of synthetic fuels that have been, or are being, used is given in Figure 2.9.

2.6 Electricity – A Clean Secondary Fuel

It has long been recognized that electricity is a clean form of energy for heating and cooking purposes. It is also a most versatile form of energy in that it powers electric motors and a wide variety of devices upon which our civilization now depends. Electricity is, however, a secondary form of energy, always derived from a primary form of energy, which is often fossil fuel or nuclear fuel, although it can be one of the renewable energy sources such as hydro, geothermal, solar, and wind. Indeed, as will be discussed in Chapter 5, many of the renewable forms of energy are only sensibly deployed *via* the medium of electricity as the energy vector. For this reason, and also to set the scene for a discussion of renewable energy, it is appropriate in the following Chapter to describe electricity generation by conventional means.

2.7 References

1 *BP Statistical Review of World Energy 2003*, June 2003; see: www.bp.com.
2 Shell Company website: www.Shell.com (November 2001).
3 Liquefied Petroleum Gas website: www.LPGA.co.uk (October 2003).

CHAPTER 3

Electricity Generation

Electricity is, without doubt, the most versatile form of energy. It is also one of the cleanest, although it has to be remembered that this is only on account of having transferred any pollution from the point of use to the power station smokestack. There, at least, it is discharged at a considerable height and dissipated over a wide area. Most electricity is generated using fossil fuels, especially coal and gas, although nuclear energy and hydroelectric energy, both make sizable contributions to the world's electricity supply. These latter two energy sources are clean in the sense of not liberating carbon dioxide or other pollutants into the atmosphere.

3.1 Electricity Generation Statistics

Before examining the scale of electricity generation, a word about measurement units is appropriate. The standard unit of domestic electricity supply is the kilowatt-hour (10^3 Wh). When discussing national or international electricity generation, it is conventional to use larger units, *i.e.* the megawatt-hour, MWh (10^6 Wh); the gigawatt-hour, GWh (10^9 Wh); the terawatt-hour, TWh (10^{12} Wh); or even the petawatt-hour, PWh (10^{15} Wh). The installed generating capacity of a nation is expressed in megawatts (MW) or gigawatts (GW). For the major countries, the quantity of electricity generated is expressed in TWh.

The total world electricity generation by fuel in 1973 and 2001 is given in Table 3.1.[1] Electricity production is 2.5 times more than what it was 28 years ago and now represents 15.6% of the world's energy consumption.[1] It has been predicted that by 2020 the demand will have risen from 15.4 to 20 PWh.[2] There has also been a significant change in the distribution of the fuels used. Coal has remained about the same in percentage terms. Nuclear has increased dramatically at the expense of oil. Gas has increased and hydro has decreased in percentage terms, though in absolute terms more hydro-capacity has been installed and electricity generation from gas has increased from 0.74 to 2.83 PWh. As might be expected, the distribution among the fuels varies greatly from one country to another as determined by the indigenous energy resources. Distribution issues will be reviewed in later discussion.

Table 3.1 *Percentage of fuel shares of world electricity generation in 1973 and 2001*[1]

Fuel	1973	2001
Coal	38.2	38.7
Oil	24.7	7.5
Gas	12.1	18.3
Hydro	21.0	16.6
Nuclear	3.3	17.1
Other[a]	0.7	1.8
Total (%)	100.0	100.0
Total (TWh)	6121	15 476

[a] Other includes combustible renewables and waste, geothermal, solar, wind.

The overall quantity of electricity generated by the top-10 countries is presented in Table 3.2. There is an enormous disparity between one country and another in the amount of electricity used per capita, as shown in Figure 3.1.[3] Clearly, there is a huge scope for increased electricity usage in much of the world.

The International Energy Agency prefers to work in terms of megatonnes of oil equivalent (Mtoe) in order to facilitate comparison with the supply and consumption of fossil fuels for all purposes. Indeed, the past and prospective contributions of fossil fuels to total primary energy supplies shown in Tables 1.1

Table 3.2 *Major producers of electricity in 2001*[1]

	TWh	% of world total
USA	3864	25.0
PR China	1472	9.5
Japan	1033	6.7
Russian Federation	889	5.7
Canada	588	3.8
Germany	580	3.7
India	577	3.7
France	546	3.5
UK	383	2.4
Brazil	328	2.1
Rest of world	5216	33.8
World total	15 476	100.0

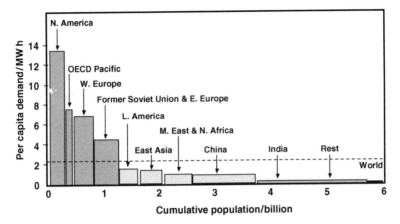

Figure 3.1 *Per capita electricity demand (see Ref. 3).*

and 1.2, Chapter 1, include the amounts supplied to the electricity-generating industry, which is why electricity, a secondary fuel, is omitted (apart from nuclear and hydro, which are primary forms of energy). For the purposes of analyzing overall energy *consumption*, it is necessary to relate the electrical units of Table 3.1 to Mtoe. The conversion factor is 1 TWh = 0.086 Mtoe (or 1 Mtoe = 11.63 TWh).[*] This is in pure energy terms and takes no account of conversion losses. Thus, in order to include hydro and nuclear in the total primary energy *supply*, it is necessary to convert backwards from electrical units and take account of any conversion losses. In the case of hydro, where no significant conversion losses are incurred, it is the convention simply to use the generation data. For nuclear electricity, however, where a thermal (steam) cycle is involved, the supply of nuclear heat energy is obtained by dividing the electrical energy generated by the average efficiency of the nuclear power plant (33%). This has the effect of making the nuclear contribution in Tables 1.1 and 1.2, Chapter 1, seem much larger than that of hydro, whereas in terms of *electricity generated* (Table 3.1) their contributions for the entire world in 2001 were of similar magnitude. In the OECD countries, where much of the nuclear generation is concentrated, its contribution to electricity supplies is indeed greater than hydro, although not by as much as might be inferred (erroneously) from Table 1.1.

3.2 Coal-fired Electricity Plant

As mentioned in Chapter 1, about half of the coal mined in the world is used in electricity generation. Countries that are large producers of coal but have little oil or gas (India, Poland, South Africa) are almost totally dependent upon coal for their electricity supplies. As shown in Table 3.1, about 39% of the world's electricity supply in 2001 was generated in coal-fired plant. The USA was the greatest producer of electricity from coal-fired plant (1983 TWh), then followed by the

---

[*] The equivalent of electrical energy units with other energy units is given in the Appendix.

People's Republic of China (1122 TWh) and India (452 TWh). In Europe, Germany is the largest user of coal for electricity (301 TWh).[1] It is unlikely that this order will have changed much since 2001.

Modern coal-fired power stations are often, although not always, large in size, *i.e.* 1–4 GW output. Burners can be designed to accept any type of coal; lignite (brown coal), sub-bituminous or bituminous coals are generally used. The economics of generating the electricity depends very much upon the cost of the coal, which, in turn, depends upon fuel quality, on the height and the depth of the coal seams, and the distance over which the coal must be transported. Open-cut mining is usually the cheapest. An example of such a mine in Victoria, Australia, with a coal-fired electricity plant adjacent to it, is shown in Figure 3.2. After removing the topsoil, coal (lignite) is simply scooped up with dredgers and deposited on a conveyor belt that takes it directly into the power station. This station has an output capacity of 2.0 GW.

By contrast, the UK has plentiful coal reserves, but much of these are deep underground in narrow seams, which makes them expensive to mine. This is the basic reason why many of the coal mines have been closed in favor of importing cheaper coal from South Africa and elsewhere. Moreover, coal from South Africa is low in sulfur content.

Britain has a number of large coal-fired power stations (2 GW output, or greater). A schematic layout of the 2-GW Didcot 'A' plant is given in Figure 3.3. The station has four parallel lines, each of which is capable of generating 500 MW, and the schematic shows just one of these. Coal is transported on a conveyor belt (01) from a stockpile to a pulverizing mill (02) where it is ground to a fine powder, picked up by a stream of hot air and blown into one of the boilers (03) to burn

Figure 3.2 *Open-cut coal mine in Victoria, Australia.*
(Courtesy of Loy Yang Power)

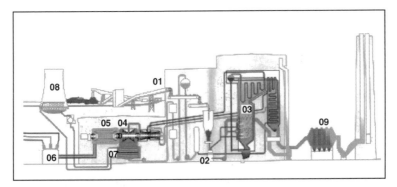

Figure 3.3 *Schematic of a modern, coal-fired, 2-GW plant (Didcot 'A'):* 01: *coal supply;* 02: *pulverizing mill;* 03: *boiler;* 04: *turbine;* 05: *generator;* 06: *generator trans-former;* 07: *condenser;* 08: *cooling towers;* 09: *electrostatic precipitators.* (Courtesy of RWE nPower)

like a gas. There are four parallel boilers. The heat produced converts extremely pure boiler water into steam in tubing that forms the boiler walls. The steam leaves the boiler at 568 °C and at ~16.5 MPa (~165 atm) pressure and passes through the high-pressure stage of a steam turbine (04) to turn the blades and shaft at 3000 rpm. The steam returns to the boiler for reheating and then is directed back to the intermediate pressure stage and three low-pressure stages of the turbine. The turbine shaft is linked to a 500-MW generator (05) whose rotor, a large electromagnet, weighs 74 t. Electricity from the generator at 23.5 kV is fed to a transformer (06) where it is raised to 400 kV for transmission along the national grid.

Spent steam from the turbine goes to a condenser (07) where it is cooled by river water and the pure condensate is pumped back to the boiler for re-use. The warmed river water from the condensers passes to six large cooling towers (08) where it is sprayed over packing in the base of the tower and cooled by evaporation in the natural up-draught of air. The tower is mounted on stilts that provide a large gap for the air to enter at the base. Before the boiler gases are discharged from the main chimney-stack, 99% of the fine dust is removed from the flue gases by electrostatic precipitators (09). Much of the ash from the coal is sold for use in the construction industry. The remainder serves to fill disused gravel pits, which are later restored to agricultural use. The SO_2 and NO_x contents of the exhaust gases are closely monitored.

In the UK, each power station is granted a license for the amount of pollutants that may be discharged per year. In order to maximize the quantity of electricity generated and still stay within the license, there is an incentive for the station manager to burn low-sulfur coal. Some power stations have been equipped with flue-gas desulfurization units that serve to absorb the SO_2 in limestone, where it is converted to calcium sulfate (gypsum). Although this permits the use of higher sulfur contents, it is an expensive option, both in terms of capital and running costs. Flue-gas desulfurization units were fitted to the largest coal-fired station in Europe, the 3.975-GW plant at Drax in North Yorkshire, UK. This station has handling facilities for 10 million tonnes of coal per year.

Table 3.3 *Advantages and disadvantages of electricity generation by coal*

Advantages	Disadvantages
Large reserves of low-cost coal available	Carbon-rich fuel leads to extensive liberation of CO_2
Well-established industry	High sulfur content of many coals leads to pollution by SO_2 and acid rain
Indigenous fuel source for many countries	

Coal-fired power stations generally operate with a thermal efficiency (ratio of electrical output to heat input) in the range 35–39%. Some of these generating plants, including Didcot 'A', have been adapted to burn gas as well as coal when this is economically justified, *i.e.* they are 'dual fuel' stations. At peak output, Didcot 'A' produces sufficient electricity to supply the needs of nearly two million people.

The advantages and disadvantages of using coal to generate electricity are summarized in Table 3.3.

3.3 Oil-fired and Gas-fired Electricity Plant

As shown in Table 3.1, the burning of oil produced only 7.5% of the world's electricity in 2001.[1] In fact, the annual percentage is declining and this is largely due to the increasing use of gas as old oil-fired plant becomes obsolete. The country that burnt most oil was the USA (134 TWh), followed closely by Japan (117 TWh) that has no indigenous coal or gas. Most oil-producing countries have gas associated with the oil and thus have little reason to use oil, a more valuable commodity, to generate electricity. Nevertheless, oil-based generation is unlikely to fade completely; it is ideally suited to small states and communities where the electricity requirement is modest and there is no native gas or coal. Oil-based generation is also generally used for small-scale combined heat and power (CHP) schemes and for fast start-up (stand-by operation) when no gas is available.

In recent years, there has been a major shift towards natural gas for electricity generation. The USA heads the league table (646 TWh) and is followed by the Russian Federation (377 TWh), Japan (using LNG, 257 TWh), and the UK (143 TWh).[1] There are several reasons for this, namely:

- the discovery and exploitation of major new gas fields
- the cleanliness of natural gas as a fuel
- the development of high-efficiency combined-cycle gas turbines (CCGTs)
- the construction of pipelines and LNG carriers to convey the gas to market.

A modern CCGT can raise the thermal efficiency of electricity generation to over 55%, a major improvement on the efficiency of coal-fired stations. A new 1.36-GW CCGT station was built at Didcot ('B' station) and commissioned in 1998. The station has two electricity-generating modules. Each module consists of two industrial gas turbines (similar in principle to a jet engine) and one steam turbine. All three turbines have associated generators. Air, at a rate of 36 t per minute is filtered, compressed to 1.6 MPa, and then directed to the combustion chamber where it is mixed with natural gas and fired. The combustion gases enter both power turbines at around 1160 °C. The subsequent expansion of the exhaust gas causes each turbine shaft to rotate at 3000 rpm and drive the rotor of the associated generator.

On leaving each power turbine, the exhaust gas at 545 °C passes to a heat recovery steam generator (HRSG) where up to 320 t per hour of steam is raised for the steam turbine. The cooled exhaust gas, at around 93 °C, is discharged to the atmosphere through a chimney. Because the gas contains no SO_2 and little NO_x, the chimney can be much lower than for a coal-fired station. The steam from both HRSGs enters the steam turbine where it expands and drives a generator. Each of the station's two generating modules produce 2×225 MW from the gas turbines and 230 MW from the steam turbine, making a total of 680 MW per module and 1.36 GW for the station.

The exhaust steam goes to a condenser to be cooled back to pure water for re-use in the HRSGs. River water is used for cooling and this, in turn, is cooled in a series of low-level cooling towers. An aerial view of the Didcot site is shown in Figure 3.4. In the background (top, right) is the coal-fired 'A' station with its tall chimney and two of its six cooling towers (left), while in the foreground is the gas-fired 'B' station with its two shorter chimneys and two long rows of low level, forced draught, cooling towers. A gas turbine of the type typically used in a CCGT station is shown in a dismantled state in Figure 3.5. The maintenance workers provide a clear indication of the size of the turbine.

Not only is gas generation much cleaner than coal in terms of pollutants (no SO_2 release and greatly reduced emissions of NO_x), but also the quantity of CO_2 greenhouse gas is more than halved, *i.e.* gas and coal produce 0.306 and 0.750 t of CO_2 per MWh in Didcot 'B' and Didcot 'A', respectively. Coal and gas will continue to be the feedstock for base-load electricity supplies well into the future. Over time, however, newer and more-efficient technologies will be introduced with attendant greenhouse gas benefits. With coal as the fuel, the widespread introduction of integrated gasification combined-cycle (IGCC) technology, with its inherent higher thermal efficiency over conventional pulverized fuel injection, would result in lower emissions of carbon dioxide per unit of power produced (Figure 3.6). Over a longer period, emissions could be reduced even further by coupling coal gasification with fuel cells (Figure 2.6, Chapter 2). Emissions from gas-based electricity are on a different, lower, curve (Figure 3.6) due to the lower carbon intensity of natural gas. Nevertheless, the same emissions–efficiency relationship applies. It should be noted, however, that the data are only applicable to the generation site. Other factors, such as the energy consumed in the extraction of coal and gas, fugitive emissions, transport and further processing, must all be considered on a full life-cycle basis to give the complete inventory of emissions for electricity leaving the power plant. Further-more, the curves in Figure 3.6 flatten out after about 55% efficiency, so that the law of diminishing returns prevails. Thus, parallel efforts must be made to capture and sequester the carbon dioxide that inevitably accompanies the generation of coal/gas-fired electricity (Figure 1.12, Chapter 1).

Figure 3.4   *Aerial view of Didcot 'A' (2* GW) *and Didcot 'B' (1.35* GW) *power stations in Oxfordshire, UK.*
(Courtesy of RWE nPower)

A comparison of the advantages and disadvantages of using natural gas to generate electricity is given in Table 3.4.

The gas turbines used in mains electricity-generating stations are huge, with outputs in the range of 200–250 MW. For dispersed generation, or for CHP applications (see Section 3.4), much smaller turbines are required. Gas microturbines typically have power outputs in the range of 25–300 kW. The basic system incorporates a compressor, recuperator, combustor, turbine and permanent magnet generator. The rotating components are mounted on a single shaft, which is the only moving part. A microturbine is uncomplicated in that it uses no lubricants and has no pumps, gearbox or other mechanical sub-systems. The relatively small size is also a major advantage in that it allows microturbines to be situated conveniently at the source of electricity demand (see Section 6.4, Chapter 6). This eliminates transmission losses. Special features of microturbines include: quiet operation with little vibration (few moving parts); high speeds (up to ~90 000 rpm); thermal efficiencies of 25–30%; low maintenance and high reliability; low emission levels. At present, the main disadvantage of microturbines is the limit to the number of

Figure 3.5 *Dismantled gas turbine from CCGT power station.*
(Courtesy of RWE nPower)

times that they can be started up and shutdown. As a result, it is normal practice to keep them running continuously. Microturbines are becoming the preferred choice of power generation in a wide variety of applications, as follows.

- *CHP generation.* Waste heat from the microturbine can be transferred by a heat-exchanger to produce steam or hot water that can be used, for example, to provide central heating in buildings in winter. Thermal hosts are easier to find because the heat produced by a microturbine is so much smaller than that by a large power station.
- *Distributed power generation.* Hospitals, hotels, factories, farms, *etc.* can install microturbines on-site to supplement power supplied by the grid. Microturbines can also be used in remote areas where there is no access to electricity.

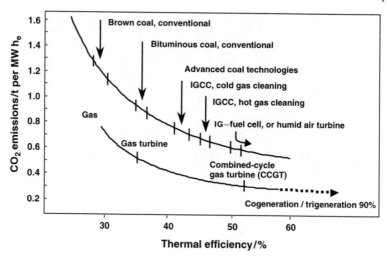

Figure 3.6 *Progressive decrease in greenhouse gas emissions from electricity-generating plants with increasing efficiency of the process when using coal and gas technologies.*

- *Off-shore oil rigs.* Microturbines can utilize the waste flue gas from the oil extraction. This removes the cost of acquiring and storing large amounts of fuel for traditional diesel engines. Moreover, whereas diesel engines are bulky and heavy and therefore can only be transported by ship, microturbines are lightweight and can be carried on-board a helicopter for fast and effective delivery.

Table 3.4 *Advantages and disadvantages of electricity generation by natural gas*

Advantages	Disadvantages
Natural gas is widely available	It may be argued that, where supplies are limited, natural gas is too valuable a fuel to be used for electricity generation and should be reserved for space heating and chemical manufacture
Natural gas has the highest H:C ratio and therefore the least CO_2 generation per kWh	
Methane is distributed by buried pipeline or LNG tanker with no environmental impact	For a country with indigenous coal, but no natural gas, the use of imported gas for electricity generation will place a strain on that nation's balance-of-payments
By using a gas turbine combined with a steam turbine high efficiencies are achieved	Liberation of CO_2
Natural gas contains almost no polluting impurities	

3.4 Combined Heat and Power (Co-generation)

CHP, or 'co-generation', is a simple means of increasing the overall efficiency of a power plant by utilizing the waste heat in the exhaust gases rather than discharging them to atmosphere while still warm. Thus, fuel is used at high efficiencies and overall emissions of carbon dioxide are minimized. In order to obtain heat at a useful temperature, it is normally necessary to raise the temperature of the exhaust gases, which thereby reduces the efficiency of electricity generation but raises the overall efficiency. The latter may be defined as the ratio of the sum of the electrical and heat output energy to the input heat of combustion. The overall efficiency for CHP is typically in the range 70–90%, compared with 35–55% for conventional power plants. At maximum efficiency, the ratio of heat output to electricity output is around three. Co-generation is particularly appropriate for uses that can accept this ratio, *e.g.* in factories or farms that operate drying processes, or in hotels where the demand for hot water is high.

Co-generation plants are usually small compared with conventional power stations and are often sized to suit an industrial or commercial application that requires a fairly constant quantity of electricity and heat for its operation. In the UK, almost half of the CHP schemes have an electrical output of less than 100 kW_e, although those larger than 10 MW_e account for over 80% of the total CHP installed capacity. Industrial applications for larger CHP units (*i.e.* megawatt size) are found in various industries, *e.g.* chemical and petrochemical, paper and board, food and drink, iron and steel industries. Such a unit might, for example, generate around 40 MW_e of electricity and 120 MW_{th} of heat to support the operation of a petrochemical plant. Smaller units (20 kW upwards) are used to supply heat and power to hotels, hospitals, apartment blocks, *etc.* In 2002, the UK had an installed CHP electrical capacity of 4.74 GW_e that generated 24.2 TWh_e of electricity and 60.7 TWh_{th} of heat.[4] Just over 6% of the total electricity generated in the UK came from CHP plants. The government aims to have 10 GW_e of installed CHP capacity by 2010 as a contribution towards meeting its Kyoto target for emissions of greenhouse gases.

The temperature at which heat is produced has to be matched to the application. For process steam in industry, a temperature of 300 °C may be needed, whereas for central-heating purposes steam at about 120 °C will suffice. Several different types of generator are possible; the most usual are gas engines or micro-gas-turbines that, together, have largely replaced steam turbines. In future, it is possible that fuel cells will be used for CHP purposes (see Section 8.3, Chapter 8). Natural gas is the most common fuel, although biogas, landfill gas and refinery waste gas may also be used.

Communal heating schemes have existed in parts of Scandinavia for many years. A local power station, operating in CHP mode, supplies both electricity and heating to a town or district. Steam circulates around the district in well-lagged pipes and customers may be metered for the heat received as well as for the electricity. It is generally necessary to include a heat-storage component to help smooth fluctuations in the demand for electricity. As more electricity is required, the amount of heat that can be delivered falls off. Consequently, the supply to the customers is maintained by the storage facility (see Section 7.1, Chapter 7). Several

European countries, *e.g.* Denmark, Finland and the Netherlands, are said to produce one-third or more of their total electricity by co-generation.

To date, district heating schemes have not found much favor in the UK. One reason is the large size of most of the power stations (500–4000 MW_e). The huge quantities of heat that are produced would necessitate a very large scheme, such as a supply to a city of moderate size. To install the system in an existing city or conurbation would be a major (and costly) project. Moreover, urban areas are often too distant from large power stations for the heat to be transmitted effectively. Finally, the degradation in the electrical output of the station, associated with raising steam at a sufficiently high temperature, would make the venture non-viable. A much better option is to have many smaller and dedicated CHP units, with both the electricity and the heat used locally, *i.e.* distributed generation as discussed in Section 3.7.

Smaller district heating schemes are ideal for new housing estates, where the heating pipes are installed during construction. In the UK, the largest such scheme is in Sheffield where a 42-km network of insulated, underground pipes carries heat to 3000 homes and 88 public and commercial buildings. Many other potential markets exist for small, independent CHP plants.

Although having a local supply of electricity and heat, and being independent of the grid, sounds fine in principle, difficulties might arise when CHP plants break down or require servicing. By contrast, an electricity grid provides greater reliability. Electricity companies may not look favorably on being simply the supplier of last resort. Taking the idea to its logical conclusion, some companies in various countries are developing micro-CHP plants for use by individual households. With the large potential that CHP holds for reducing emissions of carbon dioxide and conserving fossil fuels, it is not surprising that many governments favor CHP and are encouraging its expansion.

3.5 Nuclear Power

The release and harnessing of the immense energy locked in the nucleus of the atom was one of the great scientific achievements of the 20th century. Now, at the beginning of the 21st century, a debate still rages as to whether nuclear energy should be seen as 'clean energy' or 'dirty energy'. Proponents of nuclear power argue that a nuclear reactor gives off no gaseous pollutants, liberates no greenhouse gases, and is completely contained. Also, the reserves of uranium are such that nuclear energy may be regarded as 'sustainable', if not 'renewable'. Opponents point to the radioactive fission products that are inevitably formed in the reactor and that have to be disposed of safely. Also, the 1979 near-disaster at Three Mile Island in the USA and the 1986 tragedy at Chernobyl in the Ukraine have demonstrated the very real risk of major reactor accident. In addition, there is the possibility that nuclear materials will be diverted by countries or terrorist organizations for the purpose of making nuclear weapons.

The construction of nuclear power stations was at its peak in the 1970s, but subsequently in certain countries (USA, Germany, Sweden, UK, for example) there has been growing public opposition to the construction of new nuclear power stations and radioactive waste depositories. No new nuclear generating plant has been ordered in the USA since 1978. In other countries (France, Japan, Korea), such opposition has been relatively muted and major nuclear construction programs have taken place. Can both be right, depending upon their individual circumstances, or will a consensus about the desirability of nuclear power ultimately emerge?

The world production of electricity in 2001 was 15.476 PWh (Table 3.1). Of this, 17.1% (2.646 PWh) came from nuclear power stations. Much of this was located in OECD countries, especially the USA, France and Japan, as shown in Table 3.5. The countries that are most reliant upon nuclear electricity for domestic electricity generation (expressed as a percentage of national output), and therefore most vulnerable to a closure of their nuclear programs, are: France (77%), Sweden (45%), Ukraine (44%), Korea (40%), Japan (31%), Germany (30%), and the UK (23%).[1] For these countries, especially France, it is totally unrealistic to assume that, in the short-to-medium term, nuclear electricity could be replaced by other forms. It is significant that four of these countries (France, Sweden, Korea, Japan) have very little in the way of indigenous fossil fuels. Their strong nuclear programs were therefore undoubtedly inspired by a desire to minimize imports of oil and gas as these have an adverse effect on balance-of-payments. Germany relies heavily on both coal and nuclear power for its electricity since it has no domestic oil or gas, but does have coal.

Reactor Physics

Civil nuclear power has been developed from wartime work on the 'atomic bomb'. In 1938, the German radiochemists Otto Hahn and Fritz Strassmann, following some basic studies by the Italian-American physicist Enrico Fermi, discovered that

Table 3.5 *Major producers of nuclear electricity in 2001*[1]

Country	TWh	% of world total	Installed capacity (GW)
USA	808	30.5	95
France	421	15.9	63
Japan	320	12.1	44
Germany	171	6.4	21
Russian Federation	137	5.2	21
Korea	112	4.2	13
UK	90[a]	3.4	12
Canada	77	2.9	14
Ukraine	76	2.9	11
Sweden	72	2.7	9
Rest of world	369	13.8	53
Total	2653	100.0	356

[a] Ref. 4 gives 81.1 TWh.

uranium when bombarded by neutrons could yield two medium-mass elements (barium and lanthanum) and other products, together with the release of huge amounts of energy. Shortly afterwards, the Austrian physicists Lisa Meitner and Otto Frisch explained what had happened: the uranium nucleus had split into two nearly equal fragments after absorbing a neutron. By analogy with cell division in biology, Meitner and Frisch named the new process 'nuclear fission'. Such fundamental scientific research formed the basis of the wartime program of nuclear development in the USA that led, ultimately, to the atomic bombs that were dropped on Hiroshima and Nagasaki in August 1945.

After the war, several countries immediately began to investigate the potential of harnessing nuclear energy to provide a new and controlled way of generating electricity. Large research and development programs were established, particularly in the USA, the UK, France, and the Former Soviet Union. From these, a number of different designs of civil nuclear reactor have evolved. Before describing briefly these reactors, it is useful to examine the nature of the fission process itself. This is explained in Box 3.1 where it is seen that natural uranium consists of two isotopes, only one of which undergoes fission to release nuclear energy.

Box 3.1 Fundamentals of Reactor Physics

Natural uranium consists of two isotopes, ^{235}U (0.7%) and ^{238}U (99.3%), and it is only the lighter and less abundant isotope that undergoes fission in a nuclear reactor. (Note, 235 and 238 are the atomic masses of the respective isotopes, *i.e.* the number of neutrons and protons – 'nucleons' – in the nucleus). Neutrons released in fission events are very energetic and it is necessary to slow them down to thermal energies (*i.e.* to thermal equilibrium with the surroundings) before they are capable of bringing about fission in ^{235}U. This slowing process is effected by means of multiple collisions with light atoms such as hydrogen, deuterium or carbon, which are termed 'moderators' (see discussion of Reactor Design in main text). The moderator surrounds the uranium fuel. The probability of neutron fission increases with decreasing energy of the neutron. All commercial reactors operating today are the so-called 'thermal reactors' that utilize thermalized neutrons.

For a nuclear reactor to operate and useful quantities of energy to be extracted, it is necessary to sustain a 'chain reaction'. Each ^{235}U nucleus that undergoes fission liberates more than one neutron (Figure 3.7(a)). Some of these are lost through absorption by materials in the reactor and some through the reactor walls. To sustain a chain reaction – that is the continuous fission of ^{235}U nuclei – it is necessary that for each atom undergoing fission there shall be at least one free neutron left over (*i.e.* not lost) in order to propagate the chain. Control of the neutron economy is critical as too many free thermal neutrons would lead to a runaway fission process and result in melt-down of the reactor. This control is achieved with neutron-absorbing rods that are moved into and out of the reactor to control its reactivity.

Box 3.1 (cont'd)

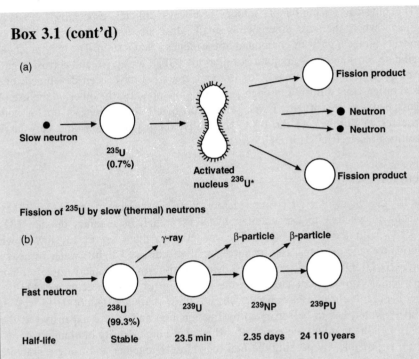

Figure 3.7 *Schematic of: (a) thermal reactor fission process; (b) neutron capture to form plutonium.*

The low concentration of ^{235}U, the fissionable isotope, in natural uranium makes it difficult to sustain a chain reaction. The majority isotope (^{238}U) reacts with fast neutrons to give, first, neptunium (^{239}Np) and then plutonium (^{239}Pu), in each case with the ejection of a β-particle (Figure 3.7(b)). This process effectively mops up many of the neutrons before they can be slowed down (moderated) to cause fission of ^{235}U. The decay of ^{239}U to ^{239}Np occurs with a half-life of 23.5 min, and ^{239}Np to ^{239}Pu with a half-life of 2.35 days. By contrast, ^{239}Pu is comparatively stable with a half-life of 24 110 years. This stability is not sufficient for plutonium to be found in nature, but it can be manufactured in a nuclear reactor and then isolated. Like ^{235}U, ^{239}Pu is a fissile isotope. In order to fabricate a nuclear weapon, it is necessary to gather together a 'critical mass' of a fissile isotope, either ^{235}U or ^{239}Pu. The ^{235}U isotope has to be 'enriched' and separated *physically* from the overwhelming majority of ^{238}U, while ^{239}Pu has to be manufactured in a nuclear reactor and separated *chemically* from the uranium host. Both routes are possible and, indeed, the Hiroshima bomb was based on ^{235}U and the Nagasaki bomb on ^{239}Pu.

The physical separation of isotopes is always difficult, especially for heavy elements where the mass difference is small. Most successful processes for heavy elements utilize a gaseous compound of the element in question. The only practical candidate for uranium is uranium hexafluoride (UF_6), a reactive and corrosive gas. Two, quite different, processes for isotope separation have been developed: one based on gaseous diffusion through permeable membranes, the other on the use of high-speed gas centrifuges. Large production plants have been set up for both processes in Europe and the USA. Ironically, these plants consume very large quantities of electricity for their operation.

Reactor Design

For the designers of civil power reactors, there are choices to be made. A key decision is whether to use uranium metal as a fuel, or uranium dioxide UO_2 enriched with the ^{235}U isotope. The moderator, which serves to slow down (thermalize) the neutrons, can be pure graphite, ordinary ('light') water or heavy water (^2D_2O). Ordinary water captures neutrons to form ^2D_2O and is therefore not a particularly good moderator from the standpoint of neutron economy. On the other hand, it is cheap. Heavy water, which is present to the extent of 0.015 wt% in ordinary water, has to be separated by fractionation and is also expensive. If the fuel is metallic uranium (un-enriched), which has a high density of uranium atoms, then it is possible, by packing the fuel rods closely together and using a graphite moderator, to sustain a fission chain reaction. This is the basis of the 'Magnox' reactors in the UK, 11 of which were built with a total design capacity of 4000 MW_e. These are slowly being phased out after 30–40 years of successful operation. The reactors use pressurized carbon dioxide gas as the coolant. The uranium fuel rods are contained in a housing made from magnesium alloy ('Magnox'). The design outlet temperature for the coolant is 400 °C and the efficiency of the steam cycle is 31%.

The Canadian 'Candu' reactor also uses natural uranium metal as the fuel, but with heavy water as both the moderator and the coolant. Ordinary water is not able to sustain a chain reaction with natural uranium metal since it absorbs too many neutrons. Heavy water does not suffer from this problem. The design of the Candu reactor has proved successful and almost 40 power plants have been constructed, including exports to India, Pakistan, Argentina, Korea and Romania, with a total installed capacity approaching 20 GW_e.

The alternative nuclear fuel, UO_2, is too dilute in uranium atoms to sustain a chain reaction unless it is enriched with ^{235}U. Most of the world's reactors utilize UO_2 fuel, enriched to the point that it is possible to use ordinary (light) water as both the moderator and the coolant. The fuel is manufactured by a complex route that involves converting all the natural uranium to UF_6, which is subsequently passed through a gaseous enrichment plant until the desired level of ^{235}U is attained, and then converted back to UO_2. The latter is fabricated into ceramic pellets, which are stacked vertically in thin metal tubes known as 'fuel rods'. After close inspection, the rods are bundled into fuel assemblies at the fuel fabrication plant and then sent to a nuclear power plant. The most common reactor types are the pressurized water reactor (PWR) and the boiling water reactor (BWR), both of which use light water as both moderator and coolant. The fuel rods are made of 'Zircaloy', an alloy of zirconium, tin and iron that possesses great strength and high resistance to corrosion from water and steam.

The advanced gas-cooled reactor (AGR) developed in the UK houses the enriched UO_2 fuel pellets in stainless-steel tubes. The coolant is pressurized carbon dioxide gas, as with the earlier Magnox reactors, and graphite blocks form the moderator. The AGR is operated at higher temperatures than either Magnox or water reactors. The outlet temperature of the coolant is 650 °C and this leads to a high steam-cycle efficiency (42%). Seven AGR power stations were built in the 1970s and 1980s and had a total nominal capacity of 9 GW_e.

There is still scope for debate over the relative merits of water reactors and AGRs; the higher operating temperature of the latter is a significant potential advantage. The reason that most electric utilities have chosen water-cooled reactors has more to do with commercial drive and enterprise on the part of the USA manufacturers, rather than with any intrinsic advantage that these reactors might possess over gas-cooled alternatives. Indeed, if the USA had chosen in the 1950s to develop gas-cooled, graphite-moderated reactors, as opposed to water-cooled and water-moderated reactors, it is quite possible that the former would now be the technology of first choice. Water-cooled reactors do, however, have a good record of reliability and safety, and also benefit from the economics-of-scale in manufacturing a series of reactors of one design. (Note, the seven AGRs mentioned above all differ in design.)

Water-Cooled Reactors

PWR has a primary coolant circuit that uses ordinary water at high temperature (300 °C) and high pressure (15 MPa). This acts as both coolant and moderator. The hot water passes from the core of the reactor to a heat-exchanger that is contained

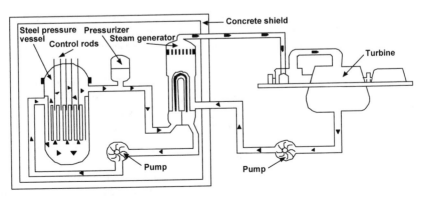

Figure 3.8 *Schematic of pressurized water reactor.*

within the same concrete shield (Figure 3.8). The secondary circuit is not under high pressure and so generates steam to drive the turbine. Condensed water from the turbine is fed back to the heat-exchanger, which then serves as a steam generator. The combined heat-exchanger and steam-producing unit is housed within the concrete shield so that radioactivity is contained should a fracture occur in the primary circuit.

Cylindrical fuel pellets of UO_2, about 10 mm in diameter and enriched to around 4 wt% in ^{235}U, are packed end-to-end in long tubes of zircaloy to form the fuel rods. These rods are typically 3.5 m in length. The fuel assembly contains 250–300 fuel rods, and each PWR has 150–200 fuel assemblies in its core. A typical PWR has a design power output of 1 GW and contains around 100 t of enriched uranium. The 'control rods', which manage the reactivity of the core, are made of a material such as cadmium or boron that is highly absorptive for neutrons. To reduce the reactivity, the control rods are introduced further into the reactor; conversely, to increase the reactivity, they are withdrawn. If the reactivity gets out of control, the rods drop into the reactor under gravity.

The second most common type of reactor is the BWR. This is similar to the PWR in terms of the choice of fuel (UO_2 enriched with ^{235}U), as well as the moderator and coolant (ordinary water). The BWR, however, is fitted with a single coolant circuit and the cooling water boils inside the reactor. The temperature of the coolant is the same (300 °C), but the pressure is lower (~7.5 MPa). At this temperature and pressure, water boils in the reactor core. A mixture of water and steam leaves the core and is separated; the water is pumped back to the bottom of the core and recirculated upwards through it, while the steam is used to operate the turbine. Because the BWR has only one coolant circuit and no heat-exchanger, the steam that passes to the turbine can be radioactive. Shielding is therefore required in the vicinity of the turbine.

There are said to be around 430 land-based power reactors in the world, of which 103 are in the USA. These generate 17% of the world's electricity and are mainly PWRs and BWRs. Several hundred power reactors, mostly PWRs, are also used to power submarines. Electricity-producing reactors in the UK are unusual in being gas cooled and graphite moderated, both Magnox and the AGRs. There is just one PWR in Britain (the Sizewell 'B' station in Suffolk). The Canadian reactors are unique in using uranium metal fuel and a heavy water moderator. Other countries (France, Japan, Korea, *etc.*) have mostly adopted designs from the USA. The Eastern European countries have tended to build reactors developed in the Russian Federation.

Several countries are engaged in co-operative programs to develop advanced designs of water reactors. These tend to focus on smaller units (100–350 MW_e) with the steam generator enclosed within the pressure vessel. With this configuration, the primary system cannot suffer a major loss of coolant, even if one of the large pipes were to break. Some advanced designs operate at higher temperatures and pressures, *i.e.* in the supercritical water range (374 °C, 221 atm), in order to improve the thermal efficiency to values approaching 45%. This introduces a major problem due to the increased corrosiveness of supercritical water.

High-Temperature Reactors

Another type of thermal reactor, still under development, is the high-temperature gas-cooled reactor (HTGCR), which operates around 900 °C and is cooled by helium gas. Work on various versions of this reactor was undertaken in the 1960s and 1970s. The most successful was that investigated in Germany, the so-called 'pebble-bed reactor'. A 300-MW demonstration reactor was built, but the program was abandoned in 1990, in part for political reasons. This concept is now being taken forward by engineering teams in South Africa, the USA, and the People's Republic of China. South Africa plans to build and operate a full-size prototype plant.

The fuel is in the form of many thousands of spheres, each 60 mm in diameter (Figure 3.9). Every sphere contains 15 000 small kernels of uranium dioxide (0.5 mm in diameter) that are coated with several impermeable barrier layers of pyrolytic graphite and silicon carbide; the latter serves as a micro-pressure vessel to contain fission products. The fueled graphite spheres are loaded at the top of

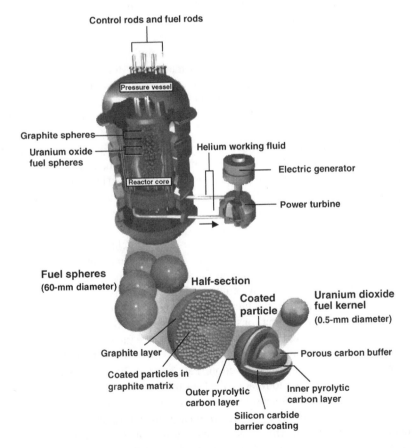

Figure 3.9 *Schematic of pebble-bed nuclear reactor.*
(Courtesy of Scientific American)

the reactor and are allowed to travel downwards under gravity, to be discharged at the bottom. The reactor design is modular, has an electrical output around 120 MW, can be factory built, and is suitable for distributed generation (see Section 3.7). The concept has several advantages, namely: operation at high temperature makes for high efficiency in electricity generation; it has passive safety features (unlike water reactors where high-pressure steam is involved); it can be refueled continuously while on load. Although this design of reactor is very attractive technically, there are divided views on the prospects for its commercial success.

Fuel Reprocessing and Fast-Breeder Reactors

At the end of their useful life and after removal from the reactor core, the fuel elements are highly radioactive. They are handled remotely and contained in shielded flasks. At this stage, there are two options: either they may be stored indefinitely in their intact form while the activity decays (open cycle), or they may be chemically reprocessed (closed cycle). The latter option involves dissolving the fuel in acid, separating the fission products from the unreacted uranium, and recovering the plutonium formed in the reactor. The highly reactive fission products may then be stored in solution, dried and entombed in concrete, or incorporated into an inert glass matrix for storage underground. The USA uses the open-cycle approach and stores all of its irradiated fuel elements, whereas France, Japan and the UK have favored the closed cycle. Reprocessing plants are in operation in France (Cap La Hague) and the UK (Sellafield).

After reprocessing, the recovered uranium is depleted in ^{235}U. This may either be converted back to UF_6 and re-enriched, or mixed with ^{239}Pu, which is also fissile, and converted to a blend of uranium and plutonium oxides $(U, Pu)O_2$, which may once again be used as a thermal reactor fuel. This is referred to as 'mixed oxide fuel' (MOX). In effect, MOX provides a means for reducing the world's inventory of plutonium.

Originally, the intention in reprocessing irradiated nuclear fuel was to recover plutonium, both for military purposes and as a fuel for use in fast-breeder reactors (FBRs). The designation 'fast' refers to the speed of the neutrons involved in the fission process ('fast fission'), as contrasted with 'thermal' (or slow) neutrons employed in the thermal reactors mentioned above. FBRs are based upon fission of the ^{239}Pu nucleus with fast neutrons. These reactors have the remarkable property of generating more plutonium than they consume, hence the name 'breeder'. This is not contradictory to the laws of thermodynamics since we are here dealing with a nuclear process, not a chemical reaction. Fast reactors have an extremely high-power density and, in order to remove the heat from the reactor core, liquid sodium is employed as a coolant. Because of their breeding potential, FBRs constitute, in principle, an almost unlimited resource of energy for the future. By progressively producing plutonium in thermal reactors and burning it in fast reactors, estimates show that it is theoretically possible to extract 50–60 times as much energy from a given quantity of uranium. Prototype FBRs have been built at Dounreay in

Scotland, as well as in France, the USA, and the Former Soviet Union. All have been shown to operate satisfactorily, albeit with some teething problems. Problems encountered with early models were more concerned with the engineering of the sodium coolant circuits than with the basic reactor physics. To date, the demonstration reactors have not led to full-scale power stations, mostly because of the costs involved. FBRs are now seen as a possible energy option for the second half of the 21st century, but much depends upon the overall future of nuclear energy. A major concern is that by the time interest in FRBs revives, much of the technical expertise will have been lost.

The Future of Nuclear Energy

What is the likely future of nuclear energy in the shorter term, until 2020? The one certain fact is that many of the older nuclear stations, including the UK Magnox stations and many of the Russian-designed stations, will reach the end of their planned life. Although, in certain cases, this may be extended by a few years, by 2020 there will be a marked decline in the world's nuclear generating capacity unless some of these stations are replaced. The factors militating against building new nuclear power stations are the widespread anti-nuclear sentiment and the high capital cost and long time-scale for constructing these plants compared with gas-fired alternatives. There is also the popular support for renewable sources of electricity to replace nuclear.

For the nuclear industry to have a viable future, it must meet three targets:

- reduce the capital costs of construction by ~30%
- build a power plant within about three years (to keep financing costs to a acceptable level)
- improve generating efficiency to ~40%

The best prospect for achieving these targets appears to lie with smaller, higher temperature reactors, especially the HTGCR. It will be necessary, whichever reactor type is chosen, to simplify sub-systems and components, to employ a standardized reactor design with factory fabrication of smaller modules, and to streamline assembly techniques.

Before leaving the subject of nuclear fission, it is worth pointing out that plutonium constitutes one of the world's large potential resources of stored energy. Plutonium is inevitably formed in a thermal reactor, whether we like it or not. Large quantities of plutonium already exist in the world and the amount will steadily accumulate with the continued production of nuclear power. The only choice facing nations is to recover and re-use the plutonium, by reprocessing the irradiated fuel (closed cycle), or to store radioactive fuel elements in their spent state (open cycle). The recovery of plutonium would provide a huge resource of non-fossil and sustainable fuel that may be burnt in either thermal reactors or fast reactors.

The advantages and disadvantages of generating electricity from nuclear power (fission) are summarized in Table 3.6.

Table 3.6 *Advantages and disadvantages of nuclear power*

Advantages	Disadvantages
An almost inexhaustible form of energy	High capital cost and long time-scale for construction
Concentrated source of energy	Problem of radioactive waste disposal
Well-established industry	Public opposition in many countries
Reliable and safe reactor designs	
Base-load generation	
Low fuel costs	
No liberation of greenhouse gases	

In the later chapters, we shall explore in detail the prospects for renewable energy as compared with the energy that is derived from fossil and nuclear sources, and attempt to assess realistically the contribution that renewable energy might make in the foreseeable future.

3.6 Nuclear Fusion

Nuclear fusion involves the combining of two light nuclei to form a heavier nucleus (see Box 3.2). Because the mass of the final nucleus is less than the combined masses of the original nuclei, there is a loss of mass accompanied by the release of energy. Very high temperatures are required to drive these energy-liberating reactions, and they are therefore called 'thermonuclear fusion reactions'. Fusion occurs continuously in the sun, liberating light and heat, and has been demonstrated on Earth in the 'hydrogen bomb', first exploded in 1952, and similar fusion weapons. In a thermonuclear explosion, the energy is released in a very small fraction of a second with catastrophic consequences. This is of limited value for the generation of a continuous source of electricity.

The major obstacle in obtaining useful energy from fusion is the large (coulombic) repulsive force between the charged nuclei at close separations. Sufficient energy must be supplied to overcome this barrier. The temperature required to produce fusion is in excess of 10^8 °C! At such high temperatures, all matter is in the form of a 'plasma', *i.e.* the atoms are stripped of all their electrons to give a mixture of atomic nuclei and free electrons. In a fusion reactor, the plasma temperature must reach at least the 'critical ignition temperature', *i.e.* the temperature at which the power generated by the fusion reactions exceeds the power lost in the system. Moreover, this temperature must be held for a 'confinement time' that is sufficiently long to ensure that more fusion energy will be released than is required to heat the plasma. Therefore, the objective of research into plasma physics and thermonuclear fusion is to design technologies whereby the fusion process can be initiated, controlled and sustained for a sufficiently long time. The

Box 3.2 Fundamentals of Nuclear Fusion

Isotopes of several of the light elements can, in principle, undergo fusion reactions, but the most promising of these involves the isotopes of the lightest element, hydrogen. Hydrogen has three isotopes, namely, protium (^1P), deuterium (^2D), and tritium (^3T). At a sufficiently high temperature, fusion may be made to take place between a deuteron (the nucleus of a deuterium atom) and a triton (the nucleus of a tritium atom), *i.e.*

$$^2D + {}^3T \longrightarrow {}^4He + n + \text{energy} \qquad (3.1)$$

The fused nuclei are unstable and fly apart liberating a helium nucleus (alpha particle) and a neutron, n, with the release of energy. Of the two isotopes that constitute the 'fuel' for the fusion reaction, deuterium is prepared from heavy water, which is made by the irradiation of ordinary water in a nuclear reactor, *i.e.*

$$^1H_2O + 2n \longrightarrow {}^2D_2O \qquad (3.2)$$

or even by the fractionation of ordinary water. Tritium is made by the irradiation of lithium in a nuclear reactor, *i.e.*

$$^6Li + n \longrightarrow {}^7Li^* \longrightarrow {}^4He + {}^3T \qquad (3.3)$$

where $^7Li^*$ is a highly energetic and unstable form of ^7Li that decomposes almost instantaneously. In nature, lithium occurs as ^6Li (7.5%) and ^7Li (92.5%); it is only the ^6Li isotope that absorbs neutrons to give tritium. In a fully functional fusion reactor, the tritium would be made *in situ* from neutron-induced reactions in a lithium 'blanket' that surrounds the reactor core; this tritium is then recirculated into the reactor for fuel. Since there are almost unlimited quantities of water and lithium available, there is no resource problem as regards fuel for a future fusion reactor.

target for a successful outcome is to hold the plasma in a stable configuration for more than a second. Clearly, there is no solid material that can be used to contain the plasma at these temperatures. Accordingly, two separate *physical* containment processes are being investigated: 'magnetic confinement' and 'inertial confinement'. Research and development has focused mainly on magnetic confinement.

Because a plasma consists of charged particles, it may be confined by a magnetic field in a 'magnetic bottle'. The classical confinement configuration is in the form of a toroid – a vacuum vessel in the form of a circular closed ring or loop (resembling a doughnut). There are two magnetic fields, one around the major

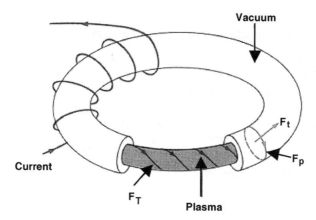

Figure 3.10   *Schematic of tokamak used for the magnetic confinement of plasmas. The total magnetic field F_T is the superposition of the toroidal field F_t and the poloidal field F_p. The plasma is trapped within the spiraling field lines, as shown.*

circumference (the toroidal field) and the other around the minor circumference (the poloidal field) (Figure 3.10). These two magnetic fields interact to give a resultant helical field that confines the plasma to the central axis of the toroid and keeps it from touching the walls of the vacuum vessel. The trick is to stabilize the plasma in this configuration so that it never comes into contact with these walls; otherwise, it would cool and also become 'poisoned' with heavy impurities sputtered from the walls, both of which would lead to large power losses. This design of fusion containment, which was invented by the Russians in the late 1960s, is known by its Russian name 'tokamak'. The plasma is heated by a very large electrical current passing through it, supplemented by microwave heating and by the injection into the plasma of energetic neutral particles. A beam of charged particles, *i.e.* hydrogen, deuterium or helium ions, is produced in an ion source. The ions are accelerated to 140 kV to increase their energy and then neutralized. They can now penetrate the magnetic fields that surround the plasma. In the plasma, the energetic atoms become ionized again and, by collisions, give up their kinetic energy to the plasma. The Joint European Torus (JET) – a tokamak constructed in the early 1980s at the Culham Science Centre in the UK – has two such neutral-beam injection units that, together, supply up to 20 MW of heating. A cut-away representation of JET is shown in Figure 3.11. To give an idea of scale, a figure of a man is included in the lower right corner. Until recently, JET was used for experiments in plasma physics.

The first studies of deuterium–tritium fusion with JET began in 1991 and by 1997, a power-level of ~21 MW had been attained. The success of subsequent JET experiments has encouraged the development of a much larger tokamak to test the plasma physics, the components and the technologies that will be required to establish a complete fusion power plant. The engineering challenges involved, the time-scale and the cost of this successor tokamak are such that full international collaboration is required to carry the project forward. This 'next step' is termed the

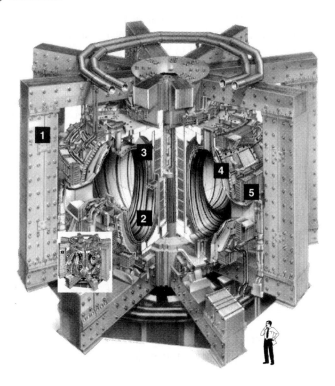

Figure 3.11   *Cut-away representation of the Joint European Torus (JET): (1) massive limbs of transformer core; (2) one of eight sections of the JET vacuum vessel; (3) one of 32 toroidal field coils; (4) the field coils are housed in a mechanical shell that surrounds the vacuum vessel; (5) one of six outer poloidal field coils.* (Courtesy of the JET project)

international thermonuclear experimental reactor (ITER) and is likely to involve scientists from the European Union, Canada, Japan, the Russian Federation, and the USA. Construction and operation will take many years. It is not envisaged that commercial fusion power plants will be available, if at all, for at least 50 years.

A toroid is by no means the only possible configuration for containing a high-temperature plasma in a magnetic bottle. Another promising configuration under active investigation is a spherical tokamak. Two spherical tokamaks have been built at Culham in recent years and have given promising experimental results. The design offers the prospect of more compact and more economic power plants than with the toroidal alternative. Various other approaches to develop magnetically controlled fusion are being studied in several European countries, Japan, the Russia Federation, and the USA.

A quite separate approach to fusion is that of 'inertial confinement'. The basic principle is to use an external driver to compress and heat a small pellet of fuel. If the compression raises the density further to a sufficiently high level in a very short confinement time (typically, 10^{-9}–10^{-11} s), while heating the fuel to a sufficiently high temperature, then economically attractive amounts of fusion energy can be produced. The resulting confinement is called 'inertial' because it depends on the inertia of the fuel itself to maintain high densities and temperatures for a time that is sufficient to burn the fuel before it rebounds and flies apart. Modern high-power lasers are capable of being focused to a spot with an intensity of 10^{19} W cm^{-2}. Consequently, it has been demonstrated that a large pulse of energy can be obtained when a small pellet of frozen deuterium+tritium (at below 14 K) in an evacuated chamber is irradiated simultaneously by several powerful lasers, *e.g.* six arranged octahedrally, as shown in Figure 3.12. The energy from the combined laser irradiation causes the surface of the fuel pellet to evaporate explosively. The escaping particles produce a reaction force on the core of the pellet that creates a strong, inwardly moving, compressive shock wave. This shock wave increases the pressure and density of the core and gives rise to a corresponding increase in temperature that is sufficient to ignite a propagating fusion reaction. By releasing 70 or more times the energy needed to compress and heat the fuel, this dynamic process provides the basis for generating inertial fusion energy. It is foreseen that a fusion reactor of this type would involve sequential bursts of energy as fuel pellets are introduced, one at a time, and the lasers fired. Experiments along these lines have been conducted, but only at a preliminary and small scale.

Many countries continue to fund research into fusion as it is seen as the ultimate means of generating electricity on a large scale in the post fossil-fuel age. The advantages cited are:

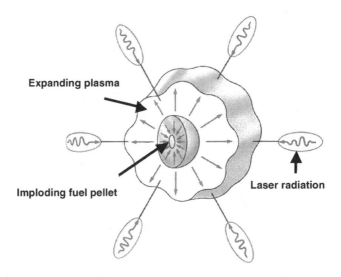

Expanding plasma

Imploding fuel pellet

Laser radiation

Figure 3.12   *In the inertial confinement of plasmas, a pellet of deuterium + tritium fuel undergoes fusion when struck by several high-intensity laser beams simultaneously.*

- a virtually limitless supply of low-cost fuel is available
- no carbon dioxide or gaseous pollutants are formed
- the fusion process is very safe because the amounts of deuterium and tritium fuels in the reaction zone are so small there is no possibility of a large and uncontrolled release of energy
- much less radioactive waste will be formed compared with a fission reactor and, moreover, the waste will have shorter half-life
- the absence of plutonium and other long-life radio-nuclides.

There is, however, a general recognition that, after almost 40 years of research, commercial fusion reactors are still several decades away from practicality, and that the economics of electricity production at that distant time can hardly be foreseen. Fusion research is an act of faith rather than of certainty – faith that it offers the prospect of providing the world's electricity without causing environmental degradation or climate instability.

Finally, it should be mentioned that the phenomenon of the so-called 'cold fusion' was reported in the early 1990s by a number of scientists. This involved a simple electrolysis process at ambient temperature during which, it was claimed, the release of heat and even neutrons were detected. Despite considerable effort in many laboratories around the world those findings have not been substantiated unequivocally.

3.7 Distributed Generation

Traditionally, in industrialized nations, most electricity has been generated in very large power plants. In some countries, the plants are owned by private companies; in other countries, they are state monopolies. Power stations are usually located at sites that are not too remote from large consumer markets and where there is easy access to fossil fuels or hydroelectric power. In the case of nuclear power, siting is also determined by safety and political considerations.

High-voltage electricity is transmitted across the country by cables mounted on pylons and the cables are interconnected to form a network, the so-called 'grid'. This ensures a high security of supply. If any one station fails or is shut down for maintenance, other stations take up the load. Similarly, if part of the grid is affected by lightning strikes or cables are brought down by snowstorms, the interconnected mesh provides alternative routing. The grid may be owned by the state utility or by a private company that is responsible for its integrity and for providing advanced protection schemes. At strategic regional centres, the grid cascades into medium-voltage distribution networks and, for the final kilometer or two, into a low-voltage network that supplies the consumer. The 'utility' is generally regarded as the company that supplies and bills the customer, and maintains the local distribution network. This may be the same organization that generates and transmits the electricity, or it may purchase electricity in bulk for sale to retail customers. This, then, is the classical electricity-supply system.

The concept of local, small-scale generation of electricity is by no means new. Indeed, early in the 20th century, before the days of the cross-country grid, most electricity supply was of this nature. With the advent of the grid, local generation became confined largely to isolated communities and islands where no grids existed. There is currently a resurgence of interest in local generation, which is referred to as 'distributed generation – DG'. This began with battery back-up systems to provide uninterruptible power supplies (UPS, see Section 6.3, Chapter 6) and with small-scale CHP systems (CHP, see Section 3.4), but has grown to encompass a range of other technologies, *e.g.* engine-generator sets, small hydroelectric plants, microturbines, wind generators, biomass-based generators, landfill-gas generators, photovoltaic systems and fuel cells, together with various storage technologies (see Chapters 4–9). Typically, generating plant is dispersed across the network, near to the consumers, rather than concentrated at one location, hence: 'distributed generation'. The generating systems may operate independently of the grid, but there is an increasing trend for them to be grid-connected, *i.e.* 'embedded generation'. The growing interest in the decentralization of electricity generation is driven by considerations of overall energy efficiency, by an increasing desire to introduce renewable energy sources and, more recently, by de-regulation of many electricity markets.

Distributed generation is considered to offer the following benefits:

- enhanced system reliability when stand-by generators are installed
- the high energy efficiency of CHP plants
- cost-effective peak-shaving
- lower grid losses, typically 10–15% saving
- ability to provide voltage support and power-factor correction
- deferred investment in upgrading the transmission and distribution system
- allows the introduction of renewable forms of electricity.

Evidently, this is a powerful set of incentives for introducing distributed generation.

The size of DG units varies widely, from less than a kW to tens of MW. Small units of a few kW are, typically, domestic CHP installations; these are likely to become more popular as fuel cells are brought to market. Large MW-sized units are often industrial CHP plant that supplies process steam and hot water, as well as electricity, to factories. The problem is how to integrate such dispersed generating units into the national grid. This is a complex issue that involves both technical factors and commercial considerations. Here, we consider briefly the technical factors.

A diagrammatic representation of how embedded generators are incorporated among the consumers and supply their neighbors *via* low-voltage distribution lines is given in Figure 3.13. So long as the amount of power generated is less than that consumed locally, no problems arise; the balance in demand is made up from the grid. Problems may surface, however, when there is a surfeit of electricity to be fed back into the grid. An inappropriately connected unit could compromise the integrity of the grid, disturb the power quality, or even become a safety hazard. The grid was designed for unidirectional flow of electricity from higher to lower voltage levels, and network protection schemes were designed for this purpose and not for

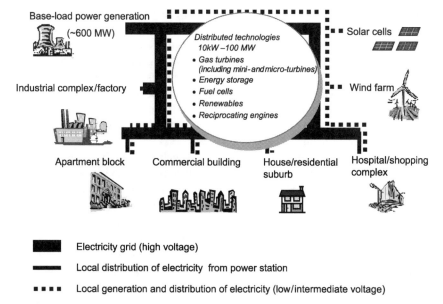

Figure 3.13 *Distributed power generation.*

reverse flows. A large number of DG installations within a low-voltage area would lead to power feedback to the medium-voltage network, and this would require different protection schemes at both voltage levels. Finally, there are technical issues concerned with short-circuit currents, overload protection, the distribution distances involved, and whether underground cables or overhead lines are to be used. And, of course, there is the commercial issue of what financial contribution the embedded generator should make towards the cost of modifying and maintaining the local distribution network into which it is exporting electricity. All of these issues will assume increasing importance as distributed generation increases in application. At present, there are no well-defined national or international regulations or standards within Europe or the USA for the integration of distributed generation within grid networks. This issue is presently being addressed.

3.8 Renewables and Energy Storage

Renewable forms of energy that are sustainable indefinitely are a major theme of this book. Many and diverse are the forms of renewable energy that are being exploited today, albeit on a comparatively small scale, and that have the potential to contribute a greater share of the world's primary energy. It is convenient to describe them under two separate headings. Chapter 4 is concerned with thermal processes, that is, processes that give rise to heat – whether through combustion (biomass), through solar heating, or through extraction of energy from the Earth (geothermal). When this heat is available at a high temperature, it may be used for process heat in industry, or for raising steam and generating electricity. Where the heat is of low quality, it may be used for heating water (domestic hot water, swimming pools), or for the space heating of buildings. Chapter 5 discusses other renewable forms of energy that may be harnessed to generate electricity directly without first going through a steam cycle. These sources include hydro, wind, tidal and wave energy. Solar energy falls into both camps, so solar heating is discussed in Chapter 4 and solar photovoltaics in Chapter 5. It may be noted that most forms of renewable energy, with the exceptions of geothermal and tidal energy, are derived indirectly from solar energy.

One of the problems with renewable energy, which has long been recognized, is its non-uniform or erratic availability. There is little correlation between where and when energy is available and where and when it is needed. For this reason, most forms of renewable energy, especially those used for the direct generation of electricity, require a storage component to be provided as an integral part of the system. This is by no means a straightforward matter. There are numerous possible options, as summarized in Figure 3.14, but few are entirely satisfactory. Certainly, it will be necessary to tailor the storage option to the type and magnitude of the primary energy form and to the size and nature of the application. The rates at which energy may be fed into or extracted from storage are also important

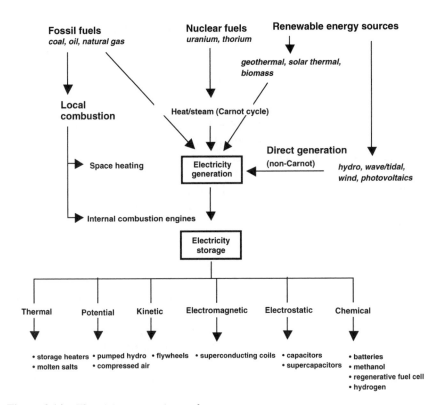

Figure 3.14 *Electricity generation and storage.*

considerations. Detailed discussion of the various options and applications for energy storage is given in Chapters 6–10.

In order to make a quantitative comparison with established fossil-fuel technology, it should be noted that the reference energy-storage system is the filled oil tank. This has a very high specific energy (11.7 kWh kg^{-1}) and energy density (10.3 MWh m^{-3}). Rates of filling are also very high (~30 MW) and losses are almost zero. Moreover, storage takes place at ambient temperatures and pressures and the tanks are cheap to purchase. These are targets that it will be very difficult to match by any other storage medium.

In the early phase of introducing renewable sources of electricity into nations with well-established electricity grids, the need for storage technologies is not so acute. This is because the grid itself and the conventional power stations that feed it act effectively as the storage facility. Fossil-fuel stations can, in principle and commercial considerations aside, serve as standbys for when hydro-reservoirs run dry or the wind does not blow. In the UK context, the government has set a target of 10% of its electricity from renewable sources by 2010, and 20% by 2020. It is likely that the grid can accommodate this much electricity without the need for dedicated storage facilities. The extent to which renewables can be substituted for fossil-derived electricity without discrete storage facilities requires further study in terms of both technical and commercial considerations.

3.9 De-regulation of Electricity Markets

Traditionally, many countries have had a state monopoly for the generation and distribution of electricity. In recent years, there has been a move towards privatization of this industry and the introduction of competition in order to promote efficiency and lower the price of electricity – so-called 'de-regulation' or 'liberalization' of the electricity market. The UK has been foremost in this field in Europe and the actions taken are therefore discussed here as a case study of de-regulation. First, the generating function was split into three major companies (two fossil and one nuclear) that all competed with one other, a transmission and distribution company, and numerous regional electricity companies who purchased electricity wholesale from the generators and marketed it retail to consumers. This situation persisted for some time before the two fossil-based generators were further sub-divided when they sold off some of their power stations. New companies tendered for and were awarded licences to build gas-fired power stations. At the same time, the retail market was de-regulated and the retail companies were permitted to supply customers nation-wide and thus were no longer confined to their regional origins. Finally, the major generators bought up some of the retail companies, thus integrating vertically. The UK now has a fully privatized and competitive electricity market in all aspects except for cross-country transmission where one company has a monopoly. Other countries are proceeding along somewhat similar paths, albeit at different rates.

Is de-regulation a good thing? From the viewpoint of the original concept of promoting efficiency and lowering the price of electricity, it has certainly been a success. Moreover, the UK government has rid itself of the chore of running a monolithic state enterprise and the need to raise taxes to invest in new power stations and transmission lines. Private capital is employed instead. In the longer term, however, problems loom large and arise from the fact that private companies tend to focus on short-term return and have little interest in making investments that will not show a profit for many years. Nor does the national interest enter much into their calculations. This regime militates strongly against nuclear power since such power stations are capital-intensive to build and have a long lead-time (typically 7–10 years) for construction. Aside from any other considerations, it is highly unlikely that, for purely financial reasons, private capital would invest in building new nuclear stations without major government incentives. Rather, private companies would all opt for CCGT technology, which is comparatively cheap and quick to build, efficient to operate, and is expected to yield profits after 4–5 years. This strategy is acceptable so long as there is an unlimited supply of natural gas available at low cost. While this may be true for some years yet (see Chapter 1), the time will come when the UK will have to import gas from far afield, just as Japan and Korea do today.

In the USA, the State of California has experienced some severe power shortages that have led to load-shedding and black-outs. In part, this is attributable to the rapid growth in electricity demand that has arisen from the mushrooming of computers and industrial equipment. Nevertheless, blame can also be attached to the commercial situation that arose following de-regulation of the electricity market. Whenever there is a power shortage, the spot market price for electricity rockets. Utilities are obliged to buy at high prices and then sell to customers at the lower prices set down in the existing supply contracts. Clearly, this is an unstable commercial situation. The long-term solution may be to construct more power stations or transmission lines, but a faster and cheaper solution may be to build distributed generation plant (microturbines and/or CHP systems) or storage facilities (pumped hydro, batteries, and/or flywheels) to meet the peaks in demand. Large consumers, who buy directly from the spot market, have an incentive to install their own storage facility, or even to shut-down their operations at times of peak price. Customers on long-term, fixed-price contracts, could even sell the unused electricity back into the market at a profit! In Los Angeles, a power company is offering major financial incentives to its largest customers to lower their energy consumption at peak times by installing their own distributed energy-storage systems.

The private ownership of the electricity industry also militates against renewable forms of energy as these too are more expensive than gas-fired power stations. So long as supplies of cheap gas are available, private industry is only likely to invest in renewables in response to government subsidies or the imposition of a sizeable carbon tax. This may well prove to be a problem for off-shore wind power, where the capital investment is sizeable, and even more so for solar PV on a large scale. Distributed generation and energy storage also have to fit within this framework, and will only be adopted to the extent that the electricity industry sees it as profitable to do so and regardless of broader considerations. The essential issue is one of the interface between the government and the electricity industry and the challenge facing the government is how to legislate and regulate the industry so as to achieve long-term national sustainability without sacrificing the benefits that stem from a privatized industry.

3.10 References

1 *Key World Energy Statistics from the IEA*, 2003 Edition, International Energy Agency, Paris, France, 2003.
2 *New and Renewable Energy: Prospects in the UK for the 21st Century*, UK Department of Trade and Industry, London, UK, March 1999.
3 Shell Company website: www.Shell.com (November 2001).
4 *UK Energy in Brief*, UK Department of Trade and Industry, July 2003.

CHAPTER 4

Renewable Energy – Thermal

From earliest times, humankind has burnt wood – a prime source of renewable energy – in order to keep warm and to cook food. Long before fossil fuels were exploited, wood was the universal fuel. Even today, it is still a principal fuel for domestic purposes in many parts of the world, although animal dung is also used in regions where timber is not freely available. These two forms of energy are said to represent the primary fuel supply for more than one-third of the world's population. In western countries where there are good supplies of fossil fuels, many people in rural areas also choose to have log fires and/or closed wood-burning stoves, either as a matter of preference or because they have easy access to low-cost wood.

As we saw in Chapter 1, Table 1.1, combustible renewables (mostly wood) and waste constitute 11% of the world's primary energy supply. This represents over 1 000 Mtoe per year and makes biomass the fourth major source of primary energy (after oil, coal, and natural gas), greater than both nuclear power and hydroelectricity. Much of this renewable energy is concentrated in non-OECD countries, and it is the burning of wood, with associated smoke and ash, that is said to be largely responsible for the massive brown cloud that stretches over much of Asia at certain seasons. This is a new form of 'smog', different from the coal-based smog of the 1950s in Europe (see Section 2.3, Chapter 2), and may be attributed to the use of open fires and simple stoves, rather than to combustion furnaces. The latter technology is usually properly engineered and equipped with a dust-collection facility. Thus, at present, most wood and biomass, although renewable, is certainly not a source of 'Clean Energy'. With the correct design of combustion equipment, this need not be so.

Much of the calorific value of wood (80%) is in the form of combustible gases that are evolved on heating. These ignite only at high temperatures (800–900 °C) and thus, for maximum efficiency, a closed wood-burning stove should have a combustion chamber with a fireproof stone-lining that will withstand the high temperature. In recent years, much development work has been undertaken on the design of these stoves, particularly in Denmark (Figure 4.1(a)) where the government is encouraging the burning of wood for both industrial and domestic heating. Modern closed stoves are at least 75% efficient when operated at their

(a)   (b)

Figure 4.1   (a) *Closed wood-burning stove approved by the Danish standardization body. The stove must meet a number of minimum requirements in terms of efficiency (at least 70% efficiency at nominal output), environmental protection (less than 0.3% carbon monoxide in smoke), safety, and design (door, ash grate, ash pan, etc.); (b) cross-section of manually fed, wood-fired boiler.*
(Courtesy of the Danish Centre for Biomass Technology)

design output. At this level, a stove burning 1 kg of dry wood per hour provides 3 kW$_{th}$ of heat. By contrast, the burning of wood in an open fireplace is hopelessly inefficient. When the temperature is too low, some of the combustible gas escapes unignited. It is also impossible to control the air supply, and much of the heat that is produced escapes up the chimney by convection. Furthermore, the smoke given off contains poisonous carbon monoxide and carcinogenic polyaromatic hydrocarbons. Unfortunately, this is how most wood is utilized today, especially in the developing world where the high cost of engineered stoves puts them beyond the reach of ordinary people. In future, it is to be hoped that, with mass production, the cost of closed stoves will fall and the world will derive substantially more useful energy from its plentiful wood supply while, at the same time, turning wood from a 'dirty' to a 'clean', non-polluting fuel. There is a historic analogy with soft coal, which today is rarely burnt on an open fire, but in closed stoves.

There have been similar developments in the design of biomass-fueled boilers for raising hot water or steam. Industrial boilers burn wood or straw. Domestic boilers, to feed radiators and supply hot water, burn either wood logs or wood chips. A schematic of a manually fed, wood-fired boiler is shown in Figure 4.1(b). This operates on the principle of downdraught combustion in which the fire burns downwards through a lined chamber at a very high temperature. This process ensures environmentally friendly combustion (carbon monoxide emissions as low as 0.1%) at high efficiency (80–90%).

The world resource of biomass is vastly greater than that used for primary energy. Estimates are that only around 2% of the biomass generated each year is used as a fuel.[1] Biomass – which includes agricultural, animal, industrial and municipal wastes, energy crops (*e.g.* fast-growing trees and grasses), standing forests, and aquatic and ocean plants – may be viewed as a useful mode of storage for solar energy, since it is formed by photosynthesis. In this process, plants combine carbon dioxide from the air and water from the ground under the action of sunlight to produce sugars and carbohydrates $[CH_2O]_n$ – which build up the plant bio-structure – and oxygen, *i.e.*

$$nCO_2 + nH_2O \longrightarrow [CH_2O]_n + nO_2 \tag{4.1}$$

The stored energy may be recovered by burning the plant or by fermenting the starches and sugars that it contains to produce alcohol. Certain plants or seeds contain oil that may be extracted and used as a fuel.

Biomass and combustible waste is certainly 'renewable', and hopefully 'sustainable'. Whether or not it contributes to the global greenhouse effect is debatable and depends upon the time-frame under consideration. All vegetation is renewable by photosynthesis, using carbon dioxide from the atmosphere, over a period of just a few years or (in the case of trees) a few decades at most. In an equilibrium situation, the carbon dioxide released through combustion is re-absorbed by photosynthesis and the net effect on the atmosphere of growing and burning vegetation is zero, (*i.e.* biomass is 'CO_2-neutral'). In the short-term, however, before equilibrium is reached, it is undeniable that all combustion processes increase the atmospheric burden of carbon dioxide. If the world could survive on other forms of non-fossil energy (nuclear, wind, solar, *etc.*), then the natural vegetation would gradually deplete the atmosphere of carbon dioxide until it was restored to the level that existed before human beings disrupted the ecosystem.

The question whether burning biomass contributes to enhanced levels of carbon dioxide in the atmosphere merits fuller consideration. One perception is that biomass is no different from fossil fuels since both combust to liberate carbon dioxide. A contrary view is that biomass may be treated as a 'renewable' form of energy because it is formed by photosynthesis from carbon dioxide and therefore combustion does not add to the global burden of greenhouse gases. In essence, both arguments are correct – the key factor is whether the short-term or the equilibrium situation is under consideration.

The view that biomass is renewable is only true, of course, if the crop from which it is derived is replanted after harvesting. Consider, for example, the planting of trees. Forests that are not harvested do not continue to accumulate carbon indefinitely. They eventually approach maturity and achieve, over time, a balance between the carbon taken up in photosynthesis and the carbon released back to the atmosphere from respiration, oxidation of dead organic matter, fires and pests. The system is complex and poses the question: is it better to use trees for energy production and recycle the carbon than to store carbon in forests while continuing to burn fossil fuels? The best strategy will differ from location to location, and will be determined by the quality of the land, its present uses, and the demands for energy and other products. Moreover, the most advantageous use of land for confronting the carbon balance may not necessarily involve trees. If the primary intent is to store carbon on site, the obvious choice is a high-density forest. On the other hand, if production of biomass energy is the goal, then a fast-growing herbaceous crop may be the best choice for some biomass energy technologies and some types of land.

Finally, it should be remembered that it takes some energy, much of it now provided by fossil fuels, to grow and harvest biomass crops and to haul the fuel to power plants. Thus, the use of biomass fuels does result in some discharge of extra carbon dioxide derived from the fossil fuel used incidentally. The extent to which biomass fuels can displace net emissions of carbon dioxide will depend on the efficiency with which they can be produced and used.

In summary, the issue of whether to grow energy crops is a complex one and involves considerations beyond those of renewable energy and carbon dioxide emissions. These include alternative uses for the land (what other crops can be grown?), sustainability (impoverishment of the soil, requirement for fertilizers), and location (accessibility, and costs of harvesting and haulage to a power plant). Each case needs to be assessed on its merits.

Apart from bio-energy (the generic name for energy derived from combustible waste materials or crops), there are other sources of renewable thermal energy. The chief among these are solar heat and geothermal heat. In this chapter, we review all the different forms of heat energy that are being harnessed or investigated.

4.1 Biomass: Agricultural and Forestry Wastes

Agricultural and forestry wastes fall into two main categories: 'dry wastes' such as cereal straw, wood wastes and poultry litter, and 'wet wastes' such as farmyard slurry, garden or green agricultural refuse, and water weeds. The two categories need to be treated quite separately. Waste paper is generally recycled and may either be pulped to manufacture lower-grade paper or burned in an incinerator to generate heat.

Dry wastes may be treated by one of the following four processes.

- *Combustion.* This is by far the most common practice. Dry wastes are compacted and then combusted to give heat. Combustion technology is well advanced and is widely used for industrial processes, for domestic purposes (hot water and space heating), and for the generation of electricity. Where applicable, the greatest overall efficiency is achieved by operating a combined heat and power scheme – a CHP scheme (see Section 3.4, Chapter 3). Combustion plant is available in a range of sizes from household boilers to 50-MW industrial furnaces.
- *Gasification.* This involves incomplete combustion in a limited supply of air and in the presence of steam to yield a mixture of combustible gases, mainly carbon monoxide and hydrogen, diluted in nitrogen. The process is similar to

the producer/water-gas reaction for the gasification of coal (see Section 2.3, Chapter 2). The gas is used for the production of electricity *via* an integrated gasification combined-cycle (IGCC) turbine (see Figure 2.6, Chapter 2).

- *Pyrolysis.* When a bio-material is heated in the absence of air, it decomposes ('pyrolyses') to give a mixture of combustible gases, organic liquids, and a residual char. The ratio of the three types of product can be adjusted according to the conditions of pyrolysis. Fast heating to moderate temperatures can give yields of up to 80 wt% liquid. This liquid ('bio-crude oil') can be used as fuel in boilers, engines, or turbines. It has a calorific value of 20–25 MJ kg^{-1}. Pyrolysis of wood and coal was extensively employed in the 19th century for the manufacture of chemicals and coal gas, but this industry has largely died out.
- *Liquefaction.* This process is conducted at relatively low temperatures (250–500 °C) in hydrogen at high pressure (up to 15 MPa or 150 atm). The product is a complex organic liquid with a heating value of 35–40 MJ kg^{-1}. Liquefaction is not extensively employed because of the high cost and inconvenience of pressure reactors.

Wet (green) wastes are best allowed to undergo anaerobic bacterial attack to produce 'biogas', a mixture of methane and carbon dioxide, which has a significant calorific value (*v.i.*). The residual solid matter (compost) may be used as a garden improver. In the UK, some Local Authorities are encouraging householders to compost their own kitchen and garden waste so as to reduce the amount going to landfill sites, which are in short supply. Whether this is a good idea in the long term is questionable. Landfill sites have the ability to utilize the biogas produced, whereas domestic composting leads to the greenhouse gases being released to the atmosphere.

It should also be borne in mind that biomass can be a store of nutrients for use in agriculture, horticulture, or even for the manufacture of chemicals. The common practice, in some parts of the world, of burning animal dung is a particular example of non-optimal use – though understandable where other fuels are in short supply. Ideally, biomass should be part of a carefully balanced ecosystem in which agriculture, silviculture, animal husbandry and bio-energy all have a role to play.

Straw

In Europe, the principal agricultural dry waste is cereal straw. Elsewhere in the world, there are other waste products in substantial quantities such as coconut shells and rice husks. Dry cereal straw has an energy content of around 18 MJ kg^{-1}, while freshly harvested straw, with a water content of around 15 wt%, is nearer 15 MJ kg^{-1}. By comparison, wood and coal yield 19 and 27 MJ kg^{-1}, respectively.[2] In its harvested state, straw burns quickly and is therefore not the most practical of fuel sources. Briquetting is a possible means to overcome this problem. This involves compacting the raw material into 'bricks' so that the energy content, in volumetric terms, is raised to a level approximately equal to that of wood fuel. The process requires the addition of a binder. Combustible binding

Figure 4.2 *Straw is abundantly available where grain is intensively cultivated. Technologies for obtaining heat and power from surplus straw are being developed in several European countries.*

agents include starches from cereal crops, molasses from sugar cane, tars, and glues from fish waste; non-combustible binders are ash, clay, or mud. Inevitably, the successful introduction of briquetting depends on the development of an appropriate technology to perform the process economically.

Meanwhile, the usual practice has been to compact the straw into high-density bales (Figure 4.2), which are conveyed by lorry to the point of use. Much of the straw is burnt on the farm, if not required for animal bedding purposes, and the heat produced is used for space heating of buildings and/or for grain drying. The resulting ash, which contains potash, is used by the farmers to fertilize their fields. This is most common in Denmark – a country with extensive cropland but few large forests – where larger straw-fired units (up to 8 MW_{th}) are used in district heating. These are often CHP schemes, with the electricity used locally or exported to the grid. Denmark has a growing number of CHP plants that use straw as fuel; together, these produce over 200 MW_{th} of heat and 85 MW_e of electricity.[2] It is estimated that straw can meet more than 7% of Denmark's energy needs. The UK has one of the world's largest straw-burning power stations at Ely in Cambridgeshire. This is a 31-MW_e installation that is designed to consume around 0.2 Mt of straw annually and generate about 270 GWh of electricity, which is sufficient to satisfy the electricity needs of some 80 000 homes. Around 12.5 Mt of straw is produced annually in the UK. Since it is estimated that 8.5 Mt is used in agriculture, around 4 Mt is potentially available for bio-energy.

Poultry Litter

Poultry litter is bedding material from broiler houses. It comprises wood shavings, shredded paper or straw that are mixed with chicken droppings. Two power stations, each of around 13 MW$_e$, have operated for a number of years in the UK using this material as fuel for a conventional steam turbine. Recently, a larger installation has been commissioned at Thetford in Norfolk that generates up to 38.5 MW$_e$ from poultry litter. As with straw, it is clearly uneconomic to convey poultry litter for any great distance.

Forestry Wastes

Only about 20–30% of the wood in a forest finishes up as useful timber ('lumber'). The remainder – brushwood, bark, and residues from the sawmill – is used in part to manufacture chipboard and paper, but much is used only as a fuel. Traditional forestry is directed towards obtaining a crop of quality timber for constructional purposes. Early in the life of a forest, it is necessary to thin out the understorey of some trees and remove others ('coppicing'). This gives rise to waste wood. Later, in the course of forestry management and the harvesting of tree trunks, a great deal of brushwood and sawdust is left behind. All of this material constitutes a source of biomass.

Hitherto, it has been a common practice to leave brushwood to compost *in situ*. This has merit from the standpoint of soil improvement and enrichment. On the other hand, organic decay leads to the release of methane – a potent greenhouse gas – as well as carbon dioxide. Moreover, the calorific value of the wood is then lost. Collecting and burning the brushwood to obtain useful heat is therefore an attractive proposition, especially as this yields only carbon dioxide and completely negates the methane-producing potential of decaying wood. In recent years, however, machines have been developed to convert the coppice-wood and brushwood into wood chips. Large wood-chipping machines enter the forest and carry out the chipping *in situ*. This converts the wood to a form that is more readily transported and utilized to make chipboard and paper. The chips may also serve as a dry fuel for combustion or pyrolysis (*v.s.*). Wood pellets are now increasingly used as fuel in closed stoves and boilers of all sizes, which include multi-MW units for district-heating schemes and for combined heat and power generation, notably in Scandinavia.

In countries with relatively few forests, the pruning of trees in urban and semi-rural locations may constitute a significant source of woodchips. The main problem encountered in utilizing waste wood from forests lies in the logistics and cost, in both energy and financial terms, of collection and chipping. Audits need to be conducted to establish whether a particular project is viable. In the UK, very little use has been made to date of forestry waste as an energy source, whereas in Scandinavia (especially, Sweden) this biomass resource is much used for district heating. The development of district-heating plants based on the use of straw or wood chips in Denmark over the period 1981–1999 is shown in Figure 4.3. The number of straw-fired plants reached a plateau in the mid 1990s, while the number of plants burning wood chips continues to grow. In North America, the generation of electricity from waste wood is commonplace and there are some 8 000 MW$_e$ of installed capacity. Australia has abundant resources of waste wood in forests and at

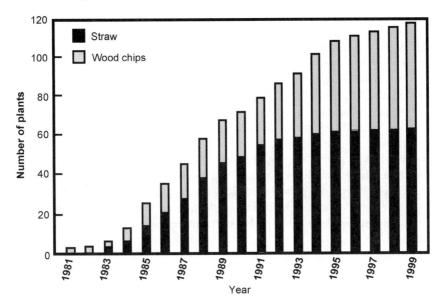

Figure 4.3 *Development of district-heating plants fired by straw and wood chips in Denmark.*
(Courtesy of the Danish Centre for Biomass Technology)

sawmills. This fuel may be used for de-centralized electricity generation to power rural communities and local industries that require chilling facilities, *e.g.* dairying, meat production, and food processing. In the UK, a 2-GW coal-fired power station is conducting a trial of co-feeding sawdust mixed in with the coal.

In summary, the technology to burn wood at all scales from the domestic room heater up to quite large power stations is well established. The factor that limits the size of the power plant is the distance over which it is economic to transport the forest waste. Many such generating units will lie in the range 5–20 MW_e.

Wet Agricultural Wastes

Farmyard slurries are highly polluting and must be carefully managed. Most farmers collect the slurry during the winter, when the animals are in sheds or stables, store it in tanks until summer, and then spread it on the land as a fertilizer during the growing season. An alternative is to store the slurry in an anaerobic digestion tank where it undergoes bacterial attack to yield biogas. Between 40 and 60% of the wet biomass is converted to biogas in 10–25 days. This gas has a composition of ~65% methane and ~35% carbon dioxide with a calorific value of about 25 MJ m^{-3}. The biogas is used in gas boilers to generate heat for farm buildings or, with larger-scale operations, in gas engines for combined heat and power. Biogas is also produced in sewage works and by the anaerobic digestion of wastes from abattoirs and food-processing plants.

In Denmark, an agricultural country, some 20 centralized biogas plants are in operation, as well as smaller plants on individual farms. Manure from dairy farms is conveyed by road to these central facilities, each of which processes up to 500 t of biomass per day – generally, a mixture of manure and other biological waste from slaughter-houses, fish-processing plants, *etc.* The production of biogas ranges from 1 000 to 15 000 m^3 per day. The disadvantage of these centralized plants is that the transport of manure is costly. There are, however, several advantages. After digestion, the residual solid/liquid is returned to farms as a fertilizer. This effects a redistribution of nutrient value from dairy farms to arable farms. Pathogens are mostly removed during the digestion process and the product does not have the obnoxious odor of raw slurry when sprayed on the fields. Although environmentally sound, these centralized plants are difficult to operate economically and depend on having zero-cost organic or biological waste material to mix with the farm manure. Fortunately, some producers of such waste are prepared to pay to have the material removed.

Most green agricultural waste is simply composted as a soil improver or goes to landfill. Farmers often harvest green grass and store in sealed silos to make winter feed ('silage') for cattle. There is little activity in collecting and drying green waste for use as a fuel.

In many developing countries (India, China, Nepal), family-sized bio-digesters have been introduced to generate biogas for cooking purposes. In India and China, several million such bio-digesters are in regular use, while Nepal expected to have 100 000 units by the middle of 2003. These are not without their problems, however; subsistence farmers often do not have an adequate supply of dung, while better-off farmers prefer to purchase fossil fuels, rather than to spend time collecting dung and managing often-temperamental bio-digesters. The digestion process also requires a considerable amount of water and this introduces difficulties in those regions where water is in short supply.

4.2 Biomass: Energy Crops

Energy crops fall into several broad categories, namely: ligno-cellulosic raw material (wood, grasses, *etc.*); sugar and starch crops (sugar cane, sugar beet, sweet sorghum); oil crops (*e.g.* cottonseed, flaxseed, rapeseed, soyabean, sunflower seed). These are treated in different ways to produce heat, electricity, or liquid bio-fuels.

Ligno-cellulosic Crops

Wood

In recent years, there has been a developing interest in growing trees specifically for use as fuel. In order to make sylviculture economic, it is necessary to select fast-growing species, such as willow, poplar or eucalyptus, and to harvest the saplings every few years. This is known as short rotation coppicing (SRC). In a typical operation, 10 000–15 000 willow cuttings are planted per hectare. After a year or so, the willows are trimmed so that they form multiple shoots, *i.e.* they are 'coppiced'.

Between two and four years later, the first harvest is taken by cutting the saplings back to near ground level. They then spring up again and can be harvested once more after a few years, and so on. The saplings, when harvested, are 25–50 mm in diameter and 3–4 m long. These dimensions keep the stems in the most productive part of the tree's growth-cycle and are ideal for passing through a wood-chipping machine. The yield is 10–15 dry-tonnes per hectare per year, with the prospect of increasing this through crop improvement.

One attractive feature of SRC as a renewable form of energy is that the wood, once harvested, constitutes an energy store (like fossil fuels) and can be used when required; other renewables (wind, solar, *etc.*) have no such in-built storage function. It is important to note, however, that the sustainability of a SRC plantation in the absence of organic nutrients or fertilizers must first be established. A 'green' solution to this problem could be the use of effluent from cattle farms, or treated sewage and wastewaters from nearby towns or villages, to fertilize the plantation.

If a power station or CHP plant is constructed to burn forestry waste (see Section 4.1, above), then it may make sense to have an auxiliary SRC plantation nearby to provide additional fuel. Because of the low density and low calorific value of wood compared with coal, the energy content per unit volume is low and transport over long distances is not economically justified. This limits the size of a wood-burning power station to about 20–30 MW_e.

In 1998, the arable biomass renewable energy (ARBRE) project in Yorkshire, UK commenced the building of a wood-burning power station of 8 MW_e installed capacity. The fuel for this was to come from around 2 000 ha of SRC, supplemented by forestry waste. The technology employed at the power station was that of a fluidized-bed gasifier, to produce a combustible gas, followed by a combined-cycle gas turbine to generate the electricity (see Section 2.3, Chapter 2 on coal gasification, where attention is drawn to problems that arise from impurities in the gas that may affect turbine operation and life, or result in air pollution). Experience has shown that under UK economic conditions, and despite a government grant, the operation of a SRC wood-burning power station is not commercially viable, when allowance is made for an adequate return to the farmers who grow the wood. The only way to attain commercial viability is to utilize the waste heat in a CHP scheme. This would require a nearby market for, say, 10–12 MW of heat, which may not exist. One possibility, yet to be developed, is to use this heat to dry wood chips that could then be sold to a coal-fueled power station that would be prepared to operate co-firing. In the near term, the co-firing of wood in an existing coal station may be the most economic option. Elsewhere, plans are being considered to build a much larger (60 MW_e) biomass-fueled power station in South Australia. This will consume 0.6 Mt of softwood per year, mostly plantation waste.

It should be noted that although the plant in Yorkshire was of considerable size (8 MW_e), it would equate to only four modern wind turbines and would produce less than 1% of the electricity from a large fossil-fueled or nuclear power station (1 000 MW_e). It has been calculated that to replace such a station with a base-load station fueled by biomass at 40% efficiency would require 230 000–380 000 ha of SRC, which hardly seems feasible in a European situation. To set the potential of

SRC in a rather different context, another calculation has shown that if Australia were to plant fast-growing *eucalyptus* on 900 000 ha of ex-grazing land (*i.e.*, an area less than 2% of that presently devoted to pasture crops), the trees would sequester about 600 Mt of carbon dioxide per year. This would equate to a 10% reduction in the nation's greenhouse gas emissions.

Grasses

Miscanthus is a hardy perennial grass that produces very high yields of a bamboo-like cane up to 3 m tall. The grass multiplies rapidly, is environmentally benign, and requires little or no pesticide or fertilizer. It is a tropical plant that is harvested annually in the autumn or winter and should grow well in parts of Europe. It holds considerable potential as an energy crop. In particular, the annual yield, at 15–30 dry tonnes per hectare, is considerably higher than that from SRC schemes. Its calorific value is similar to that of wood. Commercial experience of growing this grass for fuel is strictly limited, but it is being evaluated in Europe. It should be noted that perennial grasses (*e.g.* miscanthus, elephant grass, switch grass, bluestem) offer significant advantages in that their roots hold erodible soil and, once established, they are highly resistant to weeds. On the other hand, perennial species may involve higher production costs compared with cultivated annual species, such as sorghum, that are grown in a manner similar to corn.

Sugar and Starch Crops

Another, not dissimilar, grass is sugar cane. In sugar factories, the harvested cane is shredded and crushed to retrieve the juice that contains 10–20% sucrose. The juice is heated in a series of evaporators to form crystals of raw brown sugar and, with further refinement, white sugar. The sugar is separated in centrifuges from the molasses, a residual syrup from which no more crystalline sucrose can be obtained by simple techniques. Molasses can be fermented to produce ethanol, a liquid fuel. This process is carried out on a large scale in Brazil where the product is added to petroleum to give 'gasohol', a motor fuel (see Section 2.4, Chapter 2). Another useful by-product of sugar production is 'bagasse', the fibrous material that remains after the juices are extracted from the cane. Bagasse is the main source of fuel for heating boilers in sugar factories. It can also be used in making paper, cardboard, and fiber board.

Sugar is also obtained from sugar beet, which (unlike cane) grows in temperate climates. After extracting the sugar, the residue may be used as cattle feed, as a fuel, or in the manufacture of fiberboard. Starch-containing crops, such as potatoes, may also be hydrolyzed to sugars and then fermented to ethanol.

Oil Crops

Bio-fuels prepared from crops such as rapeseed have been discussed in Section 2.4, Chapter 2.

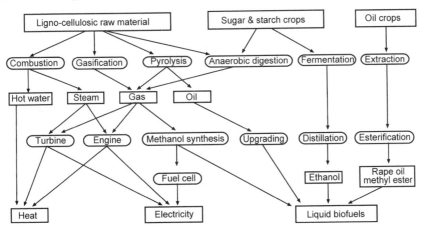

Figure 4.4 *Different routes from energy crops to end-products.*
(Courtesy of the James & James, Science Publishers Ltd.)

A summary of the various useful forms of energy that can be obtained from agricultural crops is given in Figure 4.4. In general, the economics of producing liquid fuels from sugar/starch crops or oil crops in zones with a moderate climate are poor and are unlikely to compete with fossil fuels, tax considerations aside. In some sub-tropical countries (*e.g.* Brazil, Zimbabwe), ethanol from sugar cane, added at the 10–15% level to petrol, does seem to be viable as an automotive fuel, although with government encouragement. Ligno-cellulosic biomass, on the other hand, is proving a practical and economic source of energy in many temperate countries, due to the implementation of modern combustion appliances and advanced gasification and pyrolysis technologies.

4.3 Biomass: Municipal Solid Waste

There are several possible disposal routes for domestic waste. The first, and preferred, option (wherever possible) is recycling. Most municipalities now collect newspapers and glass bottles for recycling. Plastic items pose more of a problem because different plastics require different treatments and this necessitates sorting. Nevertheless, some local authorities have undertaken to do this. Many metals are recycled, *e.g.* steel, aluminum cans, and lead-acid batteries. Over 80% of the lead used in new batteries is recycled. The second approach to the treatment of municipal solid waste (MSW) is combustion or high-temperature pyrolysis. A third route is aerobic digestion of biologically degradable material. Finally, the least favored (but most common), option is disposal by landfill.

Combustion, Gasification and Pyrolysis

Several different technologies are possible for extracting energy from waste, as follows.

- Feeding untreated MSW directly into a boiler where it is burned on a grate; 90% of the ferrous material is recovered from the bottom ash for recycling.
- Preliminary sorting to remove glass and ferrous metal for recycling, and wet putrescibles for composting. The combustible material is either shredded and burned directly, or compressed into pellets prior to combustion. These pellets are referred to as refuse derived fuel (RDF).
- Feeding the combustible material into a bed of hot sand that is fluidized by blowing air upwards through it (fluidized-bed combustion).
- Gasifying in a low-oxygen atmosphere to generate syngas, for combustion in an engine or turbine.
- Heating the waste to a high temperature in the absence of oxygen to produce secondary fuel products such as char, oil and syngas. This pyrolysis process is particularly applicable to old vehicle tires.

Full combustion is the most usual process, with gasification and pyrolysis being less extensively employed. In 1997, the world-installed capacity for electricity generation from MSW was $3\,260\ MW_e$.[2] In the UK, $200\ MW_e$ is currently generated from the combustion of 2.5 million tonnes of MSW per year at 12 plants. A flowsheet of a typical process is shown in Figure 4.5, together with a photograph of an incinerator plant in operation at Birmingham. The UK government has set demanding targets for the reduction of landfill and has also imposed a 'landfill tax'. Through such initiatives, replicated around the world, it is likely that MSW combustion will grow during this decade. Modern, technologically advanced, energy-from-wastes plants conform to extremely stringent environmental standards and achieve emission levels well within the limits set by the legislative authorities. Nevertheless, experience has shown that there is usually considerable local opposition to the proposed construction of a MSW combustion plant, and this may be a limitation to the technology.

Much of the calorific value of MSW stems from paper, cardboard and plastics. To the extent that these materials are increasingly being recycled, the scope for heat recovery from the incineration of MSW is reduced and this could restrict the implementation of incineration technology. In the medium-term, it is hoped that the separation, recycling, biological treatment and combustion of MSW will be combined into a fully integrated process in order to make best use of the resources in the waste and to minimize the environmental impact.

Bio-digestion

A relatively new process is that of large-scale bio-digestion. The MSW is first shredded and then transported *via* a conveyor belt to the top of a tall bio-digester. In one demonstration process, the digester tower consists of three sealed chambers mounted on top of each other to a total height of 16 m. Each chamber contains 67 t of shredded waste. Water is added and air circulates through the material. The air is introduced through the arms of paddles that rotate like slow-moving blades in the bottom of a blender. The MSW remains in the upper chamber for three days at a controlled temperature of 55–65 °C. After this period, a section in the floor of the

(a)

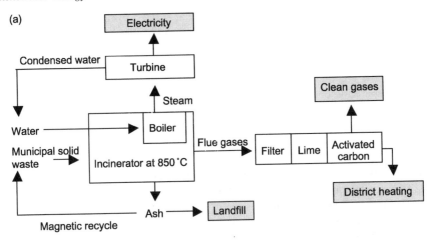

(b)

Figure 4.5 (a) *Process flow diagram for the incineration of MSW;* (b) *aerial view of energy-from-waste plant in Birmingham, UK.*
(Courtesy of the James & James, Science Publishers Ltd.)

chamber opens and the treated waste falls into the chamber below. The process is repeated in each of the chambers and, after nine days, the contents of the bottom chamber are discharged. By this time, the waste has been pasteurized and all harmful pathogens destroyed. Ferrous metals are then extracted with a powerful magnet, and plastics are removed by screening. The remaining bio-degraded material resembles dark gray soil and may be used as a soil conditioner. This process has the advantage of being odorless, relatively inexpensive to operate, and benign to the neigbourhood. Also, it is modular – each tower is able to process 8 000 t of MSW per year. On the debit side, bio-digestion does not recover the chemical energy contained in the waste and, as in all aerobic composting processes, much of the organic material is converted to carbon dioxide and released as a greenhouse gas. An interesting feature of the process is that the physical separation of metals and plastics takes place after the bio-digestion rather than before.

Landfill

Landfill sites are often worked-out gravel pits. They could also be disused open-cut coal mines, slate or stone quarries, or china-clay pits. There is an acute shortage of suitable sites in Europe and the economic value of empty pits for landfill can be greater than that of the mineral extracted from them.

A landfill site is not simply a hole in the ground, but has to be properly engineered. In particular, it is vital that pollutants from the MSW do not enter the water table. To achieve this, the pit is often lined with a thick layer of impervious clay. When the pit is being filled, it is necessary to control objectionable odors, flies, scavenging birds, *etc.* The pit is normally filled to above ground level to allow for subsidence, which occurs even when the refuse is pre-compacted. Riser pipes may be inserted at intervals in the pit to collect the landfill gas that will be evolved. When the pit has been filled with refuse, it is capped with clay so that the gas is not lost, but collected from the risers. Gas is drawn from the landfill site under suction and, after compression, is delivered to the utilization equipment under pressure. After just a few years, the site is covered with top soil and restored to agricultural use. Typically, landfill gas will continue to be produced for 15–30 years as the compostable element of the refuse undergoes anaerobic bacterial degradation. The gas has been put to numerous uses as a fuel; these uses range from heating buildings to firing brick kilns. It is also used to generate electricity using gas engines or gas turbines, but is not generally suitable for feeding into natural gas supplies without purification. For energy recovery from landfill gas to be economic, the sites must be sufficiently large to sustain gas generation for many years. The minimum size for a site is taken to be around 0.2 Mt of waste.

In the UK, about 90% of MSW is deposited in landfills. Projects for using landfill gas have been encouraged by the government, and are flourishing. Over 300 projects are being supported.[2] In 1997, the UK-installed capacity for electricity generation from landfill gas was around 150 MW_e. This was 25% of the installed capacity in the European Union, and almost 10% of that in the world. Subsequent production in the UK has increased appreciably to at least 290 MW_e. This electricity is sufficient to supply over half-a-million homes and, by substituting for fossil fuels, will save almost 0.7 Mt of carbon dioxide emissions per year.[3]

It is expected that there will be substantial growth in Europe of electricity derived from landfill gas as more sites install gas-recovery systems and generating sets. Beyond about 2 025, however, this source of electricity is expected to decline in response to an EU Directive for the diversion of organic wastes away from landfill sites. Even now, UK local authorities are being encouraged by the landfill tax to subject green waste to composting rather than disposal as landfill. This makes sense in terms of the limited landfill capacity available, but does nothing for energy conservation since the gas generated on decay is lost to the atmosphere.

The disposal of MSW is by no means easy. Aside from the issue of greenhouse gas emissions, there are numerous other issues to be taken into account. These include environmental factors such as air pollution and associated health effects, the possible release of ozone-depleting substances, contamination of water bodies, and the amount of fossil fuel used in the transport and processing of the waste. There are also socio-political-economic factors such as availability of suitable sites for landfill, national and regional government strategies, local opposition to landfill or combustion, and process costs. All of these factors need to be considered in the determination of a balanced policy for sustainable waste management.

4.4 Solar Heating

Life on Earth is totally dependent upon solar energy. All our food derives from solar energy by photosynthesis, all the fossil fuels are formed from the remains of plant and animal life, and most of the renewable forms of energy originate from the sun. Without the sun, Planet Earth would be cold and lifeless.

Fortunately, solar energy is essentially inexhaustible and environmentally benign. In the latter context, it is interesting to observe that it derives directly from nuclear fusion reactions taking place in the interior of the sun (see Section 3.6, Chapter 3). The world currently uses energy equivalent to around 10 000 Mtoe per year (see Table 1.1, Chapter 1). How does this compare with the solar radiation incident on the Earth? Expressed in electrical units, the energy consumption worldwide is 110–120 $PW_e h$ per year. In a desert area, such as the Sahara, a typical annual solar irradiation is 2.5 MWh m^{-2}. From this, and assuming a 20% collection efficiency, it may be calculated that an area of the Sahara roughly 500×500 km could, in principle, supply all the energy that the Earth uses at present. Moreover, there is no foreseeable end to the availability of this energy. Of course, this is an idealized concept that is far from being practical. We do not have the technology to collect all this solar energy, nor to store it and convert it to a usable form, nor to transport it to the world's energy markets. Nevertheless, the calculation does set the solar resource in context.

Broadly speaking, there are three possible ways to utilize solar energy:

- directly as heat
- as electricity, generated by photovoltaic cells
- as chemicals derived from solar heat, solar electricity, or directly from photon energy by photo-electrochemical processes.

In Chapter 5, we discuss photovoltaic and photo-electrochemical processes. Here, we describe the direct use of solar heat.

Passive Solar Heating

Traditionally, the most widespread use of solar heat has been to dry laundry (the current practice, in developed nations, of using electric driers is much to be deplored from an energy and environmental viewpoint). In temperate climates, greenhouses are popular to encourage the early growth of plants and to protect them from frost. In hot climates, open-air swimming pools derive some of their warmth from solar heating. These are traditional applications for solar heat.

The 'passive' solar heating of buildings can be a significant contributor to warming their interiors. In this approach, the building's structure is used to capture sunlight and store and distribute heat, which reduces the requirements for conventional heating systems. This was appreciated by the ancient Greeks, much of whose architecture took advantage of passive solar heating. Nowadays, in many temperate countries, people build conservatories or 'solaria' on the sun-facing side of their homes in order to grow tender plants and to enjoy relaxing in the warming rays that enter through the glass. Glass has the desirable property of admitting visible radiation but not re-transmitting thermal (infra-red) radiation. For this reason, in northern climes, large windows are often installed on the south side of the house, whereas in sub-tropical regions the window area is minimized on the sun-facing side in order to keep cool. Care must be taken, however, not to overheat the interior of the building in summer through solar gain as this can result in a demand for air-conditioning. Obvious strategies include the installation of sunshades and awnings (which can include in-built photovoltaic modules), and the judicious planting of trees. In winter, when the sun is not shining, a double-glazed window loses more heat by conduction than an insulated wall. Moreover, a window that is not double-glazed can set up cool down-draughts, which is why radiators are conventionally located beneath windows.

Possibilities exist for the development of three types of glass with transmission properties that vary in response to conditions. These technologies would allow dynamic control of light and solar (heat) gain to match building and occupant requirements. Photochromic glass, which has been in everyday use for some years in the form of sunglasses and spectacles, changes its transmission properties in response to prevailing radiation levels. Such glass reacts automatically, and thus does not permit a high degree of independent control. Moreover, there are considerable technical problems in scaling up to window size. Thermochromic glass varies its optical properties in response to temperature variations. Again, independent control is difficult, but as the response is tuned to variations in environmental conditions, thermochromic glass can improve thermal and lighting comfort.

Windows coated with electrochromic materials (so-called 'smart windows'), the third option, appear to offer more promise. Such a material has a color and a transparency that may be changed at will by the application of a small voltage. A photocell determines the level of incident solar radiation and automatically adjusts the voltage applied to the smart window so that it darkens when there is strong sunlight and lightens when the sky is overcast. An advantage of this technology is that the power requirement is zero while the window remains in the same optical state. Current is passed only during a change in transparency, and the total charge required is typically 20 mC cm^{-2} per unit change in optical density. The electrochemistry is similar to that in a cell of a battery. The color change is due to the change of oxidation state of the electrode, which depends on the state-of-charge of the cell. In order to make electrochromic glazing a viable product, improvements continue to be sought in performance, as well as in the manufacturing process itself. Performance targets include: greater contrast between the translucent and the transparent states; faster switching in each direction; an operating temperature suitable for use in most building situations; increased lifetime.

About one-third of global energy consumption is used in the heating, cooling and lighting of buildings. This is why it is important both to minimize heat loss by conservation measures and to utilize, to the maximum possible extent, the heat and light from the sun. Much can be done by means of architectural design to take advantage of controlled solar heating in buildings. This technology is termed 'passive solar design'. The aim is to develop low-energy buildings through a combination of insulation, other energy-conservation measures, and passive solar heating. It should be understood that with an energy system as complex as a building, passive solar design requires close attention to detail, and often there are complicated trade-offs to consider. Computer modeling and design tools have been developed to assist architects and builders in the resolution of such issues.

To benefit from passive design, effective control systems must be employed to regulate the conventional heating, lighting and ventilation systems in order to provide comfortable internal conditions whilst minimizing energy consumption. Demonstration houses have been built that have multiple windows, heavy insulation, and no source of heating other than passive solar gain and the internal heat generated by the occupants and their activities (*i.e.* cooking, washing and ironing of clothes, running electrical equipment, *etc.*). Up to 99% of the normal energy input into the building can be saved. Such houses are not necessarily more expensive to construct, since the added cost of insulation is offset by the savings effected through not requiring furnaces/boilers or radiators. With respect to cooling, commercial buildings present a more difficult challenge than residential buildings because they contain a much greater number of internal heat sources (people, computers, lighting, electrical appliances, *etc.*). Whereas more efficient lighting and equipment, building orientation and interior atria, sunshades/awnings and shade trees all assist in reducing the energy used for cooling, the practice of 'daylighting' can be especially effective. This practice involves the automatic dimming of the building's lights in proportion to available daylight, so that interiors are not overlit. Properly designed daylighting results in a smaller heat load than that experienced with even the most efficient artificial lighting, and thus makes it possible to reduce the size of air-conditioning systems.

One of the key elements of good climate-sensitive design involves the choice of appropriate building materials for energy absorption. Porous stone or brick absorbs rainwater and releases it later by evaporation, which causes cooling. (Note, the specific latent heat of evaporation of water is high, $\sim 2\,250 \text{ J g}^{-1}$). In temperate zones, where heating is more often required than air-conditioning, such evaporative cooling can be disadvantageous, particularly in poorly insulated buildings where the heat of evaporation is extracted directly from the building. In these locations, treating porous brick with an impervious surface coating, so as to keep out the rain, may save considerable energy. In hot climates, buildings may be kept cool by spraying with water, though water is usually too precious in such regions to be used for such a purpose. An exciting possibility is the use of phase-change materials (see Section 7.1, Chapter 7). When water changes from a solid (ice) to its liquid form, it absorbs heat (latent heat of fusion) before its temperature increases. Phase-change materials proposed for use in buildings operate on a similar principle, but the temperature at which they change state is close to that acceptable for thermal comfort (*i.e.* 20–30 °C). In this way, substantial amounts of heat can be absorbed to counteract temperature rise and discomfort. Of course, the reverse cycle also occurs – heat is liberated when the materials return to the solid state on cooling. Efforts are being made to develop suitable materials that can be safely incorporated into building components. One option is to impregnate plasterboard with such phase-change material, where it is held in place even when in a semi-liquid state.

Unfortunately, passive solar designs have not been adopted by the construction industry as widely as might be expected. Whereas one reason for this is a general lack of familiarity with solar design, a more serious problem is that architects and construction firms have little direct incentive to incorporate passive solar design into new buildings since the costs of heating and cooling are met by the owners. Alternatively, the cost of retrofitting existing buildings with solar-features often becomes prohibitive. For these reasons, appropriate building codes and incentive programs such as government rebates are essential for encouraging sensible energy design. In this respect, a notable initiative has been taken in Germany where legislation mandates that all new houses should consume less than seven liters of oil (or equivalent), per square meter, per year. In the longer term, it is hoped to reduce this amount to less than three liters – the so-called '3-liter house'.

Active Solar Heating and Cooling

Active solar systems are units that convert solar radiation into warm water. This may be used for such applications as domestic hot water (DHW), heated hot tubs and swimming pools, and solar-aided district heating. High-quality solar collectors can convert up to 80% of the incident radiation into useful heat. Under UK conditions, the overall system efficiency (proportion of the incident energy converted into useful energy in the house) lies in the range 40–50%.

For DHW purposes, the solar collectors are usually mounted on a south-facing roof in the Northern Hemisphere or, obviously, a north-facing roof in the Southern Hemisphere. A typical array for a single house has an area of a few square meters. For most of the year, the water leaving the collector is warm rather than hot. Therefore, integration into the conventional hot-water supply is necessary. In essence, the solar array provides base heating, which is topped up by a hot-water boiler or electric immersion heater. An example of such an installation is shown in Figure 4.6.

The solar collectors are black bodies for maximum absorption of radiation, and fall into the following two basic categories.

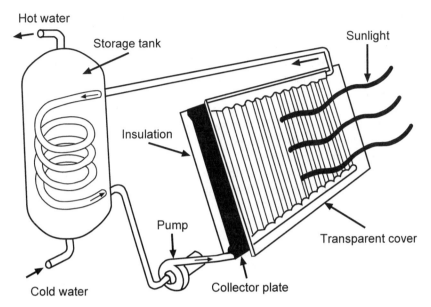

Figure 4.6 *Layout of a typical solar water-heating system for domestic applications.*

- *Flat-plate collectors.* These resemble blackened domestic radiators and are filled with water. To reduce heat loss, the collectors are insulated on the rear face and have a glass cover on top.
- *Evacuated-tube collectors.* Individual pipes are sealed into evacuated glass tubes to reduce heat loss. The pipes may contain water, or a volatile liquid that evaporates on heating, condenses at the end of the pipe, and gives up its latent heat to water in a small heat-exchanger.

Under European conditions, it is generally necessary to add antifreeze to the water to prevent damage to the collectors in winter. A water pump has to be included in the circuit if, as is usual, the hot water tank is below the level of the solar collectors. Some manufacturers market the 'all-solar' design in which the solar array incorporates a photovoltaic panel to provide the electricity to operate the pump. Another innovative design relies on a 'thermo-siphon effect', to transfer heat from the solar collector to the storage medium. The collector boils an alcohol mixture, and the bubbles push the fluid above them to the top of the collector. From there, the fluid falls down to a heat-exchanger that is coupled to the storage tank. This flow, in turn, pushes fluid that has already passed through the heat-exchanger back up to the collector to repeat the cycle.

There are many potential, non-domestic applications for solar warm water. These include hotels and guest houses, office buildings, hospitals, schools, and nursing homes. As with passive solar design, it is much simpler to incorporate active solar-heating systems in new buildings rather than retrofit them to existing buildings, which may explain why the take-up to date has been limited. Other limiting factors are the restricted insolation in temperate latitudes, the high manufacturing and installation costs associated with a small, undeveloped market, and the low cost of fossil fuels. Elsewhere, where insolation is greater, more enthusiasm has been shown for solar DHW. A case in point is Cyprus, where more than 150 000 solar water heaters are in operation – one unit for every five individuals – and provide the equivalent of 9% of the total electricity consumption in the country. This is a consequence of the high insolation level in the Mediterranean region and the lack of any indigenous fossil fuels. Also, in warm climates, solar ponds are sometimes used. These exploit the natural properties of salt water to collect and store heat energy (see Section 7.1, Chapter 7).

A novel hybrid system is a 'solar–thermal heat pump'. A heat pump is merely a device to extract heat from a low-temperature source, such as a lake or the ground (see Section 4.5), and reject it at a higher temperature to a sink, such as a building. This requires the input of electrical or thermal energy. In a solar–thermal heat pump, solar energy provides the necessary thermal input. A solar panel (mounted, for example, on the roof of a building) serves as the heat-exchanger to reheat the cold fluid rejected by the heat pump. Because of the intrinsic efficiency of a heat pump, this hybrid system is almost twice as efficient as a conventional solar hot-water system.

The second important application for solar warm water is the heating of swimming pools. The collectors are situated near the pool, either on the ground or on a support structure. For small pools, the pool water itself, after filtration, may be passed directly through the solar array. Larger pools may require an indirect system. A large solar array used to heat the swimming pool of a hotel in Madeira is shown in Figure 4.7.

Even larger arrays are occasionally installed to provide warm water for district-heating schemes – solar-aided district heating. Often this is operated in conjunction with a conventional scheme using combined heat and power. Typically, the addition of a solar–thermal component saves 10–15% of the annual heat load. If the hot water produced in summer can be stored in large, well-insulated reservoirs for distribution in winter (seasonal storage), then the solar component can supply up to 75–80% of the annual heat load. Around 100 solar-assisted district-heating plants are said to be operational in Europe. These are situated mainly in Scandinavia, but increasing interest in such technology is being shown in Austria, Germany, and Switzerland. Most of the plants have diurnal storage of hot water, but not seasonal storage. A tank of 4 500 m^3 capacity for seasonal storage of solar hot water was constructed and installed underground near Hamburg in Germany, but it was found that the investment cost was too high to be economic in comparison with fossil-fuel prices. Nevertheless, seasonal storage is a practical option for the future should fossil fuels become more expensive. Another 'green' option is to combine a biomass boiler with a solar–thermal collector to provide DHW to an estate of houses or apartments. Under Northern European conditions, there would be heavy reliance on the biomass boiler in winter, but the solar collectors would make a significant contribution in summer.

Since the mid 1980s, widespread efforts have been under way in a number of countries (*e.g.* Bolivia, the People's Republic of China, India, Pakistan, and many

Figure 4.7 *Solar array to heat a swimming pool at a hotel in Madeira.*

parts of Africa) to encourage people to use solar heat for cooking, as the burning of biomass for such a purpose is a major contributor to deforestation and the associated emission of carbon dioxide. In addition, dwindling supplies of wood are forcing people to spend inordinate amounts of time and energy in the search for cooking fuel. A wide variety of 'solar cookers' is available. The designs fall into three main categories: (i) the direct-focusing cooker, in which a pot is supported at the focal point of a parabolic (or closely similar in shape) reflector, usually made from aluminum (Figure 4.8(a)); (ii) the box cooker, the food is placed inside an insulated box that has a glass top and a reflector mounted in the lid (Figure 4.8(b)); (iii) the indirect cooker that uses a fluid (water or oil) to transfer heat from a solar collector to an insulated heat-storage tank, and from there to the stove (*i.e.* in the manner discussed above for DHW systems).

Each of the three designs of solar cooker has its advantages and disadvantages. Direct-focusing cookers are the least expensive to build and achieve the fastest cooking times (at least for small quantities of food). On the other hand, these cookers can only be used outdoors and during the day (and often only when the sun is high), have to be adjusted to keep pace with the moving sun, and usually can only accommodate one cooking pot. Box cookers are more expensive to build, but can cook and keep warm greater amounts of food (with protection from blowing dirt), and are easier to use and maintain. The main advantage of indirect cookers is the

(a)   (b)

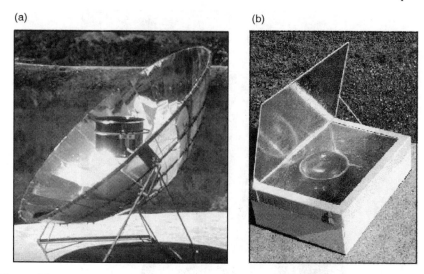

Figure 4.8   (a) *Direct-focusing and* (b) *box designs of solar cooker.*

possibility of indoor cooking. The cookers are, however, immovable units and therefore cannot be rotated to track the sun, which restricts their use to only part of the day. If steam at atmospheric pressure is the heating medium, indirect cookers cannot boil water or fry foods and are suitable only for slow cooking and stewing. It should be noted that box-type collectors are also being used for water desalination and pasteurization, as well as for the drying of fruits, vegetables and herbs. For example, an Australian company is developing a solar–thermal energy system for the cooling and drying of grain. This allows grain to be stored without deterioration, mould, or insect attack.

Although it may seem counter-intuitive, solar heat may also be used to *cool* (air-condition) buildings. The principle utilized is the same as that of the gas refrigerator. The essential point is that an input of energy is required to extract heat from ambient temperature air and then reject it, to leave behind cooled air. In a conventional electric refrigerator, the energy input is in the form of electricity to drive a compressor. In a gas refrigerator, a vapor absorption process is employed rather than a vapor compression process. The heat from the gas flame serves to liberate the refrigerant, often pressurized ammonia, from an aqueous solution. The warm ammonia gas then passes through a heat-exchanger where it cools, liquefies, and gives off its latent heat to the surroundings. The liquid ammonia is re-evaporated in the refrigerator, absorbs heat in the process, and thus cools the refrigerator box. The ammonia gas is then returned to the absorption unit, so completing the cycle. An air-conditioning unit is simply a refrigerator in which the cooled air is retained within the building, rather than in a closed box, and the warm air is rejected outside of the building. In principle, an absorption-cycle refrigerator or air-conditioner can be operated by solar heat rather than by heat from gas combustion. Solar-powered refrigerators and air-conditioners are particularly useful in tropical villages and farms where mains electricity is not available. An especially important application is for the storage of vaccines and medicines in developing countries.

The concept of solar-powered air-conditioning is very attractive because there is a direct correlation between when the facility is most needed and when most solar heat is available. Air-conditioning consumes substantial amounts of electricity, so much indeed that the electrical load in most of the USA is higher in summer than in winter. A corollary of this is that if solar–thermal heat pumps were widely employed, there would be major savings in electricity consumption and therefore in carbon dioxide emissions. For air-conditioning, a driving temperature of 80 °C upwards is required. Until now, absorption-cycle machines have generally been employed to cool large buildings using surplus heat available from CHP or district-heating schemes, or waste process heat in industry. Some demonstration projects of coupling with solar collectors have, however, been conducted and this is seen to hold promise for the future.

Solar High-Temperature Processes

Large parabolic mirrors can be used to concentrate solar energy at a point so as to produce high temperatures – a 'solar furnace'. This is nothing more than a scaled-up version of the schoolchild's experiment of burning paper by focusing the sun's rays with a magnifying glass. Antoine Lavoisier, the notable French chemist, was among the first to build a solar furnace, in the 18th century. Focusing thermal energy has the potential either to allow the generation of electricity (solar–thermal electricity), or to carry out endothermic (heat absorbing) chemical reactions that, thereby, effectively convert solar energy to useful materials.

There are three basic designs of solar furnace for electricity generation, as illustrated in Figure 4.9. The simple parabolic dish (Figure 4.9(a)) focuses the sun's rays on to a thermal receiver that is mounted above the dish at its focal point. The receiver absorbs and converts sunlight into heat. The heat is then transferred *via* a heat pipe to a steam generator to operate a small turbine, or *via* a heated fluid to power a heat engine (*e.g.* Stirling-cycle or Brayton-cycle). The turbine/engine is coupled to an alternator to generate electricity. This type of solar furnace has the highest solar concentration factor, *viz.* 1 000–4 000 (defined as the ratio of dish area to receiver aperture area), and temperatures exceeding 1 000 °C can be reached. The furnace also has the highest overall conversion of solar energy to electricity, but because the diameter of parabolic mirrors is restricted to 10–20 m the electrical output of individual dish–engine systems is limited to about 25 kW_e. The technology is therefore best suited to distributed electricity generation for small stand-alone applications in remote areas away from the mains. For such RAPS applications, an electrical storage device or a hybrid solar–fossil–fuel system is needed for when there is no sunshine. Three parabolic dish and Stirling engine units, each capable of generating 9 kW_e, are shown in Figure 4.9(a). The units were constructed at the experimental solar energy facility at Almeria, Spain, and have been in operation since 1992.

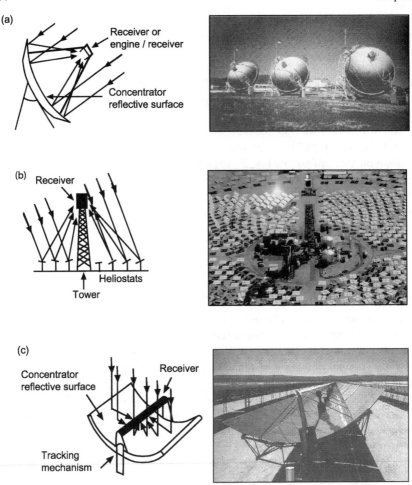

Figure 4.9 *Three main designs of solar furnace:* (a) *parabolic dish;* (b) *central receiver;*
(c) *parabolic trough.*

The second type of solar–thermal electricity generator – the 'central receiver'
type (Figure 4.9(b)) – increases the solar energy collected by arranging thousands
of individual, sun-tracking mirrors ('heliostats') around a receiver mounted on top
of a tall tower. Each heliostat has a focal length appropriate to its distance from
the receiver. The sun's heat is collected in a heat-transfer fluid (*e.g.* a molten salt)
that flows through the receiver. The heat is then used to make steam to generate
electricity in a conventional steam generator located at the foot of the tower.
The much larger distance between the heliostats and the receiver means that
the solar concentration factor is lower (200–1 000) and temperatures range from
300–1 000 °C. On the other hand, the overall installation produces considerably
more electrical power than the single-mirror type, in the range 30–200 MW_e.

In the third type of solar power generator (Figure 4.9(c)), parabolic trough mirrors track the sun as it crosses the sky. The design is parabolic only in cross-section and extends linearly, like a trough, for up to 100 m in length. The receiver is a linear heat pipe mounted at the focus of the mirror. Line focusing gives a lower concentration factor (30–80) and a lower temperature range (200–400 °C). Both trough designs and central receiver plants can incorporate thermal storage by setting aside the heat transfer fluid in its hot state. This allows for electricity generation for several hours in the evening. The attraction of this furnace is that it is modular and can be scaled up to many long parallel rows, all of which feed heat into a single generator. This lowers the unit cost of electricity. Multi-row units may generate 30–80 MW_e and many units can be combined to form a single generating station (a 'solar farm'). Plants of this type installed in California have a combined peak power output of over 350 MW_e.

The development of solar–thermal electricity generators is on-going in the areas of materials (improved mirror reflectivity and receiver absorptivity, reduced receiver thermal emissivity), heat-transfer media, engine selection, system design and engineering, scale-up, cost reduction, and market assessment. Countries that have an interest in solar–thermal electricity generation include Australia, Germany, Greece, Israel, Morocco, the Russian Federation, Spain, Switzerland, and the USA. The present position is that the concept of solar farms has been demonstrated successfully, but the economic projections are not yet favorable and the technology is still far from gaining commercial acceptance. Any future solar–thermal power stations are likely to be situated in the 'sunbelt' of Mediterranean countries, North Africa, the Middle East, Australia, or southern parts of the USA.

A more futuristic possibility is to utilize the heat from a solar furnace for the direct generation of electricity by means of photovoltaic cells, rather than by going through a steam cycle. Conventional photovoltaic ('solar') cells, which convert sunlight into electricity, are discussed in Section 5.6, Chapter 5. Here, we are concerned with the conversion of solar heat into electricity – 'thermophotovoltaics'. Because heat radiation is of much lower photon energy than light, it is necessary to employ a semiconductor with a lower band-gap (0.6 eV) than that used in conventional photovoltaic cells (1.1 eV).

Thermophotovoltaic cells have been investigated and developed to provide small-scale, portable power sources for both military and civil applications. In these applications, the source of heat is a combustion system, often based on propane or diesel fuel. Such thermal cells are particularly useful at night when conventional solar cells are inoperative, and may be seen as alternatives to batteries, fuel cells or thermoelectric generators. One application is to provide lighting and power for communications on sailing boats. Whether or not these devices might prove practical for use in conjunction with heat from solar furnaces is still highly speculative.

Another possible use of solar–thermal energy, *i.e.* the production of useful chemicals, has one practical problem: it is inconvenient to construct a chemical plant at the focus of a parabolic mirror pointing towards the sun, or at the top of a tall tower. This problem may be overcome by adopting a 'beam down' configuration that is based on the central-receiver design (Figure 4.9(b)). A field of heliostats focus the sun's rays on to a receiver at the top of the tower where there is a hyperboloidal reflector to redirect the sunlight to a solar receiver on the ground. This allows a chemical process plant to be ground-based.

Research on possible solar-assisted chemical reactions is being conducted in Australia, Switzerland, and Israel. One promising endothermic reaction is the reforming of methane using carbon dioxide, which takes place above about 700 °C, *i.e.*

$$CH_4 + CO_2 \longrightarrow 2H_2 + 2CO \quad \Delta H° = +248\,kJ\,mol^{-1} \tag{4.2}$$

The product syngas may be reacted catalytically to form methanol, CH_3OH. Alternatively, it could be reacted further with steam (the water-gas shift reaction, see Section 2.3, Chapter 2) to form more hydrogen and CO_2. A third option is to cool the syngas to a lower temperature where the reverse exothermic (heat evolving) reaction takes place.

$$2H_2 + 2CO \longrightarrow CH_4 + CO_2 \quad \Delta H° = -248\,kJ\,mol^{-1} \tag{4.3}$$

The heat evolved can be used to raise steam to drive a turbine, while the gases $(CH_4 + CO_2)$ are re-cycled back to the endothermic solar reactor so as to form a closed-reaction loop. Reaction (4.2) is effectively a chemical route for the storage of heat and syngas may be regarded as a 'solar-upgraded' fuel since its calorific value is greater than that of methane. All these possibilities are under consideration. In Australia, for example, CSIRO is using the solar collector shown in Figure 4.10 to explore the possibility of using solar–thermal energy to produce hydrogen from methane-containing gases, *e.g.* natural gas, landfill and coal-bed methane, methane derived synthetically from coal. The aim is to use the hydrogen to make electricity in a fuel cell, and to recover the carbon dioxide in a highly concentrated form for permanent disposal in CO_2 'sinks' such as subterranean reservoirs, deep coal-beds, or the ocean. A practical combination of decarbonization, hydrogen production and carbon sequestration will allow the efficient utilization of fossil fuels for electricity generation with dramatic reduction in atmospheric emissions.

Another endothermic reaction being investigated is a hybrid fossil–solar energy process, namely, the carbothermic reduction of metal oxides. Natural gas or solid coke may be used to reduce zinc oxide to metallic zinc at about 1 000 °C, a temperature that can be reached in a solar furnace. Using natural gas as the reducing agent combines, in a single step (the 'SynMet' process), the reduction of zinc oxide with the reforming of methane for the co-production of zinc and syngas. Zinc is used in many applications, which include batteries, while (as stated above) syngas finds value as fuel. Such processes are still undergoing research and development; their economic feasibility has yet to be determined.

A major disadvantage of solar heat for electricity generation or for endothermic chemical reactions is, of course, the fact that solar energy is diurnal and seasonal. Clearly, a heat source that can only generate electricity by day, and principally in summer, is of limited use. The economics of most large-scale chemical processes

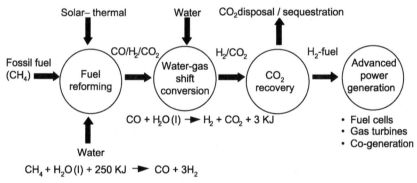

Solar– thermal   Water CO_2 disposal / sequestration

Fossil fuel (CH_4) → Fuel reforming → $CO/H_2/CO_2$ → Water-gas shift conversion → H_2/CO_2 → CO_2 recovery → H_2-fuel → Advanced power generation

Water

$$CO + H_2O\,(l) \longrightarrow H_2 + CO_2 + 3\ KJ$$

- Fuel cells
- Gas turbines
- Co-generation

Water

$$CH_4 + H_2O\,(l) + 250\ KJ \longrightarrow CO + 3H_2$$

Figure 4.10   *CSIRO solar collector composed of an array of 48 curved mirrors (area 107 m^2) that, on a clear day, can concentrate 91 kW$_{th}$ of thermal energy*

depend upon 24-h, year-round operation. Thus, supplementary heat from fossil fuels would be required much of the time to maintain constant process conditions, and this reduces greatly the value of solar–thermal heat as a useful energy source.

4.5 Geothermal Energy

An immense amount of heat is trapped within the Earth's interior, as witnessed by volcanic eruptions through the ages. Indeed, there is sufficient heat to supply the entire energy needs of the world's population indefinitely. The problem is one of gaining access to this energy source and utilizing it economically. The violent heat of volcanoes is of little practical use, but gentler forms of geothermal energy have been exploited for centuries. The ancient Greeks, Etruscans and Romans constructed public baths that used hot water from natural springs (so-called 'aquifers'). In the 18th and 19th centuries, the purported curative and prophylatic benefits of mineral-rich thermal springs made health spas popular in both Europe and the USA. Such facilities remain in common use today in countries such as

Hungary, Iceland, and Japan. The Earth's heat is also being employed directly for heating and cooling buildings, as well as in agriculture, aquaculture, and industry. For example, Iceland has pioneered large-scale district heating from geothermal energy.

Large-scale harnessing of geothermal sources for electrical energy has been a more recent activity. The generation of electricity from natural steam was in fact first demonstrated in Larderello, Italy, in 1904, but no other country followed this practice until 1958 when New Zealand built a power plant in the Wairakei area of the North Island. Two years later, a similar plant went in to service in the USA; by 2000, some 2.2 GW_e of geothermal–electric capacity had been installed, most of it at The Geysers field in Northern California. The Philippines also has around 2 GW_e. Where the temperature of the geothermal heat source is below 100 °C, and steam is therefore not available, it is possible to generate electricity using a volatile organic compound as the working fluid of a heat engine. This is the 'Organic Rankine Cycle' turbine. These turbines are widely used to generate electricity from low-temperature geothermal fluids, with over 700 MW_e of installed capacity worldwide.

Geothermal energy arises from two sources. The first source is the heat associated with molten rock ('magma'), which is present in the Earth's mantle between the crust and the core. The crust has a thickness of about 20–65 km in continental areas and about 5–6 km in oceanic areas, the mantle is roughly 2 900 km thick, and the core is about 3 500 km in radius. The temperature gradient, as one drills into the Earth, depends on the depth of the magma from the surface and on the thermal conductivity of the surface layers of rock. In most places, useful heat is available only at depths that are too great for practical drilling. Therefore, geothermal exploration focuses on those regions, often volcanic, where useful temperatures are found close to the surface.

The second source of geothermal energy arises from the radioactive decay of isotopes of uranium, thorium and potassium in the Earth's crust. Where deposits of these elements occur, the process of nuclear decay results in an accumulation of 'radiogenic' heat. Radioactive elements are often associated with hard rocks such as granite.

Because the Earth's crust conducts heat poorly, the heat remains where it is produced and dissipates only slowly. In some locations, however, geophysical activity allows molten rock to approach closer to the Earth's surface. This increases the thermal gradient and makes geothermal energy more accessible. Such hyperthermal regions exist in several parts of the world, *e.g.* Iceland, the Mediterranean, East Asia, Western USA, and New Zealand. Where surface water seeps down to come into contact with hot rock, its density is reduced and this creates sufficient buoyant pressure to bring it back to the surface, where it emerges as hot springs or steam. In some cases, pores and fractures within the rock trap this hot water or steam far below the surface to create hydrothermal reservoirs.

In summary, a geothermal system is composed of three main components: (i) a heat source, *i.e.* high-temperature (>600 °C) magmatic intrusion at relatively shallow depths, or lower temperature (100–400 °C) hyperthermal regions; (ii) a reservoir, *e.g.* hot permeable rocks from which the circulating fluids extract heat;

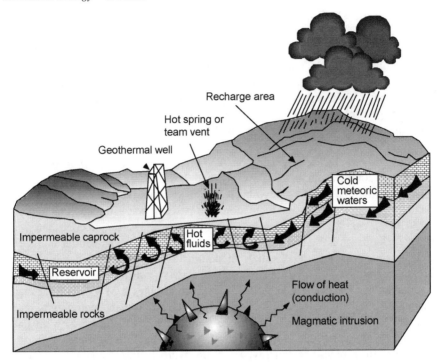

Figure 4.11   *Schematic of ideal geothermal system (see Ref. 4).*

(iii) a fluid (usually water), which is the carrier that transfers the heat. A simple representation of an ideal geothermal system is shown in Figure 4.11.[4]

Strictly speaking, geothermal energy is not a renewable resource. Because of the very low thermal conductivity of rock, it may take thousands of years to replace the heat once a significant fraction has been withdrawn from the rock. In addition, where aquifers are recharged only slowly from surface water, they can rapidly be exhausted if too much steam or hot water is removed, or if the amount replaced through injection back into the system is insufficient. Obviously, the uncertainty over the amount of energy that can be produced before a reservoir's productivity declines substantially has considerable practical significance in planning a geothermal project. It should be noted that steam output at The Geysers, the largest geothermal development in the world, has already fallen by more than 25%.

Geothermal Aquifers

Water-bearing rock strata become heated as a result of geothermal activity. In extreme cases, this can result in spectacular hot geysers, as in Iceland and New Zealand. Other places have warm water springs, for example at Bath in the UK. Most warm-water aquifers, however, have to be tapped by drilling to recover usable heat. This requires two drill holes, one to extract the warm water and another to re-inject it – unless alternative disposal sites are available.

Many aquifers are at relatively low temperature. If the temperature is below about 60 °C, the expense of drilling is not usually justified. At intermediate temperatures, the aquifer may be used for district-heating purposes and industrial applications, *e.g.* for the manufacture of chemicals, pulp and paper, or for injection into oil wells to help reduce oil viscosity and so enhance recovery. At temperatures above about 150 °C, steam may be employed for electricity generation.

Where geothermal energy is available, it is widely exploited. The seven countries with the greatest installed capacity for heat extraction are listed, in descending order, in Table 4.1 together with the seven countries with the most installed capacity for geo-electricity generation. The production of energy worldwide from aquifers in 2000 was about 15 145 MW_{th} and 7 974 MW_e.[5,6] Within the European Union, geothermal aquifers have been exploited extensively for district heating, fish farming, horticulture, and recreational purposes. It was estimated that a total of 1 850 MW_{th} and 805 MW_e was in place by 2000.[5,6] The major contributors are France, Germany, Italy, and Sweden. Paris is situated above an aquifer that serves to heat some apartment blocks. Italy predominates in electricity generation. There is little activity in the UK apart from a 12 MW_{th} district-heating scheme in Southampton to which a geothermal aquifer contributes 2 MW_{th}. Further afield, there are sizable district-heating schemes based on aquifers in Central and Eastern Europe. Higher-temperature aquifers are found in Iceland, Japan, New Zealand, and the USA. Indeed, the North Island of New Zealand derives almost all of its electricity from renewable energy, *i.e.* from hydroelectric dams and geothermal power stations.

Finally, it should be understood that the development of a new geothermal field is a costly and risky venture. Before commercial production can be considered, the field must be mapped, exploratory wells drilled, chemical analyses performed, and the reservoir tested. Moreover, district-heating projects are economical only on a substantial scale and the reservoirs must be in close proximity (*i.e.* not more than a few kilometers) to centers of heat demand to minimize piping costs.

Table 4.1 *Geothermal energy in the world in 2000. Top seven producers of non-electric (MW_{th}) and electric (MW_e) power*

Country	MW_{th}	Country	MW_e
USA	3766	USA	2228
PR China	2282	Philippines	1909
Iceland	1469	Italy	785
Japan	1167	Mexico	755
Turkey	820	Indonesia	590
Switzerland	547	Japan	547
Hungary	473	New Zealand	437

MW_{th} and MW_e data taken from Refs. 5 and 6, respectively.

Hot Dry Rocks

Geothermal hot dry rock (HDR) technology involves the extraction of heat from dry rock formations by injecting water. Two wells are drilled, one to inject cold water under pressure and the other to extract hot water. When the pressurized cold water first encounters the hot rocks it causes extensive hydraulic fracturing to create a reservoir. Water then circulates through the fractures and extracts heat from the rocks. In essence, this method creates a giant heat-exchanger. The concept was pioneered in the UK and the USA. Although the results in the UK were disappointing and work ceased in the early 1990s, efforts are continuing in Australia (Cooper Basin), the USA (New Mexico), France (near Strasbourg), and in Japan. Nevertheless, HDR technology is still at the experimental stage, and there are no operating power plants anywhere in the world.

The commercial development of HDR technology depends upon meeting three requirements. First, drilling costs have to be reduced, as they usually represent half or more of the total investment of HDR projects. Drilling through hot granite creates severe thermal and physical stresses on conventional equipment. The prospects of using advanced techniques such as thermal spallation (using a hot gas jet to break up rock) and erosive drilling are presently being evaluated. The second requirement for the successful exploitation of HDR resources is an improvement in fracturing methods to create reservoirs with greater efficiency of heat exchange. Ideally, the artificial reservoir should combine the following qualities: high initial temperature, large fracture surface-area, large connected volume, low resistance to water flow, minimal water losses to surrounding rock. Scientific research to develop a means for the accurate prediction of the useful lifetime of reservoirs is the third requirement for the practical realization of HDR technology. The thermal drawdown as the stored heat is extracted must be kept within reasonable limits (1–2% per year). Otherwise, after a few years, costly new wells will have to be drilled or the reservoir refractured to restore fluid temperatures to their original level. Taken together, these are quite substantial hurdles to overcome, but success would bring a substantial reward. For example, it has been estimated that one cubic kilometer of hot granite at 240 °C has the stored energy equivalent of 40 million barrels of oil when heat is extracted to a temperature of 140 °C. Australia alone is known to have several thousand cubic kilometers of identified high heat-producing granites, although much of this is located in remote regions where the demand for electricity is small.

For granite rocks to have a temperature that exceeds 200 °C, there has to be an overlying blanket of insulating rock of about 3 km in thickness to retain the heat. The hot granites have an internal fabric of fractures, as a result of cooling from the magma. Developing an underground heat-exchanger (Figure 4.12) involves drilling multiple holes through this blanket into the granite. The latter then has to be fractured further, by increasing the hydraulic pressure at the bottom of the drill-well, so as to open up pathways for water to circulate. Cold water is injected down one set of drill holes and, after circulating, super-heated hot water returns to the surface *via* a parallel set of holes. This water is cooled in a heat-exchanger and then re-injected. The world's first HDR power station is under construction in North-Eastern France.

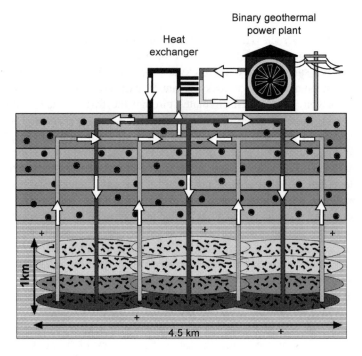

Figure 4.12   *Schematic of ideal underground heat-exchanger for the extraction of energy from hot dry rocks.*

Geothermal Heat Pumps

A heat pump is, essentially, a refrigerator with the heat-exchanger separated from the main unit. In a refrigerator, an enclosed space is cooled and the heat extracted is rejected to the surroundings. With a heat pump, heat is extracted from the external environment and rejected inside a building, thus warming it. This requires pumping heat from a colder reservoir to a hotter one by means of a closed loop that contains a heat-transfer fluid. The cold fluid is conveyed back to the external environment *via* the closed loop. The fluid then passes through a heat-exchanger where it is re-heated to the temperature of the external environment. The outdoor reservoir can be either the atmosphere, where the transfer of heat from a gas is relatively poor, or a pond or a lake, where the process is more efficient. When no pond or lake is available, the heat-exchanger is buried in the ground, which itself serves as a source of heat; this is termed a 'geothermal heat pump'. Since the Earth constitutes an almost infinite source of heat, the geothermal heat pump may be seen as a renewable and sustainable source of energy. Of course, electricity has to be used to operate the heat pump but, overall, there is a major saving in energy consumption and in CO_2 emissions compared with conventional heating. A heat pump is a versatile device that can be operated in reverse (*i.e.* like a refrigerator) when cooling (air-conditioning) is required. In this case, heat is rejected to the external reservoir rather than drawn from it.

It should be understood that geothermal heat pumps do not create heat, rather they move it from one area to another with the aid of electricity to operate the pump. The unit inside the building is essentially a refrigerator with extra valves to control the heat-exchange fluid. A major attraction of these units is that they can be used for both space heating and cooling, and thus remove the need for separate furnaces and air-conditioning units. Within the building, there is a heat-exchanger and the warmed (or cooled) air is circulated by a fan in a conventional manner.

Heat pumps are considered to offer large reductions in energy consumption in buildings and are practical engineering devices, available today. When the system is properly designed, CO_2 emissions can be cut by up to 40%. The heating and air-conditioning of buildings, both domestic and commercial, constitute a major component of overall energy consumption and this is why heat pumps have the potential to make a major contribution to energy savings.

Geothermal heat pumps have been introduced over the past decade or so. In 2000, there were estimated to be 570 000 ground-coupled units installed worldwide, with a thermal capacity of 6 850 MW_{th}. The installation rate was said to be 50 000–100 000 units per year.[3] The UK was slow to take up the technology, but now has about 150 boreholes with a thermal capacity of almost 1 MW_{th}. A number of large commercial and public buildings are already being heated by heat pumps.

Brines and Magma

Geo-pressured brine can be found in several parts of the world. These unique resources are a hybrid of geothermal energy and fossil fuel. Water is produced by the compaction and dehydration of marine sediments and into this water, methane and other hydrocarbon gases become dissolved. Thus, geo-pressurized brines contain three potentially useful forms of energy: (i) heat at temperatures between 150 and 250 °C; (ii) hydraulic pressure between 15 and 25 MPa at the wellhead; (iii) dissolved natural gas at concentrations up to 100 standard cubic feet per barrel of brine. Research on more than a dozen test wells indicates that these brines can be tapped with existing technology. This results in the hydraulic pressure forcing the fluid to the surface, and there the gases bubble out in a separator. The rejected brine is injected into another well at a sufficient depth to avoid contaminating freshwater aquifers. After filtration, the methane can be passed directly into a delivery pipeline, compressed and transported, converted to methanol, or burned on site for electricity. The heat extracted from the brine can be used either directly for space heating, for other purposes, or for the generation of electricity. As with HDR technology, researchers need to be able to predict with higher confidence the productive lifetime of a well before geo-pressurized brines can become commercially developed.

Magma – molten or partially molten rock – represents an even more vast source of energy than HDR. It has been estimated that the thermal energy in 2 km^3 of magma could run a 1 GW_e power plant for 30 years. In most areas, magma lies at least 35 km below the Earth's surface, but magma chambers or protrusions can be found closer to the surface in areas of current or past volcanic activity. Apart from thermal energy, magma can also provide chemical energy. For example, water injected directly into magma that is high in ferrous oxide could yield appreciable amounts of hydrogen.

Finally, mention should be made of the environmental issues that arise with the exploitation of geothermal energy. Obviously, air and water pollution are the two major concerns, along with safe disposal of hazardous waste, siting (effect on scenic or environmentally sensitive areas), and land subsidence. For example, open-cycle systems may generate large amounts of solid wastes as well as noxious fumes. Most of the pollution and wastes produced are gases (*e.g.* hydrogen sulfide, ammonia, methane, carbon dioxide), minerals, and metals dissolved in geothermal steam and hot water. Sludges are also usually generated when hydrothermal steam is condensed. These must be treated as hazardous wastes since they generally contain high contents of chlorides, arsenic, mercury, nickel, and other toxic heavy metals. For liquid wastes or redissolved solids, the best method of disposal is injection back into the porous strata of the geothermal well. This must be well-below freshwater aquifers so that there is no communication between the usable water and the wastewater strata. Injection may also help prevent land subsidence. Large amounts of cooling water are needed for the operation of geothermal power plants, and this can raise conflicts over water resources. Perhaps the most serious obstacle confronting the development of hydrothermal reservoirs is that they tend to be located in or near wilderness areas of great natural beauty. In this respect, some geothermal projects have been halted or abandoned through the intervention of environmental action groups. It is ironic that such activists also campaign for the reduced use of fossil fuels.

4.6 References

1 Shell Company website: www.Shell.com (November 2001).
2 *New and Renewable Energy: Prospects in the UK for the 21st Century*, UK Department of Trade and Industry, London, UK, March 1999.
3 *Guide to UK Renewable Energy Companies 2001*, James & James, Science Publishers Ltd., London, 2001.
4 M.H. Dickson and M. Fanelli, International Geothermal Association, 56122 Pisa, Italy; see: http://iga.igg.cnr.it.
5 J.W. Lund and D. Freeston, World-wide direct uses of geothermal energy 2000, *Geothermics*, 2001, **30**, 29–68.
6 G.W. Huttrer, The status of world geothermal power generation 1995–2000, *Geothermics*, 2001, **30**, 7–27.

CHAPTER 5

Renewable Energy – Electrical

By virtue of its versatility, convenience and cleanliness in use, electrical energy has become the fastest growing sector of the energy market. The range of applications for electricity is almost unlimited. In Chapter 4, we surveyed the forms of renewable energy that give rise to heat. Heat energy is usually employed for space or water heating, for industrial processes, or for conversion into electricity via a steam cycle. In this Chapter, we examine the various forms of potential energy and kinetic energy that are found in nature and that may be utilized to generate electricity via the medium of mechanical energy. It is also convenient here to include photovoltaic and photoelectrochemical systems, both of which employ solar radiation directly to generate electricity (note, thermophotovoltaic systems are discussed in Section 4.4, Chapter 4).

5.1 Hydroelectricity

The use of water power has a very long history, going back at least until Roman times. The earliest reference to an English mill was in 762 AD, in a charter issued by King Aethelbert II of Kent. In 1086, The Domesday Book (a census of English landowners and their properties) recorded that there were 5624 mills in England. Early mills (known as Norse or Greek mills) employed a horizontal waterwheel that drove a pair of grindstones directly without the intervention of gearing. Later mills were fitted with the vertically-mounted undershot wheel as it was more appropriate to the gentle landscape of England. The undershot mill was turned by the current of water striking the lower paddles of a wheel fastened to a horizontal shaft. Power was carried to the grindstones by employing gears that changed the direction of the drive by 90 degrees. In the Middle Ages, the kinetic energy of falling water streams was harnessed to grind corn to make bread. Other uses were fulling cloth (cleansing, shrinking and felting woollen fabrics), sawing wood, pumping water, crushing vegetable seeds for oil, and powering simple machines such as bellows.

During the Industrial Revolution in Britain, in the late 18th century, water power played a key role in developing the industrial base of the country. Many of the cotton mills of Lancashire and the woollen mills of Yorkshire were built in locations where water power was available to drive the machinery. With the engineering of Faraday's dynamo (see Figure 1.6, Chapter 1) into a practical device in the 1870s and the first commercial production of electric light bulbs in 1880, the idea of using water power to produce electricity took hold and many small generators were installed alongside mill-streams. These supplied electrical power to both the textile mills and the local community. In many cases, however, flowing water was an inconvenient power source because fast-running rivers were often located far from supplies of raw materials, major markets and/or trading ports. Moreover, river flows vary with the season and amount of rainfall. By contrast, the introduction of coal- and wood-fired steam engines allowed the textiles and other industries to move closer to raw materials, cities and markets. Thus, water power was largely left behind.

Water power entered a new era in 1882 when the world's first hydroelectric power plant began operation on the Fox River in Appleton, Wisconsin, USA; it had an output of 12 kW, which was sufficient to run about 250 electric lights. The generator was driven through a system of gears and belts by a waterwheel that operated under a 3-m fall of water. Soon it was realized that there were much larger sources of hydroelectricity waiting to be exploited from water stored at high altitude. Moreover, such storage could be enhanced artificially through the construction of massive dams. There are records dating back to around 2900 BC of dams being built across a stream, river or estuary to collect water for human consumption, irrigation, to control floodwaters, or to make up losses in rivers and navigation canals during lean periods of rainfall. Before the Industrial Revolution, dam construction was on only a modest scale. Then, as knowledge of the properties of materials and structures increased, engineers such as W.J.M. Rankine at Glasgow University, provided a better understanding of the principles of dam design and the performance of structures. This led, in turn, to improved construction techniques and larger dams with the result that the number and installed capacity of hydroelectric plants increased rapidly during the late 19th and early 20th centuries.

In 2001, hydroelectricity contributed around 2.2% to the world's total primary energy supply, *i.e.* approximately one-third of that provided by nuclear power (Table 1.1, Chapter 1). The contribution to the total *electricity* supply is, of course, much greater at around 17%, which is roughly equal to the present nuclear component (Table 3.1, Chapter 3).[*] In fact, hydroelectric power is presently the world's largest renewable source of electricity. World-wide, the installed capacity was 755 GW in 2000[1] and is expected to almost double (to 1435 GW) by 2010 with major new schemes in Asia and Latin America. Present statistics also show that the hydroelectricity component of the domestic electricity generation in two countries is exceptionally large, namely, 99.5% in Norway and 87.3% in Brazil. Together with Norway, the European Union has an installed capacity of at least 115 GW.[2] In the UK, the potential is much less, not least because there are relatively few

---

[*] At first sight, it is surprising that in Table 3.1 the hydro and nuclear contributions to world electricity generation in 2001 are roughly equal, whereas in Table 1.1 the nuclear contribution to total primary energy supply is much larger than that of hydro. The explanation is that in Table 1.1 the nuclear contribution is counted as heat generated and not as electricity, whereas hydroelectric power does not involve a thermal cycle. See Section 3.1, Chapter 3 for further explanation.

mountainous regions suitable for building reservoirs and dams; installed capacity is around 1.5 GW and hydroelectric power meets about 1% of the total electricity requirements of the nation.

Hydroelectricity schemes fall into two major categories:

- Large schemes with generating capacities that range from, say, 5 MW to several GW. These use water from mountain lakes or reservoirs, held back by dams. The largest of the schemes represent major works of civil engineering in their construction, and they are operated by electricity utilities that feed the electricity into the grid. A large hydroelectric dam with the power plant situated at its base is shown in Figure 5.1(a).
- Small schemes of a few kW up to, say, 5 MW. These do not normally involve reservoirs and dams, but rely on the flow of rivers and streams, often in hilly or mountainous terrain, see Figure 5.1(b). In some cases, a small pondage capacity may be added to buffer seasonal variations in river flow. These small schemes may be privately owned and operated. The smallest units typically provide power to an isolated farmhouse or a village community. Larger units may support a town, or an industrial process that is a heavy user of electricity.

All hydroelectric schemes require the same basic components:

- a hydraulic head of flowing water, which may be either a reservoir and dam constructed in the mountains (large scheme, Figure 5.1(a)), or a river cascading down a hillside (small scheme, Figure 5.1(b)) – the higher the hydraulic head, the larger is the potential energy available for transmuting into electricity

(a)   (b)

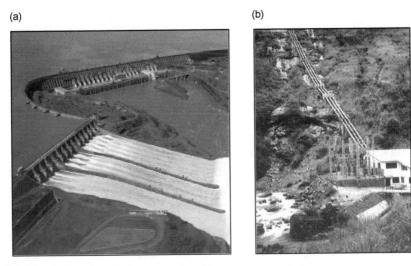

Figure 5.1   (a) *Itaipu hydroelectric dam and power plant (12.6 GW) on the River Parana at the border between Brazil and Paraguay;* (b) *hydroelectric power house (2.4 MW) in Sri Lanka.*
(Courtesy of James & James, London, UK)

- a pipeline or open channel for transporting the water from its source to the power house
- a flow control system
- a water-turbine, coupled to a generator and transformer and contained in a power house, with electrical output to the local community or to the grid
- an outflow where the water returns to the main water-course.

Some large hydroelectric schemes do not involve the construction of a high dam, but merely of a reservoir on a mountain top. The water then flows naturally down the mountain. The location of the powerhouse depends upon the average gradient of the mountain. If the slope is sufficiently steep, then the pipe conveying water from the reservoir to the turbine hall comes down the side of the hill to the powerhouse constructed at its foot. If, however, the terrain is less steep then it is necessary to increase artificially the slope of the pipes by tunneling horizontally into the base of the mountain and building the turbine hall internally. This necessitates constructing an access road, which may be a kilometer or more in length, into the base of the mountain to reach the turbine hall, as well as drilling a steeply sloping shaft to house the water pipes that descend from the reservoir.

The theoretical amount of power produced at a site depends on the potential energy of the water (which is determined by the hydraulic head), as well on the volume flow of water. In practice, there are operational losses, such as frictional losses in turbulence and viscous drag. Moreover, the water as it leave the turbines has some residual kinetic energy that is wasted. There are also energy losses in the turbines and generators. Nevertheless, the overall efficiency is very high, typically in the range 80 to 90%.

Most mountainous countries have large hydroelectric schemes. Typical examples are to be found Iceland, Norway, Pakistan, Western USA, Canada, and Tasmania in Australia. Among the best known of the large hydroelectric plants are the Aswan High Dam in Egypt and those on the large rivers of the Rocky Mountains in Western USA. The world's largest hydroelectric scheme is under construction at the Three Gorges Dam on the Yangtze River in the People's Republic of China. This will have an installed capacity of 18.2 GW, and by controlling the flow of the river's water, the dam may also provide flood protection to 15 million people living downstream. Moreover, the higher water level behind the dam will make the river navigable for over 1000 km upstream. According to the Utility Data Institute in the USA,[3] it is technically feasible to produce around 14 320 TWh per year of hydroelectric power in the world. This is close to the *total* amount of electricity that is produced from all energy sources at present (Table 3.1, Chapter 1). In practice, much of this technical potential is unlikely to be exploited for a variety of reasons, not least because of environmental concerns (see later) and the remoteness/inaccessibility of some water-catchment areas. In 2001, the total world production of hydroelectricity was around 2600 TWh; the outputs from the ten leading countries are given in Table 5.1.[1] Hydroelectricity is predicted to grow and the Utility Data Institute estimates that, over the next 10 years, 153 GW of new capacity will be installed world-wide, *i.e.* about one-fifth of the expected total new generating capacity.

Table 5.1 *Leading producers of hydroelectricity in 2001*[1]

Country	TWh	Percentage of world total	Percentage of domestic electricity
Canada	333	12.6	56.7
PR China	277	10.5	18.9
Brazil	268	10.1	81.7
USA	223	8.4	5.7
Russian Federation	176	6.7	19.7
Norway	124	4.7	99.3
Japan	94	3.6	9.0
Sweden	79	3.0	49.0
France	79	3.0	14.3
India	74	2.8	12.8
Rest of World	919	34.6	16.5
Total	2646	100.0	

Major hydroelectric power facilities often require large initial investments, but these costs are more than offset over the lifetime of the plant because the facilities require no fuel, have lower operating costs than fossil-fuel or nuclear plants, and may provide power for a long period. In most locations that are technically suitable, an economic case can be made for construction, even apart from the (notional) credit for reduced emissions and the benefit to the national balance-of-payments from less expenditure on imported fuel. These latter two factors are very significant for many countries. Another benefit of hydroelectric schemes is that the 'energy payback ratio' – the energy produced during the lifetime of a power plant compared with the energy required to construct and operate the plant – is very much higher than for all other forms of electricity generation.[3]

The industrial use of hydroelectricity has changed over the years. For example, no longer is there a large textile industry in Europe that, historically, relied upon small-scale hydroelectric schemes based on rivers to meet its power requirements. Instead, electricity from large-scale hydroelectric plants is used industrially for the extractive metallurgy and refining of metals such as aluminium, ferro-manganese and copper.

Finally, it should be understood that although large hydroelectric schemes can generate great amounts of electricity, they can also create environmental havoc. The flooding of land to fill reservoirs has often entailed the displacement of large numbers of people and the inundation of villages and sometimes towns. Historic buildings and archaeological sites may also be lost or damaged. Dams may cause widespread disruption to local ecosystems, which will affect both flora and fauna. It is also important to note that, in an age of increasing scarcity of water resources, huge amounts of fresh water are lost by evaporation from the surface of reservoirs,

Table 5.2 *Advantages and disadvantages of hydroelectric power*

Advantages	Disadvantages
Well-established process	Opportunities for hydro-power limited by availability of suitable terrain
Large-scale electricity generation at low operating cost	Involves high civil-engineering costs and may threaten historical landmarks, construction of lakes and dams
Long operating life of the station	Environmental constraints – major engineering project, often in area of outstanding natural beauty
Scope for pumped-hydro storage of electricity	A dam constitutes a danger to cities located downstream
Increased water level can promote recreational activities and provide better habitat for fish	

especially in hot countries. By contrast, small-scale hydroelectric projects have minimal environmental consequences. Moreover, in many instances, the projects lend themselves to local community involvement – in planning, construction and maintenance. To this extent, they can contribute to social cohesion and development. The People's Republic of China leads the world in small hydroelectric development, with approximately 60 000 plants providing electricity for rural communities with few or no other sources of power. Among the industralized countries, the major developers of small hydroelectric systems include France, Italy, Sweden and the USA.

A summary of the advantages and disadvantages of hydroelectric power is given in Table 5.2.

5.2 Wind Energy

Wind power, in the form of windmills, has been used since ancient times for grinding grain and to draw water for irrigation. Until the Industrial Revolution, wind was second only to wood as a source of energy. Although it is widely believed that the windmill was invented in the People's Republic of China more than 2000 years ago, the earliest known written reports of such machines were recorded in Persia during the 7th century. It is thought that Europeans discovered the use of windmills through returning merchants and crusaders in the late 12th century.

Much development took place in Holland where windmills were erected in increasing numbers to drain inland pools and lakes to reclaim new land ('polders') for an ever-growing population. Commercial application of these drainage mills began in the second half of the 16th century with the construction of large corn mills and even larger industrial mills. Notable advances in windmill construction were made in England with the invention of the fantail wheel (Edmund Lee, 1745) that kept the sails turned into the wind, and the self-regulating sail (William Cubbit, 1807) that could be adjusted according to the wind speed and, thereby, gave improvements in both safety and mechanical efficiency.

By 1850, more than 50 000 windmills were at work in Europe, but their popularity was to prove transitory given the advancements that were being made in the engineering of the Watt steam engine (see Section 1.1, Chapter 1). As steam spread, windmills that had satisfactorily provided power for small-scale industrial processes were unable to compete with the production of large-scale, steam-powered mills. Thus, for the most part, windmills were abandoned as working machines and became merely tourist attractions.

With the passing of the steam age and the birth of the electrical industry in the 1880s, renewed interest was shown in wind energy – for the conversion of kinetic energy to electrical energy via an 'aero-generator'. It is claimed that the world's first aero-generator was a 12-kW system built by Charles F. Brush in the USA in 1888. Following the pioneering work of the meteorologist Paul la Cour, a strong industry grew up in Denmark and, by 1908, several hundred machines rated between 5 and 25 kW were in operation in that country.

On-shore Wind Generation

Modern aero-generators are known as 'wind turbines' and are usually grouped together in numbers to form a 'wind farm'. Such a group of modern wind turbines is shown in Figure 5.2(a). The gearbox and generator are housed in a nacelle behind the turbine blades, which rotate to face the wind. In this way, the kinetic energy of the wind is transformed into electricity. The machines are usually spaced 5 to 10 rotor diameters apart to ensure that they do not interfere with each other. The most common form of wind turbine is the horizontal-axis machine with three blades, although the vertical-axis machine, which was invented in 1931 by the French engineer Georges Darrieus and has long, thin, curved outer blades, is sometimes employed (Figure 5.2(b)). The latter design is particularly suitable in locations where the wind noise of the horizontal-axis design may be objectionable.

It is, of course, possible to convert mechanical energy directly to heat, as shown by the experiments of Count Rumford in the 18th century and by the determination of the mechanical equivalent of heat by James Joule in the 1840s. Where there is a requirement for low-grade heat rather than electricity, as in building heating, it might be practical and economic to convert wind energy directly to heat energy using a vertical-axis machine and a paddle rotating in water. Some trials have been conducted along these lines, although often the requirement for heat will be remote from the wind turbine and then electricity is the best medium for energy transfer.

The wind's energy content is proportional to the cube of its speed, *i.e.* doubling wind speed increases power output eight-fold. Thus, small increases in average wind speed can make a large difference to the potential power output of a wind turbine. Modern turbines can work in a light breeze or a strong wind, but conditions above or below these levels are usually impractical; the operation range is usually at wind speeds of between 4 and 20 ms^{-1}. Since wind travels more slowly at ground level because of frictional pull, turbine blades should be mounted as high as possible.

(a) Horizontal-axis wind turbine

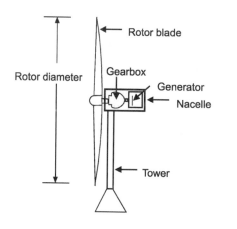

(b) Vertical-axis wind turbine

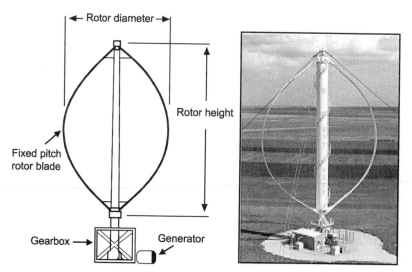

Figure 5.2 *Examples of* (a) *horizontal-axis and* (b) *vertical-axis wind turbines.*

Wind speeds can vary considerably from season to season, from year to year, and with time of day (nocturnal winds are usually lower than those in the daytime). Moreover, wind speeds also tend to increase with altitude and are location specific- neighbouring locations may experience substantially different wind speeds. Thus, determination of the best location for a wind turbine requires detailed calculations that should be based on a careful analysis of wind speeds over a long period. Another important consideration is the proximity of transmission lines to accept the electricity that is generated.

Turbulence is a further consideration in the siting of wind turbines. Where the terrain is level, winds can flow across the surface smoothly, but projections such as hills, buildings and trees, set up patterns of turbulence in the air. These subject the rotor blades of wind turbines to conflicting pressures that decrease the efficiency of power generation and shorten the life of the machinery. The output power of the turbine varies with the area swept by the rotor blades. The large blades of horizontal-axis machines can, however, be heavy and unwieldy, so that strain is placed on the hub machinery and supporting tower. It is this factor that limits the practical diameter of rotors.

Wind power differs from solar power in that it is available, in principle, for 24 hours per day, but in practice its intensity is highly variable. Large wind turbines are capable of generating 1 to 2 MW_p of electricity in a strong breeze, but the output falls off dramatically in light airs. Wind farms in Europe generally employ wind turbines with peak outputs in the range 0.5 to 2 MW_p, and often consist of around 20 turbines spread over 3 to 4 km^2 of land. A typical 0.6 MW_p machine would generate around 1.6 and 2.75 GWh per year on sites with an annual mean wind speeds of 7.0 and 10.0 m s^{-1}, respectively.[2] Several thousand such turbines would be needed to replace the quantity of electricity (measured in GWh) generated by one large fossil-fuelled or nuclear plant. Even then, there would be long periods of low or zero generation, which would necessitate the provision of back-up conventional plant or storage facilities. Because of this back-up requirement, the utilities will normally only give credit for the electricity actually supplied to them (MWh) and not for the power capability of the turbine (MW_p). This is a financial penalty that is common to most renewable forms of energy where the supply is not totally reliable. Generally, in a grid-connected system, electricity generated by wind power (or by solar photovoltaics, see Section 5.6) would not be stored, as economics would dictate that all such electricity be used at once and that conventional plant would be shut down to compensate. It should also be noted that grid systems are mostly designed to accommodate supply and demand fluctuations, and could accept a small penetration by wind energy (~10%) without changes to the existing infrastructure. Obviously, such possibilities do not apply to non-grid connected generators where either energy storage or an auxiliary diesel generator will be required.

A large, horizontal-axis turbine may have a tower height of 40 to 65 m and a rotor diameter of comparable size. The visual impact of multiple turbines in a wind farm, together with the land they occupy (often on headlands or other scenic coastal sites), may cause concern to local residents and environmentalists. Indeed, some argue that the visual intrusion on the landscape is quite disproportionate to the relatively small quantities of electricity that are generated. On the other hand, others contend that wind turbines, with their slim towers and shaped blades, are both architecturally and aesthetically pleasing, and give a powerful symbol of green energy on the skyline. Further potential problems are the aerodynamic noise from the turbines (a 'swishing' sound), possible interference with television when using steel blades, and bird kills. Moreover, wind farms frequently require the construction of overhead pylons across previously unspoilt countryside. On the credit side, not only are there no polluting emissions, but also the land on which the turbines are situated may be used for normal agricultural purposes, as the turbine bases occupy only about 1 to 2% of the land area. At the end of life, the turbines may readily be dismantled and the land returned to 'greenfield' status. Another important credit for wind power is the short lead-times for construction of a wind farm, once planning permission has been obtained.

Wind farms are being built in various countries and are already making significant contributions to national energy supply. The principal countries involved in the development of wind energy are given in Table 5.3. By the autumn of 2002, the total world installed capacity was estimated to be about 27 GW_p, of which 20 GW_p was located in Europe. The relative contributions from the individual European countries are also shown in Figure 5.3. Over the past few years, the rate of growth in Europe has been around 40% per annum, a rate equalled only by the cellular phone and computing industries. Globally, the growth rate has been 25 to 30% annually. Obviously, with such high growth rates, data rapidly become outdated.

It should be noted that a typical, large fossil-fuelled or nuclear electricity generating plant of output 1 to 2 GW_p is 1000 to 2000 times larger than a typical wind turbine of 1 MW_p output. In terms of electricity generated per annum (GWh), the comparison is even less favourable to wind since the fraction of time when peak generation is possible is much less. The 'down-time' on a fossil or nuclear station is determined only by market demand for electricity and by maintenance schedules – not by the availability of 'fuel', as is the case for wind energy.

In parts of Scandinavia and North Germany, wind accounts for more than 10% of electricity generation. On a world-wide basis, however, the electricity supplied by wind power in 1999 was under 0.5% of total world electricity generated, possibly as low as 0.2%. Nevertheless, it is now the fastest growing renewable source of electricity. In Europe, the UK is one of the windiest countries and is said to have

Table 5.3 *Top wind energy markets (by installed capacity)*

Country	Date	GW_p
Germany	autumn 2002	10.65
USA	end 2001	4.25
Spain	autumn 2002	4.08
Denmark	autumn 2002	2.52
India	end 2001	1.41
Italy	autumn 2002	0.76
Netherlands	autumn 2002	0.56
UK	autumn 2002	0.53
Sweden	autumn 2002	0.30
Greece	autumn 2002	0.28
PR China	end 2000	0.27

Figure 5.3 *Wind power installed in Europe by autumn 2002* (MW_p).

40% of the total potential wind resource for the region. The UK's 1000th wind turbine was commissioned in January 2003 and brought the total installed capacity to 0.53 GW_p. Although small compared with Germany (10.65 GW_p), it is still sufficient to service more than 300 000 homes or a large city. The savings in carbon dioxide emissions are estimated at 1.3 Mt per annum. There are also plans to build a huge wind farm (600 MW_p) on the Isle of Lewis, off the west coast of Scotland. This alone would provide 0.5% of the UK's electricity. In the Southern Hemisphere, Australia has excellent wind resources, particularly in the States of South Australia and Tasmania, which lie in the path of the prevailing westerly wind currents – the 'Roaring Forties'. The Australian Wind Energy Association has ambitious plans to construct 5 GW_p of wind turbines by 2010. All this activity stems from the fact that wind power is one of the cheapest forms of renewable electricity and under some circumstances is cost-competitive with conventional fossil-fuel generation.

A pertinent question when considering wind energy is what is the 'energy pay-back' time, *i.e.* how long does it take for a wind turbine to generate the amount of energy used in its manufacture. This is estimated at three to five months. Thus, over the course of its life (typically, 20 to 25 years), a wind turbine should produce many times more energy than was used in its construction.

Off-shore Wind Generation

Some of the aesthetic or environmental objections to land-based wind farms may be overcome by siting the turbines off-shore, on the continental shelf, where much more open space is available, the wind is often stronger, and there are no local residents to object to the construction. Here, the problems are higher capital costs of installation and cabling to bring the electricity ashore, higher maintenance costs, and possible hazards to navigation. Denmark and Sweden have pioneered such off-shore facilities. The Middelgrunden wind farm in Denmark is shown in Figure 5.4; this comprises 20 turbines, each with a generator output of 2 MW_p.

The first off-shore wind farm in the UK was commissioned in the North Sea, off the coast of Northumberland, in December 2000. Two 2-MW_p machines can provide sufficient electricity annually to service 3000 households. The total European installed off-shore wind capacity at the end of 2002 was 250 MW_p. Of this, 160 MW_p is in the new Danish Horns Rev field, which consists of 80 Vesta machines of 2 MW_p each. This recently completed facility is the largest in the

Figure 5.4 *Off-shore wind farm at Middelgrunden, Denmark.*
(Courtesy of British Wind Energy Association)

world. It is likely that many more of these wind-farms will be built. Proposals have been advanced for projects totalling 2.5 GW_p capacity in Denmark, Sweden, the Netherlands, and Ireland. In spring 2001, the UK government announced plans to build up to 18 off-shore wind farms around the coast of Britain. In mid 2003, more detailed plans were laid for the granting of licences for the first three farms. These will involve the construction of hundreds of wind turbines in the hostile environment of the North Sea. The ultimate target is to produce 10% of the UK electricity from renewables, with wind energy playing a prominent role.

Small, stand-alone, wind turbines are employed to produce electricity in places remote from mains electricity. Around the coast of Europe, small wind generators may be used, in conjunction with a storage battery, to supply electricity for navigation lights on buoys. Such generators are also employed on small craft, particularly on sailing boats when at sea and when moored in marinas.

A comparison of the advantages and disadvantages of wind power is given in Table 5.4.

5.3 Tidal Energy

Tides arise as a result of the gravitational pull of the moon on the ocean, modified by that of the sun. The height of tides varies with the time of the lunar month, with two strong 'spring tides' at full or new moon and two weak 'neap tides' during the moon's first and last quarters. There is also a high and a low tide about every 12.5 h. The influence of nearby geographical features is also important and can be particularly pronounced in river estuaries where local topography can greatly exaggerate the tidal range. In effect, tidal flows open up the possibility of generating electricity, either by means of a barrage or by extracting energy directly from marine currents.

Tidal Barrages

The available potential energy of the tide is proportional to the square of the range and therefore barrages and turbines are best located across the mouths of estuaries with large ranges. By using reversible turbines, it is possible to generate electricity during both rising and falling tides. The generator output will vary sinusoidally with a period of hours and, typically, the turbines will only supply useful power for 3 to 5 h per tide. This limitation may be of no great consequence as the power would be fed into an electricity grid. There is also the possibility of closing off the turbines and holding back the water for a while if the time of high tide corresponds with that of low electricity demand. Another option is to back-pump water from the low-level side of the barrage to the high-level side, as practised in pumped-hydroelectric schemes ('pumped hydro', see Section 7.2, Chapter 7).

The largest tidal scheme is the 240-MW_p barrage on the River Rance in northern Brittany, France (Figure 5.5). The facility has been generating electricity since 1966 and uses conventional bulb turbines. The barrage is 750 m in length and produces annually more than 500 GWh of electricity, which is sufficient to supply a

Table 5.4 *Advantages and disadvantages of wind energy*

Advantages	Disadvantages
Well established technology	Limited to sites with high average wind speed (generally, stronger winds at sea than on land)
Can generate power in remote locations and arid regions	An ephemeral energy that must be used as it is created, or stored in batteries or pumped water
Distributed electricity generation	May require back-up with other generating equipment using fossil fuels or hydro if continuous power supply is required
No pollution in the form of solid or liquid wastes, noxious or greenhouse emissions and particulates	Generating capacity of wind farms still much less than that of steam turbines in conventional power stations or nuclear plant
Manufacture and disposal of wind power equipment does not involve the use of heavy metals or toxic chemicals	Capital intensive form of energy production, especially when sea-based
Does not use up land for mining, waste disposal, or transport of fuel, as well as for construction and installation	Unattractive visual impact – public opposition to siting of large-scale wind farms close to population centres or in areas of natural beauty (sea-based turbines have less impact on population)
Fast construction times; modular designs	Other objections include noise, danger to birds, and telecommunications, radio and television interference; sea-based turbines may present obstruction to navigation
Although wind turbines are subjected to continuing stresses that vary greatly in intensity, reliability has improved; wind farms now expect 95 to 98% availability	
Low running costs (but sea-based turbines subject to salt-water corrosion)	
Short energy pay-back time	
One of the safest forms of power generation	
Wind energy schemes can be community projects: 'a friend of the democratic process'	

city the size of Rennes (population ~200 000). There are smaller barrages in Canada, the People's Republic of China, and the Former Soviet Union. The UK is well-placed with respect to potential sites for barrage installations, foremost among which is the estuary of the River Severn where topographical funnelling effects

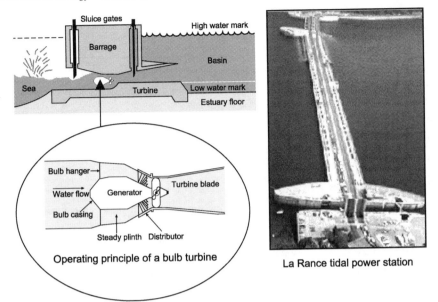

Figure 5.5   *Schematic of the design of a tidal barrage system with a bulb turbine as used in the world's first tidal power station on the estuary of the River Rance.*

produce a periodic tidal range of up to 11 m. This site was studied in depth in the late 1970s and a proposal was advanced for a 17-km barrage with 216, 40-MW turbo-generators to give a peak power output of around 8 GW_p. On a per annum basis, the generating capacity would be equal to 6% of the electricity requirements in the UK. The scheme has never been implemented for a variety of reasons, most notably: (i) concerns over the environmental impact, especially that on fish and local bird life; (ii) problems that would be caused to navigation; (iii) the huge size and capital cost of the civil engineering involved; (iv) the privatization of the electricity generating industry – private industry was simply not interested in meeting the very high capital cost given the long pay-back time.

There are several other potential sites in the UK, *e.g.* on the estuaries of the Rivers Dee, Humber, Mersey, and Solway Firth, but none of these are being developed for much the same reasons that caused the abandonment of the River Severn project. It is estimated that the total potential tidal resource in the UK, if fully exploited, would be sufficient to supply 20% of the nation's demand for electricity. Few suitable sites exist elsewhere in Europe, and at present it is doubtful whether this technology will be extended further in the region. There are, of course, other places outside of Europe where tidal energy is significant. For example, the north-west coast of Australia, off the Kimberley Region, experiences tides of 6 to 8 m. It has been estimated that the Kimberley tidal power resource could be as much as 5000 PJ per year, which is equivalent to the entire energy supply currently required in Australia. Unfortunately, the population density in this remote area is very low and the main centres of urban population are many thousands of

Table 5.5 *Advantages and disadvantages of tidal energy*

Advantages	Disadvantages
Common factors	
Predictable source of electrical power (according to tide tables)	Obstruction to navigation
No pollutants	
Long operating life of power plant	
Little visual impact	
Tidal barrage schemes	
Operation and maintenance costs are low	Few suitable sites in world's estuaries
Protection of coastline against storm surge tides	Large civil engineering undertaking and high capital cost
May provide road crossing for rivers and estuaries	Possible adverse effect on local marine ecosystem (though it is also considered that a tidal barrage will provide a new sheltered reserve for birds, fish and vegetation)
Opportunities for water-based recreational pursuits	
Marine current schemes	
Small-scale, distributed generation	New technology; little operational experience
	Restricted to areas where there are fast current flows, these are relatively few in number
	Costs uncertain

kilometres distant, which would make it uneconomic to transmit energy either as electricity or as hydrogen.

A summary of the advantages and disadvantages of tidal energy is given in Table 5.5.

Marine Currents

The ebb and flow of the tides gives rise to marine currents. In most places, these are too slow to provide practical sources of power, but often around islands and headlands, through narrow straits and in estuaries, there are fast and turbulent currents from which useful power might be obtained. The favoured concept for extracting such energy is a form of 'underwater windmill', which is analogous to a wind turbine (a schematic of a horizontal axis, axial-flow turbine is given in Figure 5.6). As with wind energy, a cube law relates instantaneous power to fluid velocity. It is here, however, that the analogy ends. The density of water is very much higher than that of air and therefore, for a given flow velocity, water has a far

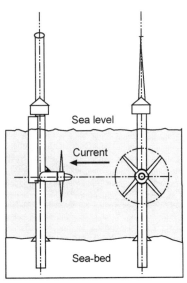

Figure 5.6 *Schematic of an axial-flow, sea-bed mounted, marine current turbine.*

greater power density per unit cross-section. Conversely, for a given power output (kW m^{-2}), the water velocity will be much smaller (only about 11%) than that needed in air. In practice, water velocities are lower than air velocities – typically, 1 to 3 m s^{-1} (about 2 to 6 knots in nautical terms), whereas wind turbines achieve their rated power at a wind velocity of about 12 m s^{-1}. The power that may be extracted from a marine turbine as a function of its rotor diameter and the speed of the current is shown in Figure 5.7.[4]

The rotor diameter of a marine-current turbine may be up to 20 m compared with 20 to 80 m for a wind turbine. The rotor is coupled through a gearbox (which may increase the rotation speed by a factor of 100) to a generator, just as in a wind turbine. The principle of electricity generation from marine currents is sound. In practice, the problems encountered are of an engineering nature and arise from the need to secure the unit against the huge forces associated with the fast flow of tens of thousands of tonnes of seawater. A water flow of at least 2 m s^{-1} is required to extract worthwhile amounts of power. The maximum flow velocity tends to be near the surface of the sea, so the turbine rotor ideally needs to intercept as much of the depth of flow as possible. Small generators may be suspended below rafts or pontoons, but large generators are likely to be housed in support structures that are fixed to the seabed. Fortunately, lessons have been learnt from the design and installation of off-shore oil platforms and the same engineering technology may be applied. The rotor would be mounted on a mono-pile, which is a steel tube of large diameter that is set in a hole drilled in the seabed. This can be done from a 'jack-up' barge that raises itself on four legs to form a stable platform, from which the hole can be drilled and the mono-pile and turbine craned into place. One design proposes the use of twin rotors of 15-m diameter that are mounted on a single

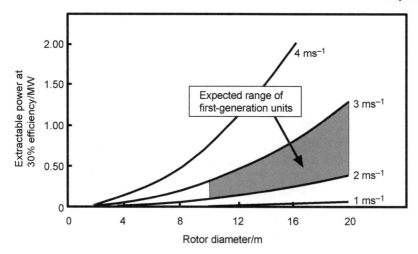

Figure 5.7   *Extractable power from marine currents.*[4]
(Courtesy of James & James, London, UK)

mono-pile and deliver up to $1\,MW_p$ of electrical power. In order to make integration with the electricity grid worthwhile, it will be necessary to have a grouping of these units close together, analogous to a wind farm.

Marine-current turbines have a number of perceived advantages over wind energy, namely:

- negligible environmental impact (no noise, little visual intrusion)
- the energy resource and its timing can be predicted confidently so that planned base-load power contributions are possible
- tidal currents offer significantly better overall capacity factors (40 to 60%) than wind, with no 'dead days'.

Marine-current turbines are much smaller than tidal-barrage schemes. The turbines do, however, avoid the need for massive and expensive constructional work. Thus, the energy pay-back time is likely to be more favourable.

To date, design and assessment studies have been conducted and a few small demonstration units (~kW size) have been built and tested, *e.g.* in Australia, Japan, and Scotland. A $3\text{-}MW_p$ prototype unit is under construction in the UK. Cost estimates suggest that a cross-flow rotor (analogous to a vertical-axis wind turbine) will be more economic than a horizontal-axis rotor. In 1992–93, a national assessment study identified sites in the UK that collectively would be capable of generating 58 TWh of electricity each year. Of this total, about 20 TWh per year was considered to be economic in the medium term (5 to 6% of electricity consumption in the UK). The largest of these tidal-current resources is in the Pentland Firth, between the Orkney Islands and the Scottish mainland. Elsewhere in the world, there is a very large resource in the waters around the Philippines. The present position is that there is no perceived technical barrier to the construction of marine-current generators, but the costs still appear to be rather high compared with conventional power stations. There is, however, scope for reducing the costs to competitive levels.

A list of the advantages and disadvantages of electricity generation from marine currents is given in Table 5.5.

5.4 Wave Energy

As winds blow over the surface of an ocean, waves are created. The stronger the wind force and the wider the expanse of ocean, the greater are the waves. Waves breaking on the shores of Western Europe from the Atlantic Ocean are generally 1 to 2 m in height, but in a storm-force wind they may reach 10 m. These waves carry a substantial quantity of energy that is dissipated at sea or on the shore, but which in principle could be harvested. It has been estimated that the waves breaking on the European Atlantic coast, from North Scotland to Portugal, have a total energy of around 1000 TWh per year. If only 10% of this could be recovered, it would be a significant contribution from renewable energy. There are also sites in the North Sea (especially around Denmark), in the Greek Islands and in the southern tip of Italy where the waves are sufficiently high to merit consideration for energy recovery.

Many designs of device have been proposed for wave energy conversion and models have been constructed for tank testing. It is a field in which there is scope for great mechanical ingenuity and it has interested a number of university engineering departments and specialist companies around the world. Broadly, the devices may be divided into three categories: shoreline, near-shore, and off-shore. Most attention has focused on the shoreline and near-shore devices, both because they will be easier to install and maintain, and because it is envisaged that the electrical power would be fed to the national grid and, thereby, would smooth the power fluctuations that arise from the variable wave intensities. On the other hand, electricity is most expensive and most needed in islands and remote areas where there may be no national grid. Under these circumstances, a wave-power system might complement a conventional CHP scheme (see Section 3.4, Chapter 3), or even wind or solar generation coupled to electricity storage. Alternatively, the wave energy might be used to operate a desalination plant to provide fresh water to communities in hot climates. There is also a growing demand for electrical power in deep-water locations for oil rigs. Assessment studies have indicated that off-shore devices based on tethered buoys could compete with diesel generators.

The best-developed shoreline system is the oscillating water column (OWC). This consists of a partially submerged, hollow structure that is open to the sea below the water line (Figure 5.8(a)). The structure encloses an air column that is alternately compressed and rarified by the moving waves. A duct allows air to flow to and from the atmosphere through a Wells turbine, which has the property of rotating the same way regardless of the direction of air-flow. Alternatively, a pair of self-rectifying, contra-rotating Wells turbines may be used. The turbine is then connected to a generator. As long ago as 1985, a 500-kW OWC system was

(a)

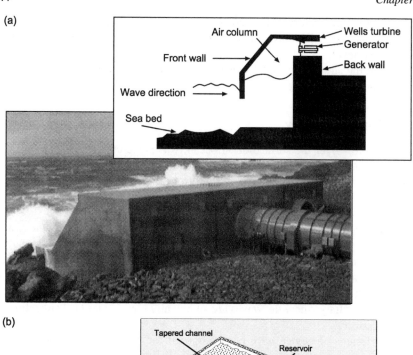

(b)

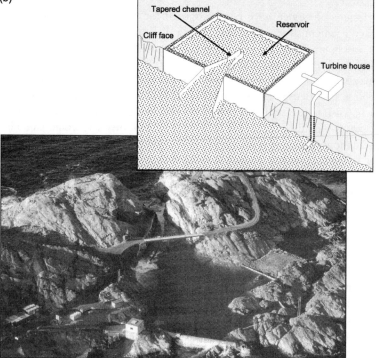

Figure 5.8   (a) *Oscillating water column device.*
(Courtesy of James & James, London, UK)
(b) *Tapered channel device.*
(Courtesy of UK Department of Trade and Industry)

installed in Norway, but this has since been abandoned after a mechanical failure. Other devices have been constructed in Japan, the People's Republic of China, and India. In 1988, a 75-kW prototype was built into a natural rock gully on the island of Islay (Scotland). The operation of this provided much valuable experience and a new 500-kW OWC has been built on the shoreline of the island (Figure 5.8(a)). There is in place an agreement to purchase the power generated over a 15-year period. A new generation of large devices is under development. For example, the European Union has sponsored the construction of a 1-MW OWC plant in the Azores, while 2-MW demonstration facilities are planned for Vancouver Island in Canada, and for the coast of Portugal.

A second design of shoreline wave energy converter is the so-called 'Tapchan', as shown in Figure 5.8(b). This comprises a gradually narrowing channel with wall heights of 3 to 5 m above the mean water level. On moving along the channel, the waves increase in amplitude until the crests spill over the walls into a lagoon. The water in the lagoon returns to the sea via a low head turbine. Because the lagoon acts as an energy store, the turbine generates a stable output. A 350-kW plant of this type was constructed in Norway in the mid 1980s and operated reasonably well for six years.

Several designs for off-shore (deep water) wave energy converters have been advanced, *e.g.* the Salter Duck, and the Pendula, hose pump, Pelamis and McCabe systems. Unfortunately, not one of these floating devices has yet been developed beyond the modelling and tank testing stage. In open water, the energy of waves during storm force conditions is many times greater than the average and device survival is a major consideration. A perceived commercial problem with such devices is their high capital cost and long pay-back times.

To summarize the present situation, demonstration shoreline devices of the OWC type are undergoing extensive trials and their future prospects will depend upon their reliability and long-term performance, as well as on the projected cost of electricity generated by them. A comparison of the advantages and disadvantages of wave energy is given in Table 5.6.

Table 5.6 *Advantages and disadvantages of wave energy*

Advantages	Disadvantages
Large potential source of renewable energy	Many competing designs of wave generator
Reliable for most of the year	Facilities must withstand high forces
Modular construction	Variable electricity output due to unpredictable nature of the sea
Little environmental impact	Comparatively little operational experience
	Facilities may cause changes to shore lines and local ecosystems
	Not cost-competitive in the near term

5.5 Ocean Thermal Energy Conversion

Electricity generation via ocean thermal energy conversion (OTEC) exploits natural temperature differences in the sea (*i.e.* between surface water and the deep) by using some form of heat engine. This is, in effect, a means to convert solar heat gained by the ocean surface water into useful energy. The theoretical efficiency of OTEC is very low (~2%) and thus the laws of thermodynamics demand as large a temperature difference as possible to deliver a technically feasible and reasonably economic system. This limits application of the technology to a few tropical regions with very deep water where temperature differences in the range 20 to 30 °C are found. A heat engine, based on the Rankine cycle (as used in conventional steam engines), can be used to work over this limited temperature range, albeit at low efficiency. There are two main processes, namely: the open-cycle system and the closed-cycle system (Figure 5.9). In the former, warm seawater is turned into vapour (at reduced pressure) and then drawn through a turbine by condensation in a unit that is cooled by cold seawater. The closed-cycle system uses warm seawater to boil a low-temperature working fluid, such as ammonia, which is drawn through a turbine by being condensed in a heat exchanger with cold seawater and is then recycled back to the boiler by a feed pump. Both approaches require enormous water flows that, in turn, involve large engineering components.

A few experimental OTEC plants have been built, notable in Hawaii. It has been suggested that OTEC may be used as a multi-purpose technology, which is considered to offer better economic prospects. For example, the nutrient-rich cold water drawn from the deep ocean has been found to be valuable for fish-farming and could also be used directly for cooling applications, such as air-conditioners, in the tropics. In addition, the production of fresh water by desalination might be usefully combined with OTEC operations.

5.6 Solar Energy

Principles and Construction of Solar (Photovoltaic) Cells, Modules and Arrays

As we discussed in Section 4.4, Chapter 4, solar energy is one of the principal sources of renewable energy potentially available to humankind. It comes in the form of both heat (infra-red radiation) and light (visible radiation). In Chapter 4, the harvesting of thermal energy was discussed; here we are concerned with the collection of light energy and its utilization as electricity.

The 'photoelectric effect' was first noted by the French physicist, Edmund Becquerel, in 1839, who found that certain materials – semiconductors – have the ability to produce low-voltage d.c. electricity when exposed to light. Later, in 1905, Albert Einstein described the nature of light and the photoelectric effect on which photovoltaic technology is now based. The scientific principles of solar cells are described in Box 5.1. When two dissimilar semiconductors are brought into

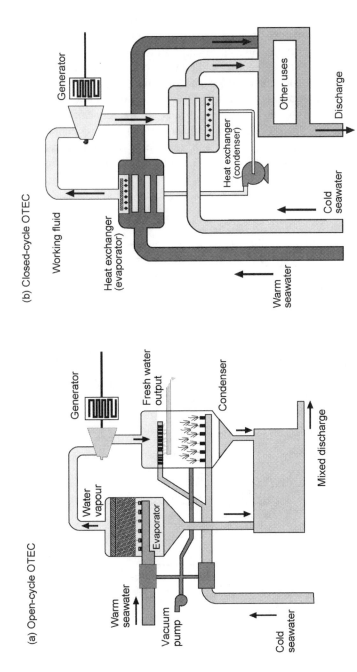

Figure 5.9 *Designs of ocean thermal energy conversion systems.*

Box 5.1 Photovoltaic Cells

Semiconductors are materials with conductivity between that of conductors (*e.g.* metals) and insulators. In solid materials, the electron energy levels form bands of allowed energy that are separated by forbidden bands (Figure 5.10(a)). The separation between the outermost filled band (the 'valence band') and next highest band (the 'conduction band') is known as the 'energy gap' or 'band-gap energy', E_g, of the material. In essence, E_g is the amount of energy required to liberate electrons from their covalent bands in the crystal lattice of the semiconductor. The energy gap is quite small (~1 eV) for a semiconductor, but is large for an insulator (~10 eV), and zero for a metal (the valence and conduction bands overlap). The energy level at which there is a 0.5 probability

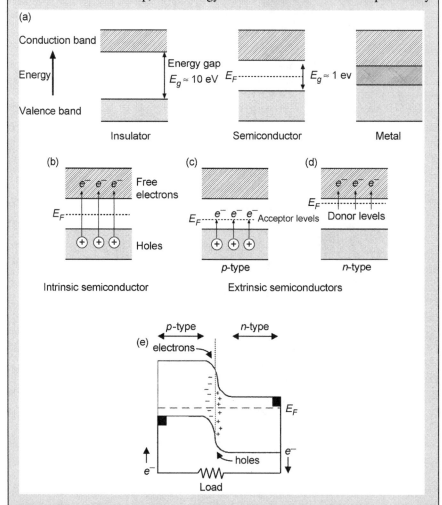

Figure 5.10 (a) *Schematics of the band structure and conductivity of insulators, semiconductors and metals.* (b)–(e) *Operating principles of semiconductor devices.*

Box 5.1 (cont'd)

of finding an electron is known as the 'Fermi level', E_F. In a metal, this level is very near the top of the valence band. In a semiconductor, however, the Fermi level is located at about the middle of the energy gap, and since this gap is small, appreciable numbers of electrons can be excited from the valence band to the conduction band, to produce a moderate current.

There are both positive and negative charge carriers in a semiconductor. When an electron moves from the valence band into the conduction band (*e.g.* by absorption of energy from a photon of sunlight), it leaves behind a vacancy, or so-called 'hole', in the otherwise filled valence band (Figure 5.10(b)). This hole (electron-deficient site) appears as a positive charge and acts as a charge carrier in the sense that a nearby valence electron can transfer into the hole, thereby filling it and leaving a hole behind in the electron's original place. Thus, the hole migrates through the material. In a pure crystal that contains only one element or compound, *e.g.* silicon, there are equal numbers of conduction electrons and holes. Such combinations of charges are called 'electron-hole pairs', and a pure semiconductor that contains such pairs is an 'intrinsic semiconductor'.

The electrons and holes in an intrinsic semiconductor recombine after a short time and their energy is wasted as heat. To obtain useful electricity, therefore, the band structure and conductivity of semiconductors must be modified. This is achieved by the addition ('doping') of controlled micro-quantities of impurities, as shown schematically in Figure 5.10(c,d). In the case of silicon semiconductors, trivalent impurities (*e.g.* boron, aluminium) that have an E_F just above the valence band, accept electrons from the silicon and so leave holes in the valence band. Such impurities are referred to as 'acceptors', and the resulting 'extrinsic' semiconductor is known as a '*p*-type' because the charge carriers are positively-charged holes. Conversely, pentavalent dopants (*e.g.* arsenic, phosphorous) that have an E_F just below the conduction band are 'donors', and provide electrons to the conduction band of silicon. This produces an '*n*-type' extrinsic semiconductor, so-called because the majority of charge carriers are now electrons, whose charge is negative.

When *p*-type and *n*-type semiconductors are brought into contact, to form a '*p-n* homojunction', the band structure at the interface is distorted so the Fermi level is the same on each side of the interface (Figure 5.10(e)). A region that is depleted of mobile charge carriers is formed at the interface (the 'depletion region') and an electric field of the order of 10^4 to 10^6 V per cm is built up across the interface – the photovoltaic effect. This internal electric field creates a potential barrier that prevents further electron-hole recombination. Thus, when photons are absorbed by the semiconductor, the electric field attracts electrons created on the *p*-side of the junction to the *n*-side, while holes are swept in the reverse direction. This charge separation induces a voltage across the device and when the two sides of the junction are connected together via an external circuit current is able to flow and, thereby, produce electrical energy (Figure 5.10(e)). The resulting device is known as a 'photovoltaic cell', or more commonly as a 'solar cell'.

conduct to form a junction, and light is absorbed by one of the pair, a voltage is developed across the junction. This is the 'photovoltaic effect'.[*]

The vast majority of modern solar cells have been made from single-crystalline silicon, although there is an increasing use of polycrystalline and amorphous silicon. Single-crystalline silicon is used in the form of thin wafers sliced from a pear-shaped crystal (boule) grown from the melt. This is an intricate and costly operation to perform. The crystals may be grown especially for photovoltaic applications, or may be a by-product from the semiconductor industry.

Silicon photovoltaic cells generally employ a *p*-type wafer. A controlled quantity of phosphorous is diffused into the surface to form a thin layer of *n*-type silicon. The time and temperature of diffusion determines the depth of the *n*-type layer. The surface layer is then given an anti-reflection coating and electrical contacts are made to both the front (*n*-type) and back of the wafer (*p*-type). A schematic representation of a basic silicon cell is shown in Figure 5.11(a). A typical, square, solar cell has dimensions of $100 \times 100 \times 0.3$ mm. The top (*n*-type) layer is only ~ 0.5 μm thick, while the base (*p*-type) layer is ~ 300 μm thick. This relatively thick layer is necessary because crystalline silicon is a poor absorber of light.

Individual solar cells generate only a low voltage (typically around 0.6 V) and power output (~ 1.5 W_p). The cells are therefore packaged together to form a 'solar module' (or 'solar panel') that provides a useful voltage and current. Modules are produced in varying sizes and power outputs. The cells are first connected electrically in series, then encapsulated and, finally, a glass front and a weatherproof backing are added. The units are shipped as sealed modules that are each fitted with a junction box. Depending upon the application, it is usually necessary to mount the modules on some form of support structure and then join them in parallel with cabling to produce a 'solar (PV) array'. The series-parallel arrangement of cells and modules multiplies the current and reduces the electrical resistance of the array. Auxiliary items include: (i) an inverter and step-up transformer to interface with the electrical load; (ii) batteries to store electricity until required and to smooth out fluctuations in module output and electricity demand; (iii) an electronic control system to match the power output of the array to the load profile, and to regulate the operating regime of the batteries so as to avoid excessive overcharge and discharge. All of these components are collectively termed the 'balance of system', BoS, and represent a significant fraction of the overall cost of a functional array.

Crystalline silicon is not the only semiconductor that may be used for photovoltaic cells. Other possible candidates include gallium arsenide (GaAs), cadmium telluride (CdTe), copper indium diselenide ($CuInSe_2$), and various organic compounds. The advantage of these materials is that they may be deposited as thin films by co-evaporation of the constituent elements. This is considerably cheaper than growing and cutting single crystals. In the 1970s and

---

[*] Interestingly, one of Edmund Becquerel's relatives – Antoine Becquerel – discovered radioactivity some years later, so photovoltaics pre-date nuclear power!

(a)

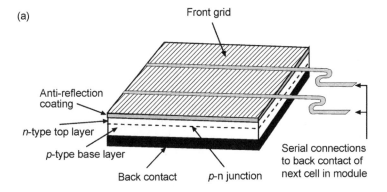

(b)

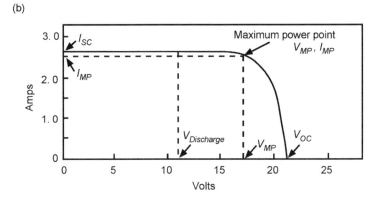

Figure 5.11 (a) *Essential features of a p-n homojunction silicon solar cell.*
(Courtesy of Imperial College Press)
(b) *Typical current-voltage characteristics of a silicon solar module at standard test conditions of 1000 W m^{-2} and 25 °C.*

1980s, the solar conversion efficiency of photovoltaic cells made from these materials was low, but over the years improvements have been made. One of the best such semiconductors for this application is $CuInSe_2$, or its alloy with gallium $Cu(In,Ga)Se_2$. Photovoltaic cells and modules made from this material are now commercially available.

Performance of Solar (Photovoltaic) Cells, Modules and Arrays

A solar cell/module can be operated at any point along its characteristic current-voltage curve, as shown in Figure 5.11(b). Two important parameters on this curve are the short-circuit current (I_{SC}, typically 20 to 40 mA cm^{-2}) and the open-circuit voltage (V_{OC}, typically 0.6 to 0.7 V per cell). To a good approximation, I_{SC} is proportional to the illumination level, whereas V_{OC} is proportional to the logarithm of the illumination level; V_{OC} is also quite sensitive to temperature. A plot of power

($V \times I$) against voltage shows that there is a unique point on the curve at which the solar cell will generate maximum power, known as the 'maximum power point' (V_{MP}, I_{MP}). In practice, the output of a cell is expressed in terms of peak watts, W_p. This is the power obtained when the cell is illuminated under standard conditions of 1000 W m^{-2} intensity, 25 °C ambient temperature, and spectrum that relates to sunlight that has passed through the atmosphere.

To extract the maximum electrical power from a solar cell, it is essential that it be operated at close to its maximum point. The value of V_{MP} is a function of cell temperature and illumination, and can vary between cells – even when these are manufactured under identical conditions. Furthermore, the load into which the cell is delivering its power, *e.g.* a battery, will have its own characteristic current-voltage curve. It is therefore highly desirable to have a d.c.-d.c. converter, together with appropriate control circuitry, that matches the solar cell to its load in such a way that maximum power is transferred. Such a device is known as a 'maximum power point tracker.' In the example shown in Figure 5.11(b), a module with a tracker will, ideally, provide a power of 43.5 W ($V_{MP} = 17.4$ V, $I_{MP} = 2.5$ A) for charging a battery at a low state-of-charge, as opposed to only 28.6 W ($V_{Discharge} = 11.0$ V, $I = 2.6$ A) for the module alone. In the former, the d.c.-d.c. converter scales down the value of V_{MP} to that required for charging the battery (typically, about 15 V for lead-acid) at a current that is proportionally higher than I_{MP}, *i.e.* $43.5/15 = 2.9$ A for the example given in Figure 5.11(b). It should be realized, however, that some power is lost in the conversion from the voltage at the maximum power point to the battery voltage. The efficiency of most trackers is usually around 93%.

Mechanical tracking systems are also used to optimize the output from a solar array and match it to the desired load profile. Such a system physically moves the array to make it point more directly to the Sun. The angle of tilt is determined largely by the latitude, which defines the path of the sun across the sky. In the tropics, the arrays will be essentially horizontal to receive maximum insolation, whereas in temperate latitudes they will be tilted at an angle towards the sun, which is lower in the sky. In reality, the situation is more complex since it is also necessary to consider the intended application and the load requirement. For example, if the load demand is small in winter and large in summer (for air-conditioning or irrigation), then it is best to have a low angle of tilt in order to maximize the output in summer. If, however, the loads are almost constant throughout the year (*e.g.* repeater stations or navigational aids), it is more appropriate to use a steeply inclined array so as to maximize the electricity generated during the winter months, while sacrificing the excess that is potentially available in summer. By choosing a tilt angle that gives the optimum match, the size and cost of the array are minimized. Optimizing the system cost is even more complex when the size and cost of the storage battery is taken into account; this is discussed further in Section 6.5, Chapter 6. It is also possible to design and build 'steerable' arrays that track the sun across the sky during the day and change the angle of inclination with the season. These are far more complex and expensive to engineer and are therefore not generally considered to be economic in terms of the extra electricity generated.

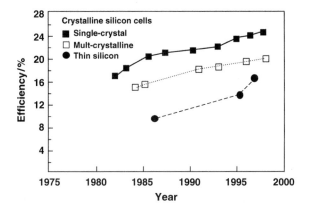

Figure 5.12 *The evolution of the efficiency of silicon solar cells.*
(Courtesy Imperial College Press)

The efficiency of silicon cells for the conversion of light energy to electricity has increased steadily over the years to present values of up to 18% for multi-crystalline silicon, and 25% for single-crystal cells (Figure 5.12).[5] At the same time, the cost of cells has fallen substantially as a result of improved manufacturing technology and greater production throughput. In the late 1970s, module prices were US$ 15 to 20 per W_p, whereas today they are around US$ 5 per W_p. In recent years, increased attention has been paid to the development of thin multi-crystalline cells and to amorphous silicon cells as these hold promise to be lower cost devices, but will be somewhat less efficient.

Different semiconductors have different band-gap energies and therefore absorb light in different parts of the electromagnetic spectrum. The efficiency of light conversion to electricity by a particular semiconductor is limited by the fraction of the solar spectrum that it absorbs and by its absorption coefficient. One way to improve the efficiency is to use two or three photovoltaic cells. These are constructed from semiconductors that absorb in different parts of the visible spectrum and are mounted on top of each other. In this way, a binary- or triple-junction cell is built up – a so-called 'tandem cell'. Naturally, each layer has to be sufficiently thick to absorb its own component of the solar spectrum, but at the same time has to transmit the other components to the underlying cells. The different layers of a tandem cell are 'grown' on top of each other, rather than the individual cells being made first and then mounted.

Tandem cells were developed initially for use in satellites, where performance is more important than cost. High solar-conversion efficiency and low mass are important parameters. A typical triple-junction cell has a gallium-indium phosphide (Ga,In)P top, a gallium arsenide (GaAs) middle, and a germanium (Ge) bottom section. These sections absorb blue, green and red light, respectively, and the device is able to convert about 30% of sunlight into electricity. As yet, tandem cells are too costly for most terrestrial applications, but this situation may well change as the technology develops and production volumes grow. It should be noted, that some solar cars have employed these triple-junction cells based on gallium arsenide (see Section 10.3, Chapter 10).

Applications of Solar (Photovoltaic) Cells, Modules and Arrays

Modern photovoltaic cells were first developed in the USA for use in the space programme. All satellites and space probes require electrical power for a variety of purposes and the only external source available stems from solar energy, converted to electricity by means of photovoltaic cells. Terrestrial applications include telecommunications and navigational aids, since these are often required in remote locations where no electricity grid is available and diesel generation is prohibitively expensive. In marine operations, solar modules are employed to provide power for navigation buoys, drilling platforms and the cathodic protection of structures, and to serve as independent electricity supplies for small boats. On land, in both developed and developing countries, photovoltaic systems operate microwave relay stations, telecommunications networks, railway signalling, road traffic monitoring, road signs, street lighting, bus-shelter lighting, irrigation equipment, and the monitoring and cathodic protection of pipelines. Such systems also supply power to remote communities, homesteads, and holiday caravans.

In tropical countries, remote homes use photovoltaic power for lighting, radio, and refrigeration. As mentioned in Section 4.4, Chapter 4, the refrigeration of medical supplies, such as vaccines, is a vital use for solar energy, either in the form of heat to operate a gas refrigerator or photovoltaic electricity to operate a compression refrigerator. All these stand-alone applications are known, collectively, as 'remote-area power-supply systems', *i.e.* RAPS systems. In some situations, it is convenient to combine solar cells with wind generators or diesel generators, or with both (so-called 'hybrid systems') to ensure a less erratic electricity supply. To some degree, wind power and solar power are complementary, especially at higher latitudes. Winds tend to be stronger in winter when sunlight is at a premium. Generally, there is more wind on a cloudy day when the insolation is reduced. This is particularly true at sea, and combined wind and solar systems have been used to power the lights and audible warnings on navigation buoys.

The importance of photovoltaic power for remote communities (via RAPS systems) is immediately clear when it is understood that some 2 billion people, mostly in the developing world, have no access to the electricity grid. This is around one-third of the entire world population! The potential market for RAPS systems is very large indeed. As the cost of modules falls through further development and mass production, this market will be progressively penetrated. A typical small RAPS installation (solar only) to supply electricity to a farm in the Australian outback is shown in Figure 5.13(a).

The second major category of photovoltaic installation is that of grid-connected systems. These are found mostly in urban areas and are mounted on buildings rather than as free-standing arrays. This practice saves on land, which is valuable in cities. Solar modules can be integrated into the building fabric, and thereby replace other

(a)

(b)

Figure 5.13   (a) *RAPS system in Australian outback;* (b) *1-MW$_p$ solar array mounted on Munich Trade Fair Centre.*

materials and reduce cost. For instance, the modules may act as sun shields, which obviates the need for blinds and reduces the electricity consumption for air-conditioning. The electricity generated is mostly used locally within the building, but at times of surplus (for instance over the week-end in office blocks), it may be fed to the grid. Because of this grid connection, there is no need for energy storage. This is a fast growing application for solar cells, with government support in several countries. Since buildings generally account for 20 to 30% of the primary energy requirements of industrialized countries, there is huge potential for solar electricity combined with active and passive solar heating.

Photovoltaics represent one of the fastest growing sectors of electricity generation. In 1998, the world-wide installed capacity was 750 to 1000 MW$_p$. Today, it is several GW$_p$. Some countries have set ambitious targets for the number of building-integrated arrays to be constructed by different dates, often with government financial support and encouragement.[6] Japan has introduced an ambitious programme to install 5000 MW$_p$ of rooftop arrays by 2010. Germany is aiming for 100 000 roofs by 2007. The Netherlands have set a target of 250 MW$_p$

by 2010. Italy intends to install 50 MW_p of grid-connected arrays on 10 000 rooftops by 2005. By contrast, the UK, with its rich resources of fossil fuels, has been relatively slow to develop building-integrated photovoltaics, but a programme is now evolving with government incentives. There are several building arrays in the range of 50 to 200 kW_p. It is interesting to note that, even under UK conditions of insolation, each kW_p installed avoids the emission of about one tonne of CO_2 per year.[5] Some very large arrays have been erected elsewhere in the world. For instance, there is a 3.3-MW_p facility at Serre in Italy, and a 1-MW_p array at the new Munich Trade Fair Centre (Figure 5.13(b)).

Previously, solar arrays have mostly been retro-fitted to existing buildings, but architects have become interested in the technology and new buildings are being designed with integrated arrays for electricity and solar thermal for heating. There is a particularly good match for offices and commercial buildings, which are mostly used during the day. In order to allow architectural freedom of expression, solar modules are now being developed in a range of sizes, colours and degrees of transparency for incorporation into building claddings. Clearly, the technology of building-integrated arrays is still at an early stage and exciting developments may be anticipated.

The downside of solar electricity lies in the diurnal and seasonal nature of sunshine, which necessitates energy storage or grid connection (see Section 6.5, Chapter 6), and in the high cost of modules and arrays. The latter issue is being addressed strenuously and there is every reason to believe that costs will fall progressively as technology improves and manufacturing volumes increase. Nevertheless, a fundamental limitation to cost reduction with silicon and similar photovoltaic devices lies in the requirement for high-temperature processing in a high-vacuum environment. This tends to restrict fabrication to batch processing on to glass substrates, with associated costs.

Organic Photovoltaic Devices

Another exciting prospect, still in the research stage but with good scope for cost reduction, is that of organic (polymer) photovoltaic devices.[5] If these can be developed successfully, then they may revolutionize the whole field of photovoltaics to bring down the cost and open up a new range of applications. The organic photovoltaic device (OPVD) is based on an organic molecular semiconductor, such as (i) a condensed aromatic polycyclic compound (*e.g.* anthracene or a perylene derivative), (ii) a metallo-porphyrin, or (iii) a polymer. The device is closely related to modern light-emitting diodes (LEDs) that also use organic molecular semiconductors. This relationship is shown in Figure 5.14. In LED mode, a voltage is applied across the organic molecular semiconductor and light is emitted through the glass substrate. In photovoltaic mode, light enters through the glass substrate and an output voltage is generated across the semiconductor. In both devices, the inner electrode adjacent to the glass substrate is a thin transparent layer of indium tin oxide, while the outer electrode is composed of metal. The organic material is placed between these two electrodes.

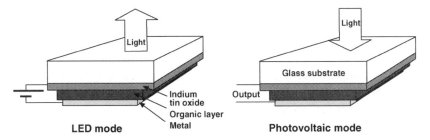

Figure 5.14 *Schematic representation of an organic photovoltaic device that, according to the selection of electrodes and semiconductor layers, can function as a light-emitting diode or as a photovoltaic diode. Fabrication is by the successive deposition on to a glass substrate of: (i) a transparent electrode, e.g. indium tin oxide; (ii) an organic semiconductor layer or layers, by vacuum sublimation and/or solution processing; (iii) a metal electrode by vacuum deposition.* (Courtesy of Imperial College Press)

The principal advantages of OPVDs are:

- comparative low-cost materials and fabrication techniques, which opens up applications in mass-produced products (such as toys) as well as making solar energy more competitive as a source of electricity
- the ability to fabricate much larger sheets of solar cells than is possible using single-crystal technology; moreover, such sheets can be made flexible so that solar cells can be moulded to fit on to curved surfaces
- the possibility to tune the colour through tailoring of the chemical structure (many of the polymers used to fabricate OPVCs are brightly coloured); this feature can be important for design and aesthetics.

The principal disadvantages are:

- poor solar conversion efficiency compared with inorganic cells
- sensitivity of the polymers to oxygen and/or water vapour leads to instability and degradation of performance.

Much research is in progress to overcome these two limitations. By using organic materials of improved purity and with better encapsulation techniques, working lives exceeding 10 000 h have already been achieved.

Dye-sensitized Solar Cells

Yet another variant on photovoltaic technology is the so-called 'dye-sensitized solar cell'. Titania itself is a semiconductor with a relatively high band-gap energy ($\sim$3.2 eV) and therefore absorbs light energy in the ultraviolet rather than in the visible part of the spectrum. By virtue of its low cost, however, titania is most attractive as a photovoltaic material and therefore considerable work has been undertaken to shift the spectral response into the visible region. Although this has not been achieved, research has now shown that it is possible to circumvent the band-gap problem by means of a subterfuge. This involves separating the optical-absorption and charge-generating functions. A dye, which is capable of being photo-excited, is adsorbed on to the surface of the titania and acts as an electron-transfer sensitizer. The coated titania serves as the negative electrode in a photoelectrochemical cell. The operating principles of such a cell are shown in Figure 5.15(a). The dye (D), after having been excited (D*) by a photon of light, transfers an electron to the conduction band of the titania

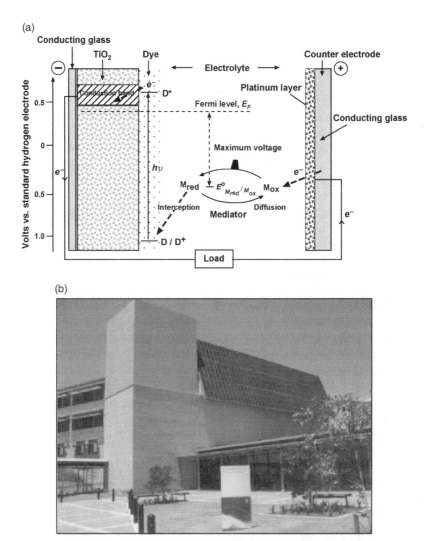

Figure 5.15   (a) *Operating principles of a dye-sensitized titania solar cell;* (b) *6-kW array mounted on CSIRO Energy Centre.*

Table 5.7 *Advantages and disadvantages of solar photovoltaic technology*

Advantages	Disadvantages
Solar energy is the world's major renewable energy resource	Diurnal and seasonal nature of sunshine
Electricity may be generated wherever the sun shines (dispersed generation)	Maximum solar insolation is limited to the tropics
Daylight is predictable (unlike wind and wave energy)	Electricity storage is needed for non-grid-connected arrays
No pollution in the form of solid or liquid wastes, noxious or greenhouse emissions and particulates	Building-integrated solar arrays require buildings to face the sun
Solar cells: – simple 'do-it-yourself' installation – have no moving parts and are silent in operation – require no cooling water – are reliable and generally require no maintenance – have a long life (thought to be 25+ years)	Silicon solar cells and modules are expensive to manufacture, although organic polymer and dye-sensitized photocells offer the prospect of reduced costs Automated steerable arrays are expensive
Solar arrays are modular in construction and factory-built	Some environmental risks from manufacturing of solar cells (though toxic substances are only used in very small quantities, large-scale manufacturing would increase the problem of emissions and waste control)
Plentiful and cheap supply of basic cell material (silicon)	

('injection process'), and itself becomes oxidized (D^+) in the process. The cell electrolyte contains a 'redox mediator', *i.e.* a substance that can be oxidized and reduced electrochemically.[*] Positive charge is transferred from the dye to the mediator (M_{red}) and the dye is returned to the reduced state ('interception process'). The oxidized mediator (M_{ox}) diffuses to the positive counter-electrode, where it is reduced again by the electrons travelling around the external circuit. The theoretical maximum voltage that such a cell can deliver is the difference between the redox potential of the mediator and the Fermi level of the semiconductor.

In practice, the titania electrode is present as a thin film of tiny crystals that have diameters of just a few nanometres and are sintered together. This film is deposited on a transparent, conducting glass substrate (generally, coated with tin oxide) that acts as a current-collector. The titania is ultra-porous and has a very high surface area. This assists the uptake of dye from solution. The dye-coated electrode is then assembled into a cell with a counter-electrode, also made from conducting glass, and the intervening space is filled with electrolyte and the mediator (typically,

---

[*] Redox-mediated reactions are very common in biochemistry, and, more specifically, photo-induced redox-mediated reactions form the basis of photosynthesis in nature.

the iodide-triiodide couple: $I^- - I_3^-$. A small amount of platinum is deposited on the counter-electrode to catalyze reduction of the mediator. Such cells have demonstrated efficiencies of 10% in the laboratory and 5% in the field. A 6-kW array of dye-sensitized solar cells that has been installed at the CSIRO Energy Centre in Newcastle, Australia is shown in Figure 5.15(b).

Compared with silicon cells, the dye-sensitized titania cell has lower efficiency under conditions of strong solar radiation, but shows far superior performance at low illumination levels. Again, this cell technology is still under development. There are, in fact, many different possibilities for photoelectrochemical cells and this appears to be an area of research that holds great promise for cost-effective *Clean Energy* in future.

A summary of the advantages and disadvantages of solar photovoltaic technology is given in Table 5.7.

5.7 References

1 Key World Energy Statistics from the IEA, 2003 Edition, International Energy Agency, Paris, France, 2003.
2 New and Renewable Energy: Prospects in the UK for the 21st Century, UK Department of Trade and Industry, London, UK, March 1999.
3 David Williams, Hydro-power from Britain, Guide to UK Renewable Energy Companies 2001, James & James, London, UK, pp. 31–33.
4 The Future of Renewable Energy – Prospects and Directions, EUREC Agency, James & James, London, 1996, p. 50.
5 M.D. Archer and R. Hill (eds), Clean Electricity from Photovoltaics, Imperial College Press, London, UK, 2001.
6 J. Plastow, Photovoltaics, in: Guide to UK Renewable Energy Companies 2001, James & James, London, UK, pp. 17–23.

CHAPTER 6

Why Store Electricity?

Energy-storage technologies can provide a vital link between the primary source of energy and its actual use. In particular, the inclusion of an energy-storage system allows flexibility in matching the availability of an energy source to the demand profile in terms of where and when power is required and at what level.

This chapter discusses the need for energy storage on the medium to very-large scale (kWh–GWh) in centralized electricity supply networks and also, at a more local level, for uninterruptible power supplies and for use in conjunction with renewable energy sources. The storage of electricity for portable devices is discussed in Chapter 9, and for electric propulsion in Chapter 10.

6.1 Candidate Energy-storage Systems

As noted in Chapter 3 (Figure 3.14), several types of energy-storage system are under investigation. These may be classified according to the form in which the energy is stored, namely:

- thermal energy: storage heaters, molten salts
- potential energy: pumped hydroelectric, compressed air
- kinetic energy: flywheels
- electromagnetic energy: superconducting coils
- electrostatic energy: capacitors, supercapacitors
- chemical energy: batteries, methanol, hydrogen.

The scale of these systems ranges from miniscule elements on integrated circuits (*e.g.* batteries, supercapacitors) to pumped hydroelectric reservoirs that store the equivalent of GWh of electrical energy.

The characteristics of those systems that are suitable for the medium-to-large-scale storage of electricity are summarized in Table 6.1. (Note, supercapacitors are an option only for power-quality applications in which the energy requirement is not large; further details of these devices are given in Section 7.5, Chapter 7.) Obviously, each technology is not without its limitations. For example, pumped hydro and compressed air energy storage (CAES) are location-specific and cannot

161

Table 6.1 *Comparison of energy-storage systems*

	Pumped hydro	CAES	Flywheels	SMES	Batteries
Efficiency	~75%	~70%+fuel	~90%	~95%	~75%
Maximum energy	10 GWh	5 GWh	5 MWh	1.5 GWh	50 MWh
Maximum power	3 GW	1 GW	10 MW	1 GW	100 MW
Modular	No	No	Yes	Possibly	Yes
Cycle-life	10 000	10 000	10 000	10 000	2000
Charge time	Hours	Hours	Minutes–hours	Minutes–hours	Hours
Siting ease	Poor	Poor	Good	Poor	Moderate
Lead time	Years	Years	Weeks	Years	Months
Environmental impact	Large	Large	Benign	Moderate	Moderate
Risk	Moderate	Moderate	Small	Moderate	Moderate
Thermal requirement	None	Cooling	Cooling	Liquid helium	Air-conditioning
Maturity	Mature	Available	Embryonic	Embryonic	Mature

provide instantaneous power. The feasibility of using flywheels looks favorable on paper, but has yet to be proven in full-scale demonstration programs and, in any event, the storage capacity is small compared with pumped hydro and CAES. Superconducting magnetic energy storage (SMES) is costly and is still at the research and development stage. (Note, pumped hydro, CAES, flywheels and SMES are discussed in more detail in Sections 7.2–7.4, Chapter 7.) Overall, these factors have resulted in batteries being the most widely used technology for energy storage at the present time; the inherent limitations of batteries are discussed in Chapter 9. Other forms of indirect electricity storage, in the form of chemicals such as methanol and hydrogen, must be classed as futuristic. Hydrogen storage is treated fully in Chapter 8.

6.2 Energy Storage for Electricity Supply Networks

Applications of Energy Storage within the Network

Early in the 20th century, before the advent of electricity grids, power was generated locally to serve individual communities. At that time, it was normal practice to use low-voltage, direct current (d.c.). A battery energy-storage system (BESS) was fairly commonplace to help smooth out fluctuations in demand. As supply systems matured and switched to high-voltage, alternating current (a.c.), and grids were constructed to link up power stations, the requirement for electricity storage faded into the background.

Present networks are generally highly interconnected at multiple locations and this complexity can give rise to a host of power and energy issues in ensuring the reliability and quality of the supply of electricity. The USA, for example, has experienced total system failures over large areas when the network has become over-stretched, or a key component has failed and caused a cascading series of incidents. Accordingly, in the USA and elsewhere, there is now renewed interest in the integration of stores of energy as a means of resolving such issues. The following applications have been demonstrated:

- *System regulation and power quality.* Energy storage serves to meet short-term, random fluctuations in demand and so avoid the need for voltage or frequency regulation by the main plant. It can also provide 'ride through' for momentary power outages, reduce harmonic distortions, and eliminate voltage sags and surges. Energy storage is particularly useful in providing reliable and high-quality power to sensitive loads, *e.g.* processing equipment that requires extremely clean power to operate properly. Sometimes the energy-storage systems are coupled directly to the critical equipment, and sometimes to the bus at the network sub-station, or even on the feeder line to the point of use.
- *Spinning reserve.* Energy-storage eliminates the requirement for part-loaded main plant to be held in readiness to meet sudden and unpredicted demands, as well as to accommodate power emergencies that arise from the failure of generating units and/or transmission lines. Traditionally, this reserve was supplied by a generator that was 'spinning' in synchronization with the supply network in order to be rapidly available. In recent years, however, the term 'rapid reserve' has come into use because, with the advent of stored energy interconnected to the supply network *via* power electronics, the reserve resource does not necessarily 'spin'. Moreover, it has a much faster response. Thus: (i) 'spinning reserve' is unloaded generation that must spin in synchronization with the network in order to be available within a few minutes, *e.g.* hydropowered generation; (ii) 'rapid reserve' is an electrical power/energy resource that is not required to spin in synchronization with the system and is available almost instantaneously (in ms), *e.g.* a BESS. Typically, the reserve power must equal the power output of the largest generating unit in operation.
- *Peak-shaving.* Energy storage accommodates the minute–hour peaks in the daily demand curve. When placed on the customer-side-of-the-meter, energy storage offers economic benefit to consumers, provided favorable time-of-day electricity rates are in effect. (Note, the price of electrical energy to large consumers may be determined not only by the total energy supplied, *i.e.* kWh, but also by the peak-power demand, *i.e.* kW.) The energy store is discharged during the period of the customer's peak demand to supply a portion of the load so as to reduce the power purchased from the utility. The energy store is replenished during off-peak hours when the demand on the utility is lower. Accordingly, there is often no demand charge and lower energy charges.

- *Load-leveling.* Surplus electricity generated during off-peak hours (typically, at night) is stored and then used to meet increased demand during the day. Such a strategy offers utilities a promising alternative to installing extremely costly, continuous-duty plants to balance the power-generation requirements, and will enable central power stations to match system generation with changes in demand (so-called 'load-following'). Load-leveling is distinguished from peak-shaving in that it accommodates longer periods of increased demand, typically, those of several hours duration.
- *Renewables support.* By their very nature, renewable energy sources are intermittent and thus electricity generation may not coincide with the demand cycle. Energy storage is used to match the output with any load profile. Also, storage is essential when electricity is generated by renewables in locations where there is no established utility grid, *i.e.* so-called RAPS systems, see Section 5.6, Chapter 5, and Section 6.5 below.
- *Area control and frequency responsive reserve.* 'Area control' is the ability of grid-connected utilities to prevent unplanned transfers of power between themselves and neighboring utilities. Interconnected utilities must be operated at the same frequency ('frequency control'), otherwise power will transfer between them. The remedy is for the offending utility to generate additional power or to import electricity from a neighboring network. 'Frequency responsive reserve' is the ability of isolated utilities to respond instantaneously to frequency deviations. Such deviations from the standard are a first indication of insufficient generation, *i.e.* excessive load. Remedies are either additional generation, perhaps from energy storage, or load-shedding.
- *Transmission voltage regulation.* Energy storage enables the voltages at both the generation and the load-ends of a transmission line to be maintained at values close to each other, *e.g.* to within 5%. This involves supplying relatively high levels of power at selected locations to meet load demands. When located at the end of the line, energy storage will lower the amount of power sent down the line during peak periods and, consequently, will reduce the resistive losses and provide a net energy saving.

Electricity Demand Profile

Countries that have a fully interconnected network tend to have surplus generating capacity for much of the year. Plants with low operating costs (nuclear, coal or gas stations of 1–2 GW size) supply the base-load, and smaller, higher cost plants are brought on line to meet peaks in demand. Storage capacity is strictly confined to a few pumped hydro facilities in mountainous regions, sometimes supplemented at a local level by limited battery storage. Whereas many utilities worldwide would welcome the introduction of more extensive battery energy-storage, the economic targets to be met are exceedingly stringent and, to date, such storage has not proved to be economically viable in most situations.

The demand for electricity fluctuates widely from summer to winter, from working day to holiday, and throughout the day from hour to hour and even minute to minute. Moreover, the seasonal fluctuations differ from country to country.

In Northern Europe, the peak demand is in winter for lighting (long hours of darkness) and heating. In warmer regions, for example, Southern Europe and Southern USA, the peak demand is in summer for air-conditioning. At present, there is very little that can be done about seasonal storage, except for the possible introduction of pumped hydro storage in particular favorable locations. In the long-term, the possibility of using water electrolysis to produce hydrogen has been debated. The gas could be stored seasonally in underground caverns or depleted oil/gas fields and fed to fuel cells to re-generate electricity when required (see Section 8.2, Chapter 8). This strategy is, however, neither economic nor practical at present. Batteries have nothing to offer to the seasonal storage of electricity since they are too expensive an option for a single (*i.e.* annual) charge–discharge cycle.

As regards weekly and daily fluctuations in demand, Figure 6.1 shows a typical load profile of a utility in the USA, operating either with or without energy storage. As illustrated by the upper profile, the conventional supply is divided into four areas, *viz.* base-load (about 40% of installed capacity), intermediate (35%), peaking (11%), and reserve (the remaining 14%). Although the detail may change, this type of load variation applies to most other industrialized countries in which cheap off-peak electricity rates exist. Two facts are immediately evident:

- there is a marked diurnal pattern of demand during the working week, *i.e.* the demand falls from about 23:00 h, remains at a low level until about

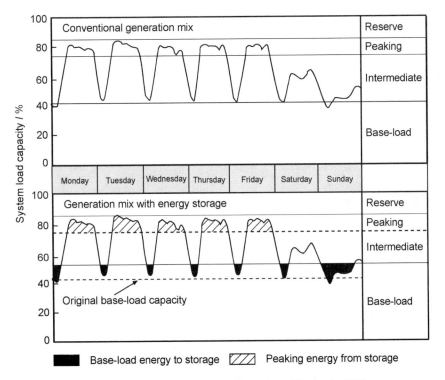

Figure 6.1 *Typical weekly load curve for an electricity utility in the USA.*

06:00 h, and then rises during the day to high levels in the afternoon and the evening;
- there is significantly less demand on Saturdays, and especially Sundays, when industrial and commercial use is low.

The solid areas of the demand profile in Figure 6.1 (lower curve) show the unused base-load that is available, if stored, to meet the peak-load (hatched areas of the profile). The base-load is supplied by large plant that is best kept running for long periods at a time and has the lowest generating costs. Intermediate demand is usually met by older, smaller plant with higher generating costs, while peak demand is furnished by reserve gas turbines or oil-fired plant at still higher costs. By storing the excess base-load electricity that is available over the week, it would be possible, in this example, to reduce the peak demand from 86 to around 75% of the installed capacity. Moreover, the higher base-load level (52 *versus* 40%) would replace part of the intermediate generation. Such a load-leveling strategy will enable central power stations to achieve better thermal efficiency, conserve fuel resources, lower maintenance costs, and may also extend the useful life of the plant and mitigate air-quality problems.

Superimposed on the daily demand curve are short-term peaks associated with the personal habits of consumers (*e.g.* times of cooking meals, watching television). Major demand 'blips' can arise quite suddenly, for example at the end of a popular television program when viewers switch on electric kettles to make cups of tea or coffee. Knowledgeable controllers of power stations can sometimes predict these incidents by studying television program schedules. The response time to meet such sudden demands should be of the order of seconds to minutes, if frequency adjustment or voltage reduction is to be avoided. Clearly, the necessity for fast action will not allow the start-up of additional generating units; the increased load must be serviced by rapid reserve or spinning reserve. Batteries respond instantaneously to load changes and are therefore considered to be the ideal strategy for coping with short-term demands, which include sudden breakdowns in the generating or transmission systems. The use of batteries would reduce or eliminate the need for spinning reserve and/or frequency control.

Storage Options and their Limitations

Through the applications listed earlier, it is considered that electricity storage can be multi-beneficial to both utilities and their customers in terms of: (i) improved power quality and reliability; (ii) reduced transmission/power losses; (iii) cost savings, *e.g.* deferral of new generation units and sub-station upgrades, and of new transmission lines and transformers (long-distance, high-voltage, *e.g.* 500 kV, 'bulk' transmission) and distribution lines (local, low-voltage, *e.g.* 23 kV, distribution); (iv) decreased environmental impact (lower emissions, diminished electric/magnetic field effects, integration of renewables); (v) strategic advantages (greater flexibility in siting and in the fuel that may be used).

The power requirements of various energy-storage applications in electricity supply networks, together with the power capabilities of the several energy-storage

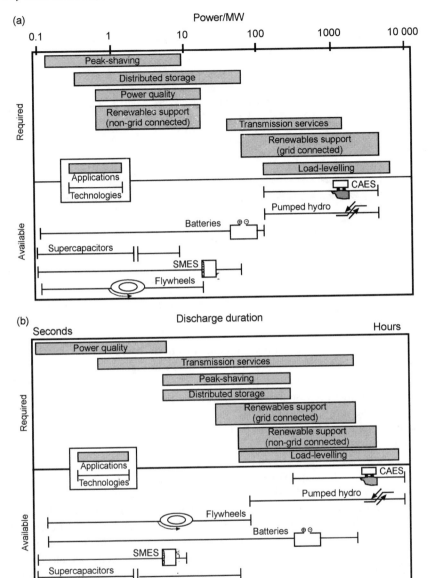

Figure 6.2 *Storage applications and technologies: (a) power; (b) discharge duration[1]
(note, the power and energy capabilities expressed in this figure relate to the
maximum size of the devices presently available and not to their intrinsic
capabilities as expressed in* W kg^{-1} *and* Wh kg^{-1}*).*

options, are illustrated schematically in Figure 6.2(a) along a common scale
(abscissa).[1] Setting aside pumped hydro and CAES, which, as mentioned above,
require special topologic formations for implementation, the energy-storage
systems rank in terms of power rating in the following descending order: batteries,

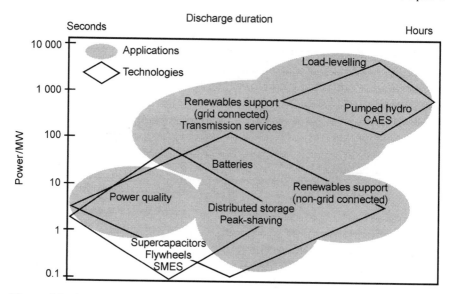

Figure 6.3   *Storage applications and technologies: energy and power (data taken from Figure 6.2(a) and (b))*[1].

SMES, flywheels, supercapacitors. A comparison of energy requirements and availability, in terms of discharge duration, is given in Figure 6.2(b). In this case, the descending-order ranking is: batteries, flywheels, supercapacitors, SMES. It should be noted that because these technologies are still evolving, the rankings are not absolute and are subject to change.

The information given in Figure 6.2(a) and (b) is combined in Figure 6.3[1] by presenting power as a function of discharge duration for both applications and energy-storage technologies. More precise details of the operational requirements of a wider range of energy-storage applications are given in Table 6.2. Super-capacitors, flywheels and SMES are shown to be primarily useful for power-quality applications, *i.e.* applications that do not require discharges of long duration. Batteries have the broadest overall range of applications, and pumped hydro and CAES are best for load-leveling. It is notable that, among these methods of energy storage, only supercapacitors and SMES store energy directly as electricity. Batteries store energy in chemical form, and flywheels, pumped hydro and CAES store energy in mechanical forms. For these latter four options, conversions are necessary both to charge the store from an electrical supply and to recover and use the stored energy as electricity.

Experience with Battery Energy-storage Systems

Since 1980, at least 17 BESSs have been employed in electricity networks. These systems, summarized in Table 6.3, encompass a wide range of capacities and a multitude of applications. They also reflect a maturing of BESS technology. Of the 10 systems dating from 1980 through 1992, two were multipurpose demonstrations

Table 6.2 *Energy-storage applications and appropriate technology (modified from Ref. 1)*

Application	Power/Energy	Storage time	Response time	Technologies	
Power quality	≤1 MW/~0.2 kW h	s	<1/4 cycle	Supercapacitor, Flywheel	Micro-SMES Battery
Transmission stabilization	up to 100's MW/20–50 kW h	s	<1/4 cycle	SMES	Battery
Very short duration					
Short duration					
Rapid reserve	1–100 MW/5–500 MW h	<30 min	<3 s	Flywheel, SMES	Battery
Spinning reserve	1–100 MW/5–500 MW h	≤30 min	<10 min	Pumped hydro, CAES, Flywheel	SMES Battery
Peak-shaving (customer side)	<1 MW/1 MW h	1 h	<1 min	Flywheel, SMES	Battery
Renewables support	up to 10 MW/0.01–10 MW h	min–h	<1 cycle	Battery	
UPS	up to ~2 MW/0.1–4 MW h	2 h	s	Flywheel, SMES	Battery
Distributed generation	0.5–5 MW/5–50 MW h	~1 h	<1 min	Flywheel, SMES	Battery
Long duration					
Load-leveling	100's MW/0.01–1 GW h	6–10 h	Min	Pumped hydro CAES	Battery SMES
Load-following	100's MW/0.01–1 GW h	several h	<cycle	SMES	Battery
Very long duration					
Emergency back-up	1 MW/24 MW h	24 h	s–min	Battery	
Renewables back-up	0.1–1 MW/20–200 MW h	up to 7 days	s–min	Pumped hydro CAES	Battery
Seasonal storage	50–300 MW/10–100 GW h	weeks	min	Pumped hydro	CAES

Note: microturbines and fuel cells are not storage devices and are therefore not included. Their application as standby generators is discussed in Section 3.3, Chapter 3 and Section 8.3, Chapter 8, respectively.

Table 6.3 *Implementation of large-scale BESSs since 1980*

BESS	Location	System capacity	Applications	Date[a]
Elektrizitätswerk	Hammermuehle, Germany	400 kW, 400 kWh	Peak-shaving	1980
BEWAG AG	Berlin, Germany	17 MW, 14 MWh	Frequency control Spinning reserve	1986
Kansai Power Co.	Tatsumi, Japan	1 MW, 4 MWh	Multipurpose demonstration	1986
Hagen Batterie AG	Soest, Germany	500 kW, 7 MWh	Load-leveling	1986
Crescent Electric Membership Corporation	Statesville, NC, USA	500 kW, 500 kWh	Peak-shaving	1987
Delco Remy Division of General Motors	Muncie, IN, USA	300 kW, 600 kWh	Peak-shaving	1987
Southern California Edison	Chino, CA, USA	10 MW, 40 MWh	Multipurpose demonstration	1988
Vaal Reefs Exploration & Mining Co.	South Africa	4 MW, 7 MWh	Peak-shaving Emergency power	1989
Johnson Controls, Inc.	Humboldt Foundry Milwaukee, WI, USA	300 kW, 600 kWh	Peak-shaving Load-leveling	1989
San Diego Gas & Electric	San Diego, CA, USA	200 kW, 400 kWh	Peak-shaving	1992
Pacific Gas & Electric (PM250)	San Ramon, CA, USA	250 kW, 167 kWh	Power management	1993
Puerto Rico Electric Power Authority (PREPA)	San Juan, Puerto Rico	20 MW, 14 MWh	Spinning reserve Frequency control Voltage regulation	1994
GNB Technologies	Vernon, CA, USA	3.5 MW, 3.5 MWh	Peak-shaving	1996
		2.45 MW, 4.9 MWh	Spinning reserve	
		1.8 MW, 5.5 MWh	Environmental	
Metlakatla Power & Light	Metlakatla, AK, USA	1.3 MW, 1.3 MWh	Utility stabilization	1997
		915 kW, 1.83 MWh	Power quality	
		700 kW, 2.1 MWh	Environmental	

(Continued)

Table 6.3 *(Continued)*

BESS	Location	System capacity	Applications	Date[a]
Grid interactive, 1-MW solar plant with BESS	Herne, Germany	1.2 MW, 1.2 MWh	Peak-shaving Power quality	1999
Grid interactive, 2-MW wind farm with BESS	Bochold, Germany	1.2 MW, 1.2 MWh	Peak-shaving	1999
			Power quality	
PQ2000	Homerville, GA, USA	2 MW, 10 s	Power quality Standby power	1997

[a] It is not always clear whether the reported date is for completion or commissioning of the BESS. Most are thought to be commissioning dates.

(Kansai Power and Southern California Edison). Three of the 10 systems were operated by supply-side (utility) interests, namely, BEWAG, Kansai Power and Southern California Edison, and the remaining seven by demand-side (customer) interests. Only BEWAG, an 'island' utility, lists frequency control and spinning (rapid) reserve as applications; these are typical needs for such isolated utilities. The San Diego Gas & Electric (SDG & E) facility was a demonstration system that provided peak-shaving for a specific load, namely, a light-rail transportation system.

Among the more recent BESSs, the Herne and Bochold installations in Germany interface renewable energy sources (wind and solar) to utility grids. The systems enable such energy to be used for peak-shaving the utility demand, and also to enhance power quality (reliability) by supplying stored energy in the event of an outage. The PQ2000 is a commercially available BESS. Following its introduction at Homerville, USA, additional units have been sold and placed in service. This is evidence of commercial maturity and economic viability. Note that its primary application is standby power (rapid reserve) for intervention in the interest of assuring power quality. In one sense, it is an industrial-scale uninterruptible power supply (UPS) system.

Most of the above BESSs have been one-of-a-kind systems built for demonstration purposes and, in some instances, costs may have been a secondary issue. The possibility of improving the performance of utility networks by means of BESSs has come about through the evolution of power electronics technologies that are capable of rapidly and seamlessly transferring electrical power of high quality between a.c. and d.c. systems. With power electronics and battery technologies becoming ever more sophisticated, and with the demand for reliable electrical power being of high priority, it is likely that BESSs will once again be more widely used.

What battery specifications are set by the electricity utilities? The prime consideration is the economics – what is the value of storage to the utility per kWh and per kW, and can battery manufacturers meet this target? For weekly or diurnal storage, the rates of charge and discharge are fairly conventional (~hours), and so the focus is on the energy stored (kWh). By contrast, the short-term peaks require rapid battery discharge, and perhaps rapid charge too. The emphasis here is on the power output of the battery (kW) rather than on the stored energy. In both types of service, the overall energy efficiency of the battery system (Wh-output: Wh-input) is crucial to the economics.

Until now, where battery storage has been demonstrated, it has invariably been in the form of lead–acid batteries. This choice has been made on the grounds of availability and comparative cheapness. Battery mass is not a major consideration for stationary applications, provided the units are floor-mounted and not stacked. There are, however, certain drawbacks with the use of lead–acid batteries. Ideally, the energy-storage system will be situated outdoors under a simple, inexpensive shelter. Lead–acid batteries do require some maintenance (particularly, flooded types, see Chapter 9), and are often housed in a dedicated room under temperature-controlled conditions. Further, the life cycle of the batteries under the conditions of use is barely adequate and their cost, when added to the other capital costs of the system (battery housing, power-conditioning equipment), is presently too high for many potential applications in the electricity-supply industry. Another problem is the fine acid mist that is liberated from flooded designs of lead–acid battery during charging; this mist is highly corrosive. (Note, in recent years, sealed batteries that require no water maintenance and that release no corrosive gases have been developed – the so-called 'valve-regulated lead–acid' (VRLA) design – see Section 9.1, Chapter 9.) Utilities consider the ideal battery to be one of large size and low cost that can be stacked outdoors, requires no maintenance, is not sensitive to ambient temperature, and has a long life in terms of both years and charge–discharge cycles.

Contrary to the hitherto strong preference for lead–acid, a BESS made up of four parallel strings each of 3440 nickel–cadmium cells will be installed in Alaska, USA, to secure electricity supply to the city of Fairbanks. Alaska poses a particular operational problem in that it is effectively an island, cut off from the main US grid, and also experiences winter temperatures as low as $-50\,^\circ C$. Spinning reserve is expensive to provide in an island situation and, anyway, does not give the instantaneous response required by much electronic equipment. The battery ordered by the Golden Valley Electric Association, Inc. in Fairbanks will be capable of supplying 40 MW of back-up power for 3 min, or 30 MW for 10 min, and is expected to become the most powerful battery in the world. Evidently, the much higher cost of nickel–cadmium is justified for the harsh Alaskan conditions.

For centralized storage at the power station, the size of the facility required is more appropriate to an electrochemical engineering plant than to an assembly of unit cells. This is the reason why some utilities have shown an interest in flow batteries, which are constructed on a plant basis with the reactant chemicals contained in large reservoirs external to the cells (see Section 9.3, Chapter 9).

6.3 Uninterruptible Power Supplies

Historically, the requirement for standby power arose first in telephone exchanges, railway carriage lighting and hospital operating theatres, but has now expanded into many other applications. Almost every public building, *e.g.* airports, hotels, railway stations and track operations, stores, supermarkets – as well as factories, office blocks, aircraft, ships and even power stations – each has its own UPS. This facility requires a battery pack to take over seamlessly when the mains supply fails, until such time as a local generator can be started. Uninterruptible power is especially vital for such modern applications as central computing facilities and nuclear power control rooms. In aircraft applications, standby batteries are used to back-up navigation computers, communications systems, and fly-by-wire controls.

In many situations, a standby diesel generator is installed and the role of the battery is then to provide power until the generator can be started. This will be quite a short period, typically 15 min or less. The requirement then is for a battery that may be discharged at a high rate (power) rather than for one with a high-energy-storage capacity. Traditionally, lead–acid batteries have been employed for this duty since they are commercially available, of relatively low cost, and may be discharged rapidly. Lives of 20 years or more are common for such batteries on standby duty. A typical large installation of lead–acid batteries as might provide back-up for a major electricity user is shown in Figure 6.4.

6.4 Storage of Distributed Electricity

By virtue of the demonstrated attributes of small size, modular construction, silent operation, negligible emissions, high efficiency and instantaneous response, battery energy-storage units can be placed practically anywhere on the distribution system at critical points in the network. Furthermore, batteries can be installed in incremental capacities that meet more closely the demand at the time. Distributed strategically within a utility network, a group of battery units, each sized between ~1 and 10 MW, yields greater benefits than a single, centrally located unit of equivalent total capacity. A convenient location for decentralized storage might be at one of the step-down transformers, with the most local of all being at the district transformer.

Figure 6.4 *Large UPS installation.*

A further area in which distributed generation plays an important role is for those communities that are too remote to form part of the grid system. These may range from isolated farms and homesteads, where electricity generated from renewable sources (*e.g.* wind and/or solar) may be appropriate, to sizable towns that are too distant to justify extension of the grid from major centers of population, or to island communities. The installation planned for Fairbanks, Alaska, mentioned earlier, is a good example. For islands and isolated towns, a local generating station and distribution network will be appropriate and auxiliary generators or electricity storage will be required to meet peaks in demand.

Another important feature of electricity storage is that it can be used not only to level the power demand placed on the generating plant, but also that placed on the transmission and distribution system. Some utilities are moving away from large, centralized, power plants back towards smaller generators (*e.g.* microturbines, wind-turbines, biomass-fueled plant, photovoltaics (PV), fuel cells) and storage units that are distributed geographically in close proximity to end-users (see Section 3.7, Chapter 3). Such decentralization would provide many advantages, namely: protection from major mishaps; lower power losses; decreased costs of transmission and distribution; greater flexibility in matching capacity additions and retirements with changes in demand; reduced concerns over possible health effects of electric and magnetic fields; shorter construction times; less disruption from digging up roads and laying cables. All of these attributes are attractive to the utilities, but they have to be balanced against the life-cycle cost of the storage batteries.

Thus, it would appear that the utility of tomorrow will be a hybrid network of traditional (but improved) central-generation, distributed-generation and energy-efficiency technologies (Figure 6.5). Such distributed power supplies are seen as an effective means whereby utilities can withstand increased competition, improve customer service, reduce costs (through increased efficiency), and achieve cleaner operation. In essence, the overall target is for electricity generation to use the least resources with the least environmental impact.

6.5 Storage of Renewable Electricity

As is evident from the earlier discussion of Clean Fuels (Chapter 2), fossil fuels have two important characteristics in addition to being concentrated sources of energy. They are *energy stores* and they are *readily transportable*. This means that the fuels may be stored until such time as they are required and may be transported by rail, road or pipeline to where they are to be used. By contrast, most renewables cannot be stored and cannot be transported to the place of use, except by first converting them to electricity. (Note, combustible biomass can be both stored and transported, while hydropower can be stored but is location-specific.) Electricity is the most versatile and preferred form of energy for many applications and therefore it is not surprising that renewable energies and electricity generation are so intimately bound together. The problem is that of matching the supply of renewable energy to the demand, which is particularly acute for RAPS applications where there is no grid to help smooth fluctuations in the supply and the demand. This calls

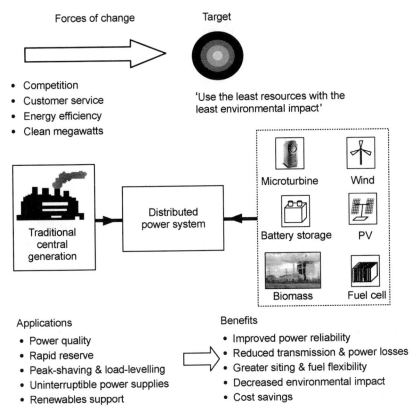

Figure 6.5 *The distributed utility concept.*

for the development and application of systems for the efficient storage of electricity at a local level.

The supply–demand problem is critical for wind energy and solar energy – two of the most promising sources of renewable electricity. Wind energy is erratic, while solar energy is subject to regular diurnal patterns of supply that necessitate electricity storage when used in an isolated (*i.e.* non-grid-connected) mode. The vast majority of the installed power from wind turbines in the world is grid-connected. At present, this contribution is only a small fraction of the total electricity supply and, hence, utilities are well equipped to handle variations in wind power (the grid itself acts as a storage facility). If, however, wind power is ever to become a major provider of electricity, then some form of storage or back-up capacity will be required to accommodate occasions when there is not a good match of wind and load. Moreover, the amount of storage or back-up will vary widely from one utility system to the next. In theory, the storage requirement can be greatly reduced by linking a number of widely separated wind farms into a single grid, so that a wind drop in one area would be offset by an increase in another. Utility systems that have relatively large amounts of hydropower capacity (which can be used for pumped storage or load-following) can support substantial levels of intermittent renewable generation with no additional storage.

The PV generation of electricity from solar energy is particularly appropriate for medium-scale applications and is therefore more commonly implemented by means of stand-alone systems (*e.g.* in RAPS systems) than by large-scale grid-connected installations. It should be noted, however, that small-scale systems are being increasingly used (even in cities and towns) as a cost-effective method of providing power to a wide variety of services, *e.g.* lighting, parking-ticket machines, UPS systems, monitoring devices, railway signaling.

Secondary batteries have a number of features that make them well suited to storing solar electricity, namely:

- system input and output is in the form of low-voltage, d.c. electricity
- modular construction allows flexible sizing and easy battery exchange
- batteries respond immediately to supply-and-load variations, and are very reliable
- it is possible to match the internal resistance of the battery to that of the load for maximum power output (see Section 5.6, Chapter 5)
- batteries have a short lead–time in manufacture.

A schematic of a solar PV array and battery combination connected to a load is shown in Figure 6.6. A particular problem is that for maximum power output, it is necessary to match the internal resistance of the battery not only to that of the load, but also to that of the array (see Section 5.6, Chapter 5). This can be difficult, as the resistance of a solar cell is, typically, many times higher than that of a battery.

In order to select a battery for use with a solar PV installation it is necessary to take into account:

- the output of the array and its variation with time of day and month of the year (variations in insolation level result in changes in the battery charging current)

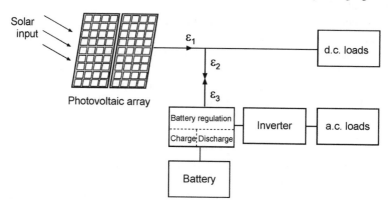

Figure 6.6   *Schematic of photovoltaic system to supply solar-generated electricity to d.c. and a.c. loads. (ε_1, ε_2 and ε_3 are array, charge and discharge efficiencies, respectively).*

- the pattern of load demand placed on the battery as a function of time over each charge–discharge cycle
- the energy efficiency of the battery (Wh-output:Wh-input)
- the rate of self-discharge of the battery on standing
- the environmental conditions, especially temperature
- the requirement for reliability and freedom from maintenance, as dictated by remoteness and difficulty of access
- the operational life of the battery
- the commercially acceptable cost for the battery.

Most present-day RAPS applications lie in the power range 0.1–10 kW and require a battery with a storage capacity of 1–100 kWh. For comparative purposes, this capacity range corresponds at the lower end to that of a starter battery for a large vehicle, and at the upper end to that of, say, a traction battery for an electric bus. In many locations, it will be necessary to use a diesel generator and, optionally, a wind turbine in conjunction with an array, either because the insolation is inadequate or because an array of the required size is too expensive.

The role of a battery in a RAPS installation depends very much on the insolation level. In a tropical zone, there is little seasonal variation in insolation, while in a desert region there is little rainfall and little cloud cover. Under these conditions, the insolation is fairly predictable and the function of the battery is usually to store electricity that is generated during the hours of daylight for use during the hours of darkness (so-called 'supply-leveling'). Only occasionally will it be necessary to store electricity for more than a day and so the extra margin of battery capacity required is quite small. The situation is quite different in temperate zones, in which there is typically a high incidence of cloud cover. Not only will it be necessary to have a much larger array, because the insolation is low and erratic, but also it will be necessary to install a considerable excess of battery capacity to allow for several consecutive days of cloud cover. Superimposed on this is the major seasonal variation at higher latitudes. Clearly, for a given application, both the array and the battery will be much larger and more expensive than the counterparts required for a RAPS system in a tropical zone. It has been calculated, for example, that the year-round solar electricity supply in Denmark (latitude 56° N) would prove to be quite uneconomic, in terms of the size of array and battery required, unless augmented by a wind turbine and/or diesel generator.

Algorithms have been prepared that allow determination of the daily load (in terms of kW and kWh) for a particular application, and also the weekly averaged load. The maximum anticipated daily load determines the size of the battery and calculation of this load will avoid installing a battery that is too large and expensive. The weekly averaged load is particularly important when the use factor is variable day-by-day. An example of this might be a school used only on weekdays, or a holiday caravan occupied only at weekends. The weekly averaged load determines the size of the array required, with the battery acting as a 'ballast' between days of low load and days of high load. Obviously, the objective of these considerations is to minimize the overall cost of the complete installation. The economics of electricity supply may be further improved through good load-management, *e.g.* by using low-energy lamps for lighting, or well-insulated and high-efficiency refrigerators. The extra capital cost of these items is likely to be small compared with that of additional solar arrays and batteries.

Seasonal variations in insolation present a greater problem as it will not generally be economic to install batteries for seasonal energy-storage. Fortunately, for some applications, electricity may be required only in summer when insolation is at a maximum; examples would be caravans, remote holiday cottages, and pleasure boats. Where a RAPS system is required all year round, it may prove practical to combine solar PV in summer with wind generation in winter since, statistically, winds are stronger and more consistent in winter when insolation is at its lowest. Another seasonal aspect of installation design, as discussed in Section 5.6, Chapter 5, is to match the tilt angle of the array to the load demand.

Most batteries, although not all, operate best at ambient temperature, say 0–40 °C. Outside this temperature range, their performance deteriorates rapidly. This may pose problems for certain applications. In polar latitudes, there may be good insolation on winter days when the ambient temperature is −20 to −40 °C. This will necessitate the selection of a battery that operates well at low temperatures, for instance, nickel–cadmium. At the opposite extreme, the temperature inside a battery enclosure could easily exceed 70 °C in tropical situations in summer. No battery with an aqueous electrolyte will perform well under these conditions and it is therefore necessary to use a thermally insulated (or ventilated) container.

A second role for the battery in a solar installation – quite distinct from its primary storage role – is to provide for surges in power demand (*i.e.* peak-shaving operation). Power surges of relatively short duration and up to six times the steady load are common when appliances are turned on and off. Without the provision of a battery, it would be necessary to have a much larger array that is sufficient to meet the maximum instantaneous demand. This would be uneconomic. Similarly, when using an inverter to provide a.c. for domestic uses, inductive loads (such as motors in washing machines and vacuum cleaners, and compressors in refrigerators) all give rise to current surges of short duration. Both the battery and the inverter must be capable of meeting these instantaneous loads. While direct-current versions of some of these appliances are commercially available, as used in caravans and boats, they do tend to be more expensive.

Third, and finally, the battery is used to smooth the swings in current and voltage output from the array. Without this buffer function, the power supply to the load would be erratic. Thus, as well as providing a diurnal storage capability – the primary role – the battery also serves as a buffer to match a fluctuating electricity supply to a fluctuating load.

6.6 Reference

1 S.M. Schoenung, *Characteristics and Technologies for Long-Vs. Short-Term Energy Storage*, A Study by the DOE Energy Storage Systems Program, Sandia Report, SAND2001-0765, March 2001, Sandia National Laboratories, Albuquerque, New Mexico, USA.

CHAPTER 7

Physical Techniques for Storing Energy

The different modes of energy storage have been itemized in Figure 3.14, Chapter 3. Here, we review the major physical modes (as thermal, potential and kinetic energy) and also, for completeness, the electromagnetic and electrostatic forms of energy storage. Storage as chemical energy is treated later, in Chapters 8 and 9.

7.1 Thermal Energy Storage

Thermal energy can be stored in three ways, as follows:

- Sensible heat storage, in which a substance absorbs heat passively and becomes warm (see, for example, passive solar heating, Section 4.4, Chapter 4).
- Latent heat storage, which makes use of the energy stored when a substance changes from one state, or 'phase', to another. For example, heat is absorbed when a substance changes from a solid to a liquid, and is released again when changing from a liquid to a solid. There is no temperature change associated with latent heat.
- Thermochemical storage, which uses the energy stored in reversible chemical reactions.

There are many everyday illustrations of the benefits to be gained by providing the means to store heat (or cold): clothes serve to retain body heat and buildings are thermally insulated to retain or exclude heat; refrigerators are used everywhere to keep food fresh and 'cold storage' is an established industry for the preservation and conveyance of perishable foodstuffs.

Sensible Heat Storage

The storage of sensible heat requires, ideally, materials of high specific heat (*i.e.* energy stored per unit weight) and low cost. In practice, it is the latter consideration that prevails and the two most commonly used materials are water and rocks (or gravel). Water has a specific heat capacity of $4.18\,kJ\,kg^{-1}\,K^{-1}$ ($4.18\,kJ\,dm^{-3}\,K^{-1}$) and rocks 0.7–$0.8\,kJ\,kg^{-1}\,K^{-1}$ (1.4–$2.2\,kJ\,dm^{-3}\,K^{-1}$). Hence, water is almost six times as effective as minerals on a gravimetric basis, and two-to-three times as effective on a volumetric basis (Table 7.1). Water, however, is limited to $100\,°C$ and for higher temperatures it is necessary to use ceramic materials. Traditionally, firebricks have been employed to store heat in fireplaces and also to line industrial furnaces and bakers' ovens. The passive solar heating of buildings (Section 4.4, Chapter 4) also involves the use of bricks to store heat. In the UK, 'night storage heaters' were widely employed for domestic space heating before the advent of central heating and cheap gas. These heavy, insulated units contained bricks or similar ceramic materials and were heated overnight by in-built resistance heaters using low-tariff electricity. The heat stored overnight was liberated slowly and progressively during daylight hours. From a thermodynamic and environmental standpoint, it is grossly inefficient to utilise electricity, a premium form of energy, for low-grade space heating. On occasions, however, it can be justified when other forms of heating are not available, or when heat is required for a short period only. If a room or building is to be used occasionally for short periods, it is more energy efficient to heat it by electricity for those periods than to maintain permanent central heating. Electrical space heating generally has the lowest capital cost but the highest running cost.

Water-based thermal stores are used domestically to ensure a constant supply of hot water. For instance, a hot-water cylinder is essential when the primary source of heat is from a boiler operated intermittently on a time-clock, from an immersion heater based on low-tariff, off-peak electricity, or even from solar heating. On a

Table 7.1 *Energy-storage capacities of various materials*

	Stored energy		*Temperature range* (°C)
	($kJ\,kg^{-1}$)	($kJ\,dm^{-3}$)	
Conventional fuels (calorific value)			
Oil	~41 000	~34 000	
Coal	29 800	45 000	
Wood	15 100	7200	
Heat-capacity storage			
Hot water	334	334	20–100
Hot rocks, bricks	55–65	~150	20–100
Phase-change storage			
Ice (heat of fusion)	334	334	0
Water (vaporization)	2270	2270	100
Salt hydrates (heat of fusion)	200–350	330–500	30–70 (typically)

It is apparent that, overwhelmingly, fossil fuels constitute the best form of energy storage – on both a gravimetric and a volumetric basis.

much larger scale, water reservoirs have been used in conjunction with district heating schemes to store hot water derived from combined heat and power units (CHP units, see Section 3.4, Chapter 3), or from solar energy sources. For instance, the city of Odense in Denmark, which has a large district-heating scheme, installed a 12 000 m³, highly insulated, hot-water storage facility adjacent to the power station in 1978. On occasions when full electricity output has been demanded from the CHP plant, this store has provided sufficient hot water to support the heating needs of the city for up to 2 h. Once district heating lines have been installed, it also makes sense to have communal water tanks to store solar-generated heat. Several such schemes are in operation in Sweden and employ large underground, well-insulated, storage reservoirs from which hot water is delivered to office buildings or apartment blocks. When necessary, *e.g.* in winter, CHP heat is added to maintain the temperature of the supply.

The loss of heat from a spherical storage vessel is proportional to the square of the radius, whereas the volume of the vessel is proportional to its radius cubed. It follows that for efficient and long-term storage a very large unit should be used. Such a facility has been installed in Germany with a view to seasonal storage of solar heat from summer to winter (see Section 4.4, Chapter 4). The technical problems have been addressed successfully, but the economics have yet to prove favourable.

In equatorial countries, where insolation is at a maximum, solar cooking is a realistic possibility (see Section 4.4, Chapter 4). To hold heat longer during periods when the sun is covered by clouds and to provide evening meals after the sun has set, it is necessary to have a storage medium that operates at 200–250 °C. If cooking is to be done indoors, a facility is required to transfer the heat from the outdoor store to the stove. The store might then take the form of hot oil, or some other high-temperature liquid, and an insulated thermo-siphon to conduct the heat indoors. Obviously, this is an added sophistication that costs more.

A salt ('solar') pond is another option for storing sensible heat. In this approach, highly saline water sits on the bottom of the pond, with less saline (and therefore less dense) layers stratified above. Solar energy penetrates the upper layers and warms the bottom layer. The 'membrane' pond is a modified design in which the salt layers are physically separated by thin transparent membranes. The salts commonly used are sodium chloride and magnesium chloride. Dark-coloured material is often used to line the pool to enhance the absorption of solar radiation, and to prevent leakage and hence possible contamination of groundwater. The concentrations of the salt layers are designed such that even when the bottom layer is heated to near boiling, it is still denser than the layer above and thus inhibits the convection of heat throughout the pond; for this reason, salt ponds are termed 'non-convecting' ponds.* Pipes carrying a heat-transfer fluid run through the bottom layer and extract the heat, which can then be used to drive a turbine. The overall

---

* Some studies have also been conducted on 'convecting' solar ponds. In one version, the pond is shallow and uses pure water that is enclosed in a large bag with a blackened bottom and plastic or glass glazing on top. This arrangement permits convection but hinders evaporation. After being heated by the sun during the day, the hot water is pumped into a large heat-storage tank at night to minimise heat loss. To date, convecting ponds have met with little practical success.

thermodynamic efficiency of converting sunlight to electricity is very small (typically around 1%) and thus, to produce practical amounts of electrical power, ponds with large areas are required. Successful power generation by this technology has been demonstrated primarily in arid and semi-arid parts of the world. A 20-hectare pond that supplied a 5-MW_e power plant operated successfully in Israel for several years, and smaller systems in the kW_e range have been demonstrated in California and Texas. Salt ponds have also been coupled to thermal desalination systems in order to provide potable water together with brine for recharge of the hot bottom layer.

Molten salts can provide sensible heat storage at very high temperatures. A mixture of sodium nitrate and potassium nitrate has been tested as both a heat transfer agent and a storage medium for capturing solar energy *via* a central-receiver solar furnace (see Section 4.4, Chapter 4). Similar mixtures have been evaluated for storing waste heat generated by industrial processes such as paper and pulp production, the smelting of metals (iron, steel, aluminium), and brick-making.

Phase-change Storage

Latent heat storage is potentially a more efficient and cost-effective approach than sensible heat storage. In addition, phase-change materials (PCMs) help to regulate the temperature, which is an especially useful characteristic for designing passive solar facilities for buildings. In principle, solid–solid PCMs absorb and release heat in the same manner as solid–liquid PCMs but, to date, have been found to be impractical for use as thermal stores. The same holds true for liquid–gas PCMs, on account of the increase in volume following transformation to the gaseous state.

For the above reasons, interest in the application of PCMs has centred primarily on solid – liquid systems. Among these, some inorganic salts, particularly fluorides and fluoride mixtures, absorb considerable heat on melting, but the temperatures involved are high, *viz.* 600–900 °C. Such salts could prove useful for industrial heat storage. Salt hydrates such as $CaCl_2 \cdot 6H_2O$, $Na_2SO_4 \cdot 10H_2O$ and $Na_2HPO_4 \cdot 12H_2O$ are more promising in that they melt below 100 °C. These salts melt incongruently (*i.e.* melting combined with dissociation) at 29–35 °C and with latent heats of 200–350 $kJ\,kg^{-1}$. For example, $Na_2SO_4 \cdot 10H_2O$ (Glauber's salt) decomposes at 32 °C to a saturated water solution of Na_2SO_4 plus a residue of the anhydrous salt. The heat evolved in this process is 250 $kJ\,kg^{-1}$ and is equivalent, in terms of heat storage, to raising the temperature of the same weight of water from 0 to 60 °C (Table 7.1). There are, however, complications of a physicochemical nature associated with incongruent melting. Reversing the reaction quantitatively, and reforming the crystal hydrate, may not be straightforward, especially in a large container. Moreover, the melting temperatures are too low for central-heating purposes, but could perhaps be useful in the passive solar heating or cooling of buildings. In fact, crystal hydrates have not been much employed for heat storage.

The integration of salt hydrates, paraffin waxes and fatty acids in plasterboard (wallboard) and other lightweight building materials is being investigated to increase heat storage capacity and, thereby, provide an effective means for moderating the thermal environment within rooms. Moreover, the use of PCM plasterboard does not require radical changes in building design or construction. Although salt hydrates are relatively inexpensive, the packaging and processing required to obtain reliable performance is both complicated and costly. The encapsulating material should be a good conductor of heat, should withstand frequent changes in volume as phase changes occur, should be impermeable to water, and should resist leakage and corrosion. Paraffin waxes and fatty acids are readily available, inexpensive and melt at different temperatures according to their chain length. Both types of material can be incorporated in plasterboard, either by direct immersion or by encapsulation in plastic pellets that are added during the manufacturing process. The plasterboard can be tailored to a specific climate by varying the PCM content. Other possible applications of solid–liquid PCMs include use in paving materials to minimise night-time icing and damage from freeze–thaw cycling on bridges and underpasses, and in outdoor clothing for protection against extremes in temperature.

Thermochemical Storage

Reversible thermochemical reactions are the third group of thermal-storage options. These include the reversible decomposition of metal hydrides, discussed in Section 8.2, Chapter 8, and the reversible reaction of methane with steam to give synthesis gas, see Section 2.1, Chapter 2. In the latter process, a nickel catalyst is employed at 900 °C for the endothermic production of synthesis gas, as well as for its exothermic conversion back to methane and steam. Since this cycle takes place at high temperature, it is of potential interest in storing and recovering process heat in industry. Other reversible processes that are theoretically possible for heat storage include splitting ammonia into nitrogen and hydrogen, reducing sulfur trioxide into sulfur dioxide and oxygen, or decomposing calcium hydroxide to calcium oxide and water. The practicality and economics of operating such systems are, however, dubious.

7.2 Potential Energy Storage

Pumped Hydro

The largest form of electricity storage practiced today is pumped hydro. This requires two water reservoirs, in the form of lakes, separated by a substantial difference in vertical distance (*i.e.* a 'head'). During the night, when surplus generating capacity is generally available, water is pumped from the lower lake to the upper. During the day, when extra electricity is required, the process is reversed and the falling water passes through a turbine to generate electricity. In some pumped-hydro schemes, two sets of equipment are used: one for generating and the other for pumping. In alternative schemes, the generator becomes an electric motor, receives electricity from a nearby power station, and operates the turbine as a pump. Both approaches are cost-effective, with round-trip efficiencies of at least 70%. An immediate benefit of pumped hydro is its fast response to energy demands. On the other hand, construction costs are high and use is limited by the availability of suitably mountainous terrain together with twin lakes of sufficient size – one at the top of a mountain and the other at the foot.

Some 300 pumped-hydro schemes are in operation world-wide with a combined power output that is approaching $100\,GW_e$. The largest facility in the UK has been constructed at Dinorwig in North Wales (Figure 7.1). This is a massive storage undertaking, with a capability of generating $1800\,MW_e$ for up to 5 h and a start-up time of only 12 s when on 'spinning reserve', *i.e.* when the generator is spinning in air. From a standing start to full power takes 100 s. Therefore, when sudden demands for extra power are anticipated, one or two generators are held on spinning reserve. Because the station is located within Snowdonia National Park, which is an area of outstanding natural beauty, the visual impact has been minimised by building the entire works within the core of the mountain. Construction necessitated first blasting out from the rock (slate) one of the largest man-made caverns ever created; it has a height of 60 m and could accommodate St Paul's Cathedral (London). The upper lake, Marchlyn Mawr, is situated at over 500 m above the lower lake Llyn Peris, and this vertical separation provides the head of water required to operate the generators. It was necessary to extend and dam the upper lake to increase its water holding capacity to 7 million cubic metres. During full operation, the water level rises and falls by 34 m daily. The lower lake was deepened to accommodate the extra water. Inside the Dinorwig power station, there are six large pump-turbine units in a line that run along the 179-m length of the cavern. These are situated on floors that are well below the level of the lower lake. The water enters and leaves through steel penstocks of about 3 m in diameter, and power is transferred to and from the pump–turbine units by means of vertical drive-shafts that connect with the motor-generators located on a higher floor. The overall construction began in the mid-1970s and the station was opened in May 1984.

Pumped hydro is particularly appropriate for a network that has a large nuclear component since, for both technical and economic reasons, nuclear reactors are best

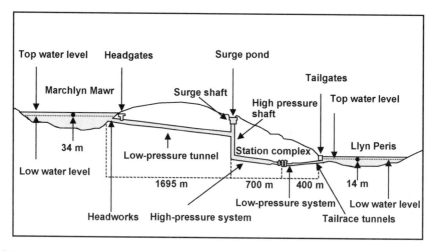

Figure 7.1   *Cross-sectional view of Dinorwig pumped-hydro scheme in North Wales, UK.*

operated on base-load. The rapid response of pumped hydro makes it suitable for peak-shaving, load-levelling, and meeting rapidly fluctuating power demands. It should be appreciated, however, that a pumped-hydro scheme is a net *consumer* of electricity – it takes more electricity to pump the water up than is recovered – and does nothing to add to a nation's overall electricity supply. Rather, it serves to redistribute generating capacity from when it is in surplus to when it is most needed. To date, this is the most practical and economic means of storing electricity on the multi-MW$_e$ scale. Unfortunately, the scope for expanding pumped hydro in many countries is limited by the availability of suitable sites and by competing concerns, such as the preservation of natural environments and wildlife habitats.

Pumped hydro is also valuable when electricity is to be derived from intermittent renewable sources of energy. Accordingly, some small systems have been constructed to store surplus energy from wind turbines in isolated communities with a local grid. Again, though, it is necessary to have a suitable hill with a small reservoir or pond at its summit. Such small-scale ('mini') hydro capacity is likely to increase as research on computerised control systems and improved turbines lowers the costs of producing such power.

Compressed Air Energy Storage (CAES)

Compressed gas also involves the storage of potential energy. When a piston is used to compress a gas, energy is stored in the gas and can be released later by reversing the movement of the piston and expanding the gas. In order to exploit this on a scale that is useful for significant electricity storage, it is necessary to have a very large container capable of holding gas at high pressure. Natural or man-made underground caverns are employed. Possible reservoirs include: caverns formed naturally or excavated from salt domes or limestone strata, caverns mined from bedrock, aquifers, exhausted oil and gas wells, and abandoned coal mines.

A schematic representation of a CAES system is shown in Figure 7.2. Air is compressed to, say, 4–7 MPa (40–70 atm.) pressure in either one or two stages before being stored in the underground cavern. Off-peak electricity is used to power the compressor. When extra electricity is demanded, the high-pressure air is pre-heated in a heat-exchanger (a 'recuperator') before being combined with a small amount of fuel, *e.g.* natural gas, and passing to a two-stage gas turbine. The latter, in turn, drives a generator for the production of electricity. The hot exhaust gases from the low-pressure turbine are returned to the recuperator to pre-heat the incoming air. Because the gas is pre-compressed, the system is much more efficient than a conventional gas turbine in which a significant fraction of the energy is consumed in compressing the gas. Note, it is possible to operate a turbine on compressed air alone, without an integral combustor fired by a fuel, but at a reduced efficiency. An attractive feature of CAES systems is that the overall cost is less than that of building and operating extra peak-generating plant. Other claimed benefits include: fast response time, lower emissions than from conventional fossil-fuelled generating plants; efficiency maintained at low capacity factors; more efficient and less expensive than pumped hydro for large-scale energy storage.

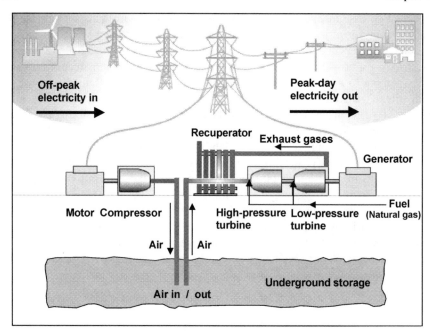

Figure 7.2 *Schematic representation of components of a CAES system.*

The first CAES plant to be built was a 290-MW$_e$ system at Huntorf in Germany in 1978 (Figure 7.3). Here, compressed air is stored in two salt caverns, with a total volume of 0.3 million cubic metres, and power can be generated for up to 4 h. The second plant (110 MW$_e$) was commissioned in McIntosh, Alabama, USA, in 1991. This also uses a salt cavern (0.54 million cubic metres) and has improved on the Huntorf design by incorporating a recuperator to preheat the air from the cavern with waste heat from the turbines (Figure 7.2). The total system has functioned with over 95% reliability. A large facility is under construction at a limestone mine near Norton, Ohio, USA. It is of modular design and plans call for nine 300-MW$_e$ turbines to be brought on line over a 5-year period.

Figure 7.3 *Aerial view of CAES plant at Huntorf, Germany.*

7.3 Kinetic Energy Storage in Flywheels

The principle of the flywheel for the short-term storage of energy has been known since prehistoric times, in the form of the potter's wheel. In the Industrial Revolution, large flywheels were adopted in steam engines to smooth out the power pulses from the pistons (see Section 1.1, Chapter 1). Today, simple steel flywheels are employed in internal-combustion engines for the same reason.

Flywheels may also be used to store electricity by converting it to kinetic energy, a method sometimes referred to as 'mechanical energy storage'. Electricity to be stored powers an electric motor that increases the speed of the flywheel, while electricity is recovered by running the motor as a generator, which causes the flywheel to slow down. The amount of energy stored is proportional to the mass of the flywheel (or more accurately, its mass moment of inertia) and to the square of its angular velocity. Thus, the rotational speed is much more important than the mass in determining the amount of energy stored.* The maximum energy that can be stored is dependent upon the tensile strength of the material from which the flywheel is constructed. The circumferential tensile stress in the rim is also proportional to the square of the angular velocity. It follows that the maximum stored energy is to be found in a flywheel of high tensile strength that is rotating at the maximum safe speed.

The flywheels with the highest tensile strength are not made of steel, but of fibre-reinforced composites such as carbon or Kevlar® fibres embedded in a matrix of epoxy resin. These composites are often laminated and thickened towards the centre in order to relieve stress concentration and increase the stored energy. As well as rotating faster and storing more energy than steel flywheels, composite wheels are much safer if the maximum safe speed is exceeded, since they tend to delaminate and disintegrate gradually from the outer circumference, to produce fine fibres rather than to explode catastrophically. Various flywheel designs are shown in Figure 7.4(a), and the typical maximum stored energy per kg for various materials used in flywheel construction is listed in Table 7.2. It is clear that composites of carbon or Kevlar® fibres with epoxy resin are, overall, substantially better than the other materials. In forming the composite, the fibres are wound circumferentially around the wheel so as to give maximum strength.

Flywheels have a number of attractive features for energy storage, particularly with respect to battery alternatives. For example, they:

- act as high-power devices, which absorb and release energy at a high rate;
- do not have the electrical inefficiencies associated with electrochemical devices;
- have a long life, which is unaffected either by the frequency of cycling (charge–discharge) or by the rates of uptake and release of energy;

---

* The mass moment of inertia provides a measure of an object's ability to resist rotational speed about a specific axis. It is the sum of all the point masses of a rotating object multiplied by the squares of the respective distances of the masses from the axis of rotation. Thus, a flywheel is more effective when its inertia is larger, *i.e.* when the mass is located farther from the centre of rotation, *e.g.* by having a more massive rim or a larger diameter.

(a)   (b)

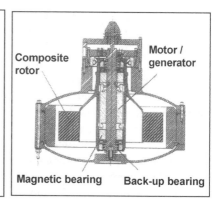

Figure 7.4   (a) *Various designs of flywheel;* (b) *schematic of flywheel assembly developed by United Technologies Corporation.*
[(a) Courtesy of Imperial College Press]

- give greater reliability than batteries;
- have flexibility in both design and unit size;
- are constructed from readily available materials;
- have a small footprint;
- require little maintenance (unlike some batteries) and, hence, life-cycle costs are low (note, capital investment is high but, in principle, flywheels can be mass-produced at reasonable cost as power devices);
- are relatively unaffected by extremes in ambient temperature associated with installation in restricted space or outdoor use;
- create no environmental impact in use or in recycling.

Apart from capital cost, the most significant limitation of flywheels lies in their relatively modest capability for energy storage. They are essentially surge-power devices rather than energy-storage devices, and are best suited to applications that involve the frequent charge and discharge of modest quantities of energy at

Table 7.2 *Some properties of flywheel materials*

Material	Design stress ($MN\,m^{-2}$)	Density ($10^3\,kg\,m^{-3}$)	Specific energy ($Wh\,kg^{-1}$)
Carbon-fibre–epoxy	750	1.55	51.5
Silica-fibre–epoxy	350	1.90	14.0
Kevlar®-fibre–epoxy	1000	1.40	76.2
Mild steel	300	7.80	8.2
Maraging steel[a]	900	8.00	24.0

[a] A special hardened steel that is strengthened by the introduction of intermetallic compounds, *e.g.* Ni_3Ti and Ni_3Mo, and by keeping the concentration of carbon as low as possible.

high power ratings. In this respect, flywheels are complementary to batteries. A typical advanced flywheel will store ~1 kWh of electricity, but may be charged–discharged at a rate of 25–50 kW. Advanced flywheels use composite materials, operate at high speeds (tens of thousands of rpm), and tend to be of the 0.5–1 kWh size.

High-speed flywheels are generally mounted in vacuum enclosures, to eliminate air drag, and on low-friction bearings or magnetic suspension systems. A schematic diagram of a flywheel assembly developed by the United Technologies Corporation in the USA is shown in Figure 7.4(b). The motor-generator is mounted on magnetic bearings, with conventional bearings as back-up. The technology is flexible in terms of the amount of energy stored as controlled by the size and speed of the rotating wheel, and the power rating as determined by the motor-generator.

Appreciable efforts were made to develop improved flywheels in the 1970s and early 1980s, but the activity declined thereafter. A recent resurgence of interest has come about as a result of the above-mentioned technological developments in fibre-composite materials and low-loss bearings, coupled with advances in power electronics. Added factors have been the growing market interest in electric and hybrid electric vehicles (see Sections 10.2, and 10.4, Chapter 10) and in uninterruptible power supply (UPS) systems (see Section 6.3, Chapter 6), as well as the increased need for improving power quality in electricity networks.

The use of a relatively small flywheel unit in electrically powered vehicles would significantly enhance the short-term power available for accelerating and hill-climbing, and would also facilitate the recuperation of energy during braking. Most batteries will not readily accept charge at the rates encountered with regenerative braking, whereas the flywheel is a surge-power unit and has no such limitation. Of course, the extra mass of the unit, the volume it occupies and its cost are factors that have to be taken into account. There is also the question of safety with a fast-rotating flywheel (up to 30 000 rpm) in close proximity to passengers. Finally, there is the problem of the processional torque when the vehicle is turning. Altogether, it seems that advanced flywheels are better suited to stationary applications although, conceivably, they could be used in large road vehicles.

Recently, flywheels have also been finding use in electric-rail applications such as metro trains. When the electricity supply is provided *via* a live rail, it is possible to mount flywheels beside the track rather than on the train. As the train accelerates and draws a heavy current, the voltage along the track falls. This is sensed at the flywheel installation and activates the flywheel to act as a generator and pass current into the system. Conversely, when a train decelerates, the voltage along the track rises. Again this is sensed and the generator system now acts as a motor and draws current from the track to speed up the flywheel. Such an arrangement has advantages compared with mounting the flywheel on-board the train.

Trackside flywheels can offer several benefits, *e.g.* voltage support at weak points of the network, capture of regenerative-braking, and provision of supplementary power to reduce current peaks. A 1-MW system has been installed by the New York Transit Authority to capture braking energy from a decelerating train and deliver it to one that is accelerating further up the track. It is also possible to run a train and accelerate it to $40 \, \text{km} \, \text{h}^{-1}$ should one of the sub-stations fail.

The installation consists of ten 100-kW flywheel units developed by Urenco Power Technologies in the UK (Figure 7.5(a,b)). A 250-kW unit has also been designed and built for trolleybus systems (Figure 7.5(c)).

Two very large, stationary, flywheel-generators are in use at the Culham Laboratory, UK to provide pulse power to the Joint European Torus experiment on nuclear fusion (see Section 3.6, Chapter 3). This experiment requires power pulses of 700 MW, which is more than can be drawn from the grid. The flywheel-generators have been constructed to accommodate this deficit (Figure 7.5(d)). Each flywheel consists of a 9-m diameter rotor, which weighs 775 t and is driven by a 8.8-MW motor. The speed of such a huge rotor is only 225–250 rpm. This design was chosen because the Culham flywheel is a power device and not an energy-storage device. When pulsed power is demanded, each generator converts its rotational energy into 2600 MJ of electrical energy and delivers 400 MW of power. Flywheel-generators of this order of size could find applications in the utility electricity market. In an ambitious research programme, the New Energy Development Organisation (NEDO) in Japan has been attempting to develop a 10-MWh commercial flywheel system for load-levelling at electricity sub-stations.

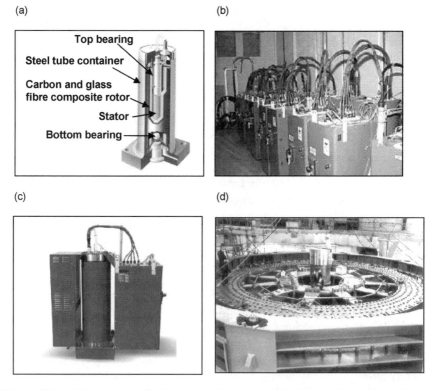

Figure 7.5 (a) *Schematic of flywheel assembly developed by Urenco Power Technologies;* (b) *bank of 100-kW Urenco flywheels;* (c) *250-kW Urenco flywheel;* (d) *400-MW flywheel–generator in the course of construction at Culham, UK.*

Flywheels are also of potential interest for the localised storage of electricity generated by wind turbines and photovoltaic arrays. Since these two technologies may exhibit large, frequent and rapid fluctuations in power output, a flywheel-based buffer store could remove the need for downstream power electronics to track such fluctuations and so improve the overall electrical efficiency. In many situations, rechargeable batteries would seem to be a more appropriate storage medium and these are widely used today, but a battery-flywheel combination is worthy of consideration. It seems that rather little development work on flywheels has been directed specifically to these renewable energy applications.

7.4 Superconducting Electromagnetic Energy Storage (SMES)

Magnetic fields can be used for energy storage. When an electromagnet is connected to a source of d.c. power, the flow of energy into the magnet builds up together with the current and the magnetic field. This energy can be released again as an electric current when the power source is disconnected. Unfortunately, when using conventional magnetic materials, the quantities of energy involved are very small, a few hundred joules at most in an electromagnet of one cubic metre in size. Moreover, when the power is switched off the stored energy immediately decays.

The situation improves dramatically when using a superconductor. This is a material that, below a certain critical temperature (T_c) offers no resistance to current flow. A typical superconductor is the niobium–tin compound Nb_3Sn. For most superconductors, T_c lies below 20 K, although materials with higher critical temperatures have been found and are being investigated. This activity commenced in 1986 when it was demonstrated that a ceramic composed of lanthanum, barium and copper oxide exhibits superconducting behaviour at 30 K. Intense research programmes were set up world-wide and resulted in the discovery of a material, $YBa_2Cu_3O_7$ (known as 'YBCO', $T_c = 93$ K) that displays superconductivity at temperatures warmer than liquid nitrogen – a commonly used coolant. Unfortunately, YBCO has poor mechanical properties (*e.g.* brittleness, low ductility), is difficult to produce in bulk and, inevitably, is expensive. Although the search for high-temperature superconductors has continued, little progress has been achieved since 1997 when a mercury cuprate ceramic under an extreme pressure of 30 GPa was shown to have a T_c of 164 K. Clearly, whilst scientifically interesting, this finding has little practical significance.

The critical temperature is not the only major factor that decides the practicality of a superconductor. When magnetic field strengths or current densities go above certain critical values, superconducting properties are again lost. The values of the magnetic field and current density that a material will withstand without losing its superconductivity are, therefore, also important in determining how much energy can be stored. The parameters of temperature, magnetic field strength and current density all trade-off against one another in superconductors – in an ideal case, a single material would maximise all three. Unfortunately, nature is not that cooperative. Furthermore, additional trade-offs exist in manufacturing wire with good stability and low a.c. loss properties. A composite of many fine filaments of superconductor embedded in copper achieves the best set of characteristics.

In order to wind coils with the many turns that are necessary to store large amounts of energy, long lengths (many metres) of superconducting wires are required. At present, such lengths of high-temperature, high-field, high-current superconductors have not even been approached in single strands! A clear example of the compromises involved in material selection is evident when comparing Nb_3Sn and NbTi. While Nb_3Sn has better thermal, electric and magnetic properties, it is brittle and therefore both difficult and expensive to form into a wire. By contrast, NbTi is a ductile metal alloy that forms wires well, while still providing acceptable thermal, electric and magnetic performance. Consequently, the material of choice for most superconducting wire applications is NbTi.

When a superconducting coil is connected to a d.c. power source, the current in the coil (that is a pure inductance) increases and the magnetic field along with it. So long as the superconductor is kept below its critical temperature, the current continues to circulate even after the power source is removed and the energy is stored until required, when it may be released in high power pulses of electricity (of millisecond duration if desired). This, then, is a SMES system.

The ability of a superconducting coil to carry very high currents allows the development of high magnetic fields. These fields are proving extremely useful in a number of important areas. Superconducting magnets are employed in magnetic resonance imaging (MRI) scanners in hospitals; the devices allow physicians to obtain detailed images of the interior of the human body without surgery or exposure to ionising radiation. In high-energy physics, superconductors produce the large magnetic fields required for particle accelerators. They are also an essential component of plasma fusion devices, such as the tokamak (Section 3.6, Chapter 3), where a very high magnetic field is required to contain the plasma in a 'magnetic bottle' and to prevent it from touching the walls of the vacuum vessel, which would result in power losses.

The major disadvantage of superconducting magnets has been the need to operate at low temperatures. In most cases, the magnet is held at 4 K in a helium cryostat. This is an expensive exercise, both in terms of the capital cost of the equipment and the running cost of liquid helium. Even MRI scanners, which have become commonplace in the world of medicine, are still very expensive. As discussed above, high-temperature superconductors that operate at liquid nitrogen temperatures have not yet reached technical maturity to be considered for SMES applications. Given their high cycle-life, SMES systems are suitable for applications that require constant full cycling and a continuous mode of operation. Moreover, compared with batteries, the energy output of an SMES system is much less dependent on the discharge rate. It is also noteworthy that SMES has the potential to be more than 90% efficient. By contrast, pumped hydro has an efficiency of around 70% and is restricted to specific locations. Although research is being conducted on large SMES systems in the range 10–100 MW, recent focus has been on smaller facilities with outputs 1–10 MW. The latter are now available commercially for assisting electric utilities to improve power quality and reliability. These so-called 'Distributed-SMES' (D-SMES) systems are compact, self-contained,

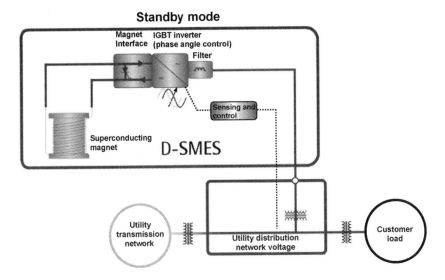

Figure 7.6 *Schematic of D-SMES system offered by American Superconductor Corporation.* (Courtesy of American Superconductor Corporation)

and can be transported on semi-trailers. A schematic of the operating principles of a 2-MW system produced by Anglo Superconductor Corporation in the USA is shown in Figure 7.6. Six such systems have been installed in northern Wisconsin to enhance the stability of the local transmission grid. Whether SMES will ever become economic for the bulk storage of electricity remains an open question.

7.5 Electrostatic Energy Storage (ESES)

A conventional 'electrostatic' capacitor consists of two conductors (usually metal 'parallel plates' of equal area) separated by an insulator (a 'dielectric', *e.g.* vacuum, air, mica, oil, paper, plastic). On the application of an applied voltage, the device stores energy by the separation of positive and negative electrostatic charge; the two plates carry equal but opposite charges. The capacitance, C, is the ratio of the magnitude of the charge, q, on either plate to the magnitude of the potential difference between them, V_c. It is also directly proportional to the dielectric constant, ε, of the insulating material and the area of each plate, A, and it is inversely proportional to the inter-plate spacing, d, *i.e.*

$$C = q/V_c = \varepsilon A/d \tag{7.1}$$

The energy stored in a charged capacitor is given by:

$$U = 1/2\, CV_c^2 \tag{7.2}$$

The energy density is very low, typically $\sim 0.05\,\mathrm{Wh\,dm^{-3}}$.

The 'electrolytic capacitor' is another well-known type of conventional capacitor. The dielectric is a very thin film of oxide that is formed electrolytically on a metal such as aluminium, tantalum, titanium, or niobium. The metal substrate acts as one conducting phase and an electrolyte solution as the other. The capacitors are divided into two general types according to the choice of electrolyte: 'wet' types employ an aqueous electrolyte (typically, a solution of boric acid and ammonium or sodium borate), whereas 'dry' types use a wide range of non-aqueous electrolytes that range from highly viscous liquids to semi-hard crystalline masses (*e.g.* glycerol–ammonium borates, ammonium lactates, amine acetates). In a wet electrolytic capacitor, the can or container serves to make electrical contact with the aqueous electrolyte. This design is not satisfactory for the dry type, due to the low conductivity of the non-aqueous electrolyte, and a plate or foil is added to provide the electrical contact. Electrolytic capacitors can only be used with a flow of current in one direction. The oxide-covered electrode must, therefore, always be connected to the positive side of the applied voltage, and the electrolyte must always be negative. If the direction of current flow is reversed, a large current will flow and the capacitor becomes useless. In other words, the system exhibits the characteristics of a rectifier. Given that the dielectric (oxide film) is extremely thin, the specific capacitances ($F\,g^{-1}$) of electrolytic capacitors are much larger (up to 1000 times) than those of electrostatic counterparts. The specific energy is higher (typically 0.06 *versus* 0.003 $Wh\,kg^{-1}$), but both types of capacitor are extremely poor energy-storage devices.

So-called 'supercapacitors' differ from conventional electrostatic and electrolytic capacitors in that they store electrostatic charge in the form of ions, rather than electrons, on the surfaces of materials with high specific areas ($m^2\,g^{-1}$). The electrodes are usually prepared as compacts of finely divided, porous carbon, which provide a much greater charge density than is possible with non-porous, planar electrodes. Much of the storage capacity is due to the charging and discharging of the electrical double-layers that are formed at the electrode–electrolyte interfaces.[*] The voltage is lower than for a conventional capacitor, while the time for charge, as well as that for discharge, is longer because ions move and re-orientate more slowly than electrons. In these respects, the supercapacitor begins to take on some of the characteristics of a battery, although no electrochemical reactions are involved in the charge and the discharge processes.

The 'ultracapacitor' moves one step closer to a battery. In a typical design, an electrode material with a large specific surface-area is combined with a material that can be reversibly oxidised and reduced over a wide potential range, *e.g.* the oxides of multivalent metals such as ruthenium and iridium. The energy is stored both by ionic capacitance and by surface (and near-surface) redox processes that occur during charge and discharge. The latter are electrochemical reactions in which surface ions are reduced and oxidised. This enhances the amount of stored

---

[*] When an electrode is immersed in an electrolyte solution charge separation occurs at the interface and a so-called 'double-layer' is formed. The excess charge on the electrode surface is compensated by an accumulation of excess ions of the opposite charge in the solution. The amount of charge is a function of the electrode potential. The double-layer behaves essentially as a capacitor and since charge separation occurs at the molecular level, very high values of capacitance are obtained.

energy. Moreover, because the ions are confined to surface layers, the redox reactions are rapid and are fully reversible many thousands of times, which therefore make for a long cycle-life. The electrolyte may be either an aqueous solution, *e.g.* sulfuric acid or potassium hydroxide, or an organic solution, *e.g.* tetraethylammonium tetrafluoroborate, $(C_2H_5)_4NBF_4$, dissolved in acetonitrile, CH_3CN. Capacitors with aqueous electrolytes have a very low resistance, but also a low breakdown voltage, (*i.e.* the voltage that when exceeded, causes the dielectric to conduct) while the converse is true for capacitors with organic electrolytes. It should be noted that there is considerable confusion in the literature over the definition of 'supercapacitor' and 'ultracapacitor', especially as groups working on the development of capacitor devices tend to use the terms interchangeably. Thus, it has been proposed that the generic name 'electrochemical capacitor' should be adopted.

Electrochemical capacitors can store vastly more energy than conventional capacitors. They vary in size from the small capacitors used for memory back-up applications in electronics, to devices with capacities up to 2700 F that form the basic module for the units used in hybrid electric vehicles. They may be discharged at rates up to 10–20 times faster than batteries and, equally important, can also be recharged at much greater rates than batteries. Moreover, they have much longer lives than batteries when subjected to such high rates of charge and discharge. A further advantage is that unlike many batteries, they will operate satisfactorily at temperatures as low as $-40\,°C$. To summarise:

- batteries store electrical energy, whereas capacitors are power devices;
- batteries may only be charged and discharged comparatively slowly, whereas capacitors may be discharged or charged in seconds, or less;
- capacitors may be cycled many thousands of times before they deteriorate (well in excess of the life performance obtained from batteries);
- capacitors operate at lower temperatures than most batteries.

The performance of batteries and capacitors as energy-storage devices is compared in Table 7.3. It will be seen that conventional capacitors contrast with batteries in almost all respects. They store comparatively little energy, but can produce high power output for very short periods. They also have exceptionally

Table 7.3 *Comparison of characteristics of energy-storage devices*

Characteristic	Battery	Capacitor Conventional	Electrochemical
Energy density (Wh dm^{-3})	50–250	0.05	1–10
Power density (W dm^{-3})	150	$>10^8$	$>10^3$
Discharge time (s)	$>4\times10^3$	<1	1–10
Cycle-life	10–1000	$>10^6$	$>10^5$

long cycle-lives. The contrast is so great that the applications for capacitors and batteries are quite different. Electrochemical capacitors are intermediate devices with improved energy density and longer discharge time than conventional capacitors, while still retaining respectable power density and cycle-life. They are ideal for meeting sudden transient power demands that are too great to be supplied by a battery. These properties make batteries and capacitors ideal partners. In a typical application, the battery and capacitor are connected in parallel and the capacitor 'floats' at the same terminal voltage as the battery. When there is a sudden load demand for a pulse of high current, the capacitor discharges preferentially to meet this requirement. After the pulse, the battery recharges the capacitor at leisure. Eliminating the need for high-current pulses from the battery prolongs its life and may permit the use of a battery of smaller capacity.

Advanced capacitors are being developed and evaluated for many possible applications. For small-scale applications, a capacitor may be used in place of a battery for memory-maintenance on a computer processor board when there are short fluctuations in mains supply. They are also finding markets as energy-storage devices in tools and toys. The advantage over a battery is the long life of the capacitor, so that it need never be changed. At a medium scale, a battery–capacitor combination may be used for vehicle engine starting, especially in very cold weather. At low temperatures, the battery may be too weak to fulfil this function,

(a)   (b)

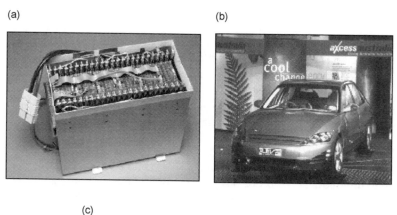

(c)

Figure 7.7   (a) *CSIRO electrochemical capacitor fitted to* (b) *'aXcessaustralia LEV' and* (c) *Holden 'ECOmmodore' hybrid electric cars.*

but will provide sufficient current to trickle-charge the capacitor. Finally, at a large scale, a battery–electrochemical capacitor combination is ideal for use as a power source in electric or hybrid electric vehicles where peak-power demands of short duration are encountered. Not only does the capacitor provide peak power for acceleration, but also it allows energy to be captured from regenerative braking since it is more efficient in accepting large currents than a battery.

Two hybrid electric cars equipped with electrochemical capacitors have been built and demonstrated in Australia (Figure 7.7). The pack of 104 capacitors can deliver 50 kW of power for 10 s, and is combined with a purpose-built lead-acid battery (the CSIRO 'Double-Impact$^{®}$' design) to provide a practicable and affordable surge-power unit for vehicle acceleration and hill-climbing.

Clearly, electrochemical capacitors are entirely complementary to batteries. Whether or not they might be of use alongside batteries in applications such as portable-power supplies and the storage of photovoltaic electricity has yet to be established, but obviously they would be employed only in situations that require transient pulses of high power. Wherever there is a rapidly fluctuating source of power, for example with wind energy or wave energy, this combination may prove useful in accepting high-current pulses and, thereby, smooth the supply output. This is an area for further exploration and development as larger electrochemical capacitor packs become available.

CHAPTER 8

Hydrogen Energy

Hydrogen, the most common element on earth, is widely seen as the ultimate form of *Clean Energy* and, as such, merits a chapter of its own. The proposition that hydrogen should be a sustainable medium of energy has become known as the 'Hydrogen Economy'. This term is thought to have been coined in 1970 by Neil Triner at the General Motors Technical Laboratory in Warren, USA.[*] The overall scheme of the Hydrogen Economy is illustrated conceptually in Figure 8.1, which outlines the many different possible routes to hydrogen from both conventional and novel primary energy sources, the storage and transportation modes for hydrogen, and its end-uses in fuel cells, engines, and industrial processes. This is a broad canvas and many authors restrict the use of the term Hydrogen Economy (or 'Hydrogen Energy') to the production of hydrogen from non-fossil sources, its distribution and storage, and its combustion in a fuel cell to generate electricity.

Hydrogen has many potential attractions as a new fuel. It may be derived from non-fossil sources, it burns cleanly to water with no pollutants being emitted, it is suitable for use in a fuel cell to generate electricity directly, and it has a high energy content expressed on a per mass basis (see Table 2.1, Chapter 2). Unfortunately, these attractive features are counter-balanced by many practical engineering and economic considerations that explain why hydrogen does not already find extensive use as a fuel.

As discussed in Section 2.5, Chapter 2, most hydrogen today is made from fossil fuels by chemical-reforming reactions and its major uses are in the refining of crude oil and the manufacture of ammonia. Lesser, non-energy, applications are found in the manufacture of other chemicals, as well as in the food, plastics, metals, electronics, glass, electric power and space industries (Table 8.1). By contrast, the present use of hydrogen for electricity generation *via* fuel cells is negligible. In the short term, however, there is environmental benefit in converting fossil fuels into hydrogen to serve as a clean fuel in fuel cells or internal-combustion engines.

---

[*] The concept of using hydrogen had in fact been suggested much earlier in such diverse publications as Jules Verne's science-fiction novel *The Mysterious Island* (1874) and J.B.S. Haldane's paper *Daedalus, or, Science and the Future* (1923). It is further notable that Haldane proposed the use of wind power to produce hydrogen *via* electrolysis of water; the gas would be liquefied and stored in vacuum-jacketed reservoirs that would probably be sunk in the ground.

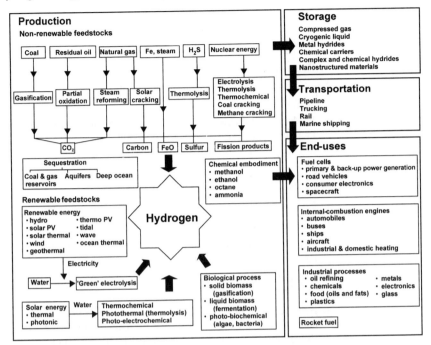

Figure 8.1   *The Hydrogen Economy: a summary diagram showing various possible means of producing, storing and transporting hydrogen, as well as possible end-uses.*

This benefit stems from the relative ease of pollution management at a central production facility compared with dispersed sites. Moreover, in principle, emissions of carbon dioxide are more easily captured and sequestered at a single plant than when fossil fuels are deployed in the field. A prototype plant for the conversion of coal to hydrogen, with sequestration of carbon dioxide, is being planned in the USA. The hydrogen will then be used for electricity generation (see Section 2.3, Chapter 2).

In this chapter, we look ahead to a very different future – possibly decades away – in which hydrogen is produced by the electrolysis of water on a large scale *via* the use of renewable electricity (solar, wind, *etc.*). It would then be stored in one of several different forms, distributed to where it is needed, and then re-converted to electricity in a fuel cell. The technology for this vision is still in embryonic form, principally because the economics of such energy production and use are not favorable. Nevertheless, for almost 30 years, much attention has been focused on the considerable challenges that would confront the practical introduction of hydrogen as an energy vector. Many conferences have been held and a specialist journal, the *International Journal of Hydrogen Energy*, is devoted to the subject.

At first, the interest in hydrogen energy arose from the shortfall in fossil fuels – especially oil – that was anticipated in the mid-1970s, and from a projected surplus of off-peak electricity from nuclear power stations. At that time, it was envisaged that nuclear power would expand much more rapidly than it has, and that surplus

Table 8.1 *Principal non-energy uses of hydrogen*

Industry	Uses
Oil	Removal of sulfur and other impurities ('hydro-desulfurization')
	Conversion of large oil molecules to fuel distillates for blending with petroleum ('hydrocracking')
Chemical	Production of ammonia (and then fertilizers), methanol, hydrogen peroxide, acetic acid, oxo alcohols, dyestuffs
Food	Conversion of sugars to polyols (bulk sweeteners)
	Conversion of edible oils (from coconuts, cotton seed, fish, peanuts, soybeans, etc.) to fats, *e.g.* margarine
	Conversion of tallow and grease to animal feed (and soap)
Plastics	Production of nylons, polyurethanes, polyesters, polyolefins
	Cracking of used plastics to produce lighter molecules that can be recycled in new polymers
Metals	Reductive atmosphere for production of iron, magnesium, molybdenum, nickel, tungsten
	Heat-treatment of ferrous metals to improve ductility and machining quality, to relieve stress, to harden, to increase tensile strength, to change magnetic or electrical characteristics
	Oxygen scavenger in metalworking
	Welding torches
Electronics	'Epitaxial' growth of polysilicon
	Manufacture of vacuum tubes, light bulbs
	Heat-bonding of materials ('brazing')
Glass	High-temperature cutting torches
	Reductive atmosphere for float-glass process
	Glass polishing
	Heat-treatment of optical fibers
Electric power	Coolant for large generators and motors
	Nuclear fuel processing
Space	Rocket fuel

night-time electricity would be available since nuclear plant normally operates on base-load and is not readily switched on and off. In parallel, the idea arose of using hydrogen to store electricity from renewable energy sources. The attractions of hydrogen as a storage medium are:

• it is universally available in the form of water, from which it may be extracted conveniently by electrolysis

- it may be transmitted over long distances in buried pipelines, which are cheaper to construct and operate than electricity grids, and have no visual impact
- the gas in the pipeline provides a storage component within the electricity-supply system
- hydrogen is the ideal fuel for use in fuel cells to regenerate electricity
- hydrogen is oxidized cleanly to water, thereby the cycle is closed and no significant pollutants are formed.

While these attractions are still true in principle, many extraneous events have conspired to delay the widespread introduction of hydrogen-based energy systems, irrespective of whether the hydrogen is to be produced from fossil fuels or from renewables. Chief among these events have been the demise of nuclear energy programs, and the discovery of far more petroleum and natural gas than was foreseen in the 1970s (see Chapter 1). Technological advances in conventional power production have also played a role, such as modern combined-cycle gas/steam turbines that are over 50% efficient in electricity production from natural gas (see Section 3.3, Chapter 3). Other factors that bear indirectly on the situation are the widespread privatization of electricity utilities and the general imperative of short-term profitability that now pervades industry (see Section 3.9, Chapter 3). Quite simply, there is as yet no economic case for hydrogen energy and the time-scale on which it may become affordable is uncertain. Despite all these difficulties, however, interest in hydrogen energy is re-awakening, especially in the USA where there is acute appreciation of the forward supply position for oil and natural gas. Many industrial companies, as well as government and academia, are becoming involved.

Several automotive companies are developing vehicles powered by fuel cells (see Section 10.5, Chapter 10). There is also considerable interest in fuel cells for the distributed generation of electricity. In fact, there is greater activity in the advancement of fuel cells than there are of electrolyzers or hydrogen stores, which are the two other key components of the hydrogen-from-renewables scenario. Since fuel cells and electrolyzers are essentially complementary, much of the technology for fuel cells is likely to be applicable also to electrolyzers.

From this discussion, it seems that the short-term prospects for electrolytic hydrogen on a global scale are poor. Nevertheless, it is necessary to look ahead beyond a decade or two, when many of the premium fossil fuels are expected to be in short supply and expensive, and when there may be even greater concern over greenhouse gases and associated global warming. At that point, electricity generation from renewable energy sources should become more generally desirable and competitive, and the requirement for energy storage on a larger scale will emerge. Electrolytic hydrogen must then be considered as one option. In the meantime, circumstances in different countries and in different locations are sufficiently diverse that, in some regions, electricity from renewable sources is becoming the preferred technology. For these reasons, it is expedient to continue the development of hydrogen energy, which should include research aimed at increasing the efficiency of hydrogen production from water electrolysis, improving methods for hydrogen storage, and perfecting electricity generation from hydrogen in fuel cells – and all at reduced cost.

8.1 Hydrogen Production

The theoretical ('reversible') voltage required to split water at 25 °C (298.15 K) and 101.325 kPa (1 atm) by electrolysis is 1.229 V; see Box 8.1 for further explanation. Hydrogen is produced at the negative electrode and oxygen at the positive electrode. The convention used to distinguish the two types of electrode polarity in electrochemical cells is discussed in Box 8.2 (Figure 8.2). The voltage for water decomposition decreases almost linearly to 1.088 V at 200 °C (473 K) (Figure 8.3). The decrease in ΔG with increasing temperature is largely offset by an increase in the entropy term $T\Delta S$, so that the enthalpy of the reaction, ΔH, is almost independent of temperature. Since electrolyzers are essentially adiabatic (*i.e.* little heat is absorbed from the surroundings), the energy corresponding to the entropy term is also supplied electrically. At 25 °C, this increases the minimum cell voltage for water electrolysis to 1.47 V, and the electrical energy consumed in the reversible reaction is almost temperature-independent. In other words, the voltage is 'thermo-neutral' (upper line in Figure 8.3). Many electrolyzers operate at enhanced pressure in order to increase the production rate of hydrogen from a given size of unit, and to avoid the necessity of employing compressors subsequent to electrolysis (without compression, the storage volumes would be enormous). Raising the operating pressure from, for example, 0.1 to 2.5 MPa (1–25 atm) leads to a further increase of about 0.7 V in the reversible voltage. This corresponds to the free energy required to compress the product gases.

In practice, the operating cell voltage exceeds the theoretical value by an amount that represents the electrical losses in the cell. These arise from four different sources: (i) resistive ('ohmic') losses in the electrolyte; (ii) overpotential at the oxygen electrode; (iii) overpotential at the hydrogen electrode; (iv) resistive losses in the electrodes.[*] The voltage–current characteristics of a cell operating at 90 °C are shown in Figure 8.4, together with the various components that give rise to the extra voltage required above and beyond the thermodynamically reversible value. It is seen that the cell voltage increases sharply with current density, predominantly because of resistive losses in the electrolyte. For this reason, it is important to choose an electrolyte of maximum conductivity. In most cases, a concentrated aqueous solution of potassium hydroxide (30–40 wt%) is employed. This must be prepared from very pure water, otherwise impurities will accumulate during electrolysis; chloride ion, which is a common impurity in water, is particularly harmful in that it causes pitting of the protective films formed on metal surfaces in alkaline solutions.

[*] For an electrode reaction to proceed at an appreciable rate, the potential of the electrode must be changed in such a direction as will sustain the flow of current. The greater the current, the more is the change in potential, and the greater is the 'irreversibility' of the electrode process. The degree of irreversibility is measured by the departure of the electrode potential from the reversible value (no current flowing) under the same experimental conditions. An irreversible electrode is said to exhibit 'overpotential'.

Box 8.1   Thermodynamics of Equilibrium and Spontaneous Processes

By providing electrical energy from an outside source, water can be decomposed ('electrolyzed') into molecules of hydrogen and oxygen. Under thermodynamically reversible conditions, *i.e.* maximum efficiency, the expenditure of electrical energy results in the performance of maximum work, that is, in the maximum amount of water decomposed. Under conditions of constant temperature and constant pressure, this maximum amount of work equates to the change in the so-called 'Gibbs free energy' of the system, ΔG. In thermodynamic theory, ΔG is defined as:

$$\Delta G = \Delta H - T\Delta S \qquad (8.1)$$

where ΔH is the change in heat content, or 'enthalpy', of the system and is calculated by adding up the amount of energy released or used to form or break chemical bonds during the reaction; T is the temperature; ΔS is the change in the amount of disorder, or 'entropy', that occurs in molecules involved during the reaction. Equation (8.1) may be expressed in words as: 'the change in free energy of a system that is acting reversibly at constant temperature and constant pressure is equal to the change in heat content of the system less the heat absorbed by the system during the reversible change'. This is the Second Law of Thermodynamics.

The change in free energy is a useful indicator of the spontaneity of electrochemical (or chemical) reactions: (i) if ΔG is positive, the process is not spontaneous, but can be driven by application of sufficient energy from the surroundings; (ii) if ΔG is negative, the process can take place spontaneously and do work on the surroundings; (iii) if ΔG is zero, the system is at equilibrium since the amounts of work being done on the process and by the process itself are equal. Clearly, category (i) applies to the electrolysis of water, and category (ii) to the reverse process that takes place in a fuel cell.

The change in Gibbs free energy may be expressed in terms of the electrical voltage of a reversible galvanic cell, V_r, *i.e.*

$$\Delta G = -nFV_r \qquad (8.2)$$

where n is the number of electrons required in either electrode during completion of a molar quantity of the overall reaction; F is the Faraday constant (the charge on one mole of electrons, *i.e.* 9.6484×10^4 C). Under the arbitrarily chosen 'standard' conditions of 25 °C (298.15 K) and 101.325 kPa (1 atm), the thermodymanic decomposition voltage of water ($V°$) is 1.229 V.

Box 8.2   Terminology Used in the Operation of Electrochemical Cells

During the operation of an electrolysis cell, *i.e.* a cell driven by the application of an external voltage, the positive electrode sustains an oxidation (or 'anodic') reaction with the liberation of electrons, while a reduction (or 'cathodic') reaction takes place at the negative electrode with the uptake of electrons (Figure 8.2(a)). For this reason, the positive electrode is often known as 'anode' and the negative electrode as the 'cathode'. The internal circuit between the two electrodes is provided by the electrolyte, in which negative ions ('anions') move towards the positive electrode, and positive ions ('cations') move towards the negative electrode.

A fuel cell operates in the reverse manner to an electrolysis cell, *i.e.* it is a 'galvanic' cell that spontaneously produces a voltage. The anode of the electrolysis cell now becomes the cathode and the cathode becomes the anode (Figure 8.2(b)). Nevertheless, the positive electrode remains a positive electrode and the negative electrode remains a negative electrode. The same also holds true for a secondary battery, whether it is being charged (*i.e.* as an electrolysis cell) or being discharged (*i.e.* as a galvanic cell) (Figure 8.2(c)).

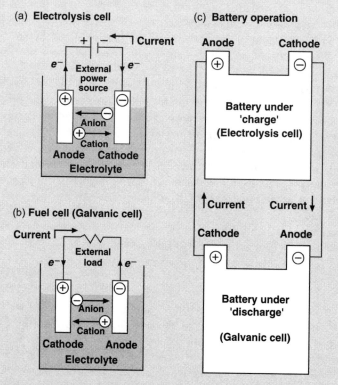

Figure 8.2   *Terminology used in the operation of electrochemical cells.*

Box 8.2 (cont'd)

Thus, to prevent confusion, it is better to avoid the use of the terms anode and cathode altogether and to adhere to 'negative electrode' and 'positive electrode' for the different types of electrochemical technology, which is the approach taken in this book.

The overpotentials at the hydrogen and oxygen electrodes may be reduced by applying coatings of electrocatalysts and by raising the temperature. Alkaline cells with base-metal electrodes (*e.g.* mild-steel negative electrodes and nickel positive electrodes) operate typically at a current density of 200 mA cm^{-2} when 2.1 V is applied (at 70 °C and atmospheric pressure). By using electrodes that are electroplated with suitable catalysts, it is possible to lower this voltage to 1.7 V and thus achieve significant savings in electricity consumption. Further reductions in operating voltage are possible by taking advantage of improved precious-metal electrocatalysts – based on platinum-group metals – that have been developed for use in fuel cells (see Section 8.3). Nevertheless, in order to reduce the capital cost of the electrolyzer plant, it is necessary to operate at as high a current density as possible, often greater than 500 mA cm^{-2}. This results in lower electrical efficiency and a trade-off must be made between capital costs and operating (electricity) costs.

Early electrolysis cells were about 60–75% efficient, but the small-scale, best-practice figure is now closer to 80–85%. Larger units are usually a little less efficient at 70–75%. It should be noted that when using electricity generated from thermal power stations, the overall efficiency is only about 25–30% for the

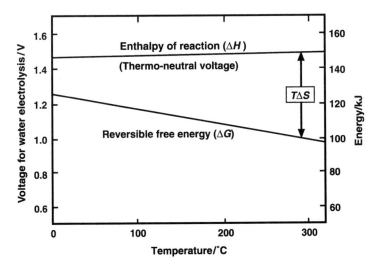

Figure 8.3 *Theoretical voltage versus temperature relationship for water electrolysis.*

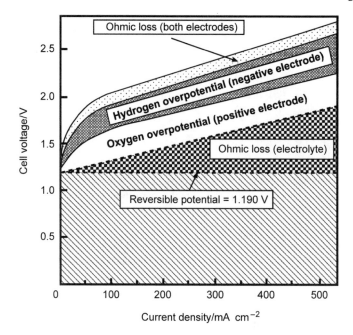

Figure 8.4 *Performance (voltage–current density) of a basic (unactivated), unipolar electrolyzer operating at 90 °C.*

conversion of fossil fuel to hydrogen. This is not a very attractive return on the invested energy.

Electrolyzers, like batteries, may be constructed in either monopolar or bipolar designs. The monopolar (or 'tank-type') electrolyzer (Figure 8.5(a)) consists of alternative positive and negative electrodes held apart by porous separators. The positives are all coupled together in parallel, as are the negatives, and the whole assembly is immersed in a single electrolyte bath ('tank') to form a unit cell. Cells based on aqueous potassium hydroxide are often of this type. A plant-scale electrolyzer is then built up by connecting these units in series electrically (this is analogous to the construction of a typical automotive lead–acid battery). Bipolar electrolyzers and batteries, on the other hand, have a metal sheet (or 'bipole') that connects adjacent cells in series electrically. As seen in Figure 8.5(b), the electrocatalyst for the negative electrode is coated on one face of the bipole, and that for the positive electrode of the adjacent cell is on the reverse face. A series-connected stack of such cells forms a module that operates at a higher voltage and lower current than the tank-type design. To meet the requirements of a large electrolysis plant, these modules are connected in parallel so as to increase the current. Bipolar cells may employ either liquid electrolyte or a solid polymer membrane.

Whenever hydrogen is formed by electrolysis, there is also by-product oxygen in the ratio, 1 mol of oxygen to 2 mol of hydrogen. If electrolysis were to be carried out on a very large scale, it may prove difficult to find a market for the oxygen.

(a) **Monopolar**

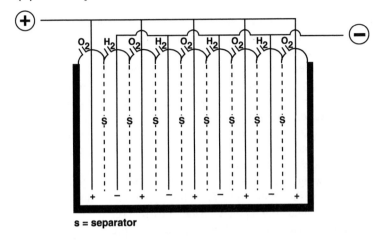

s = separator

(b) **Bipolar**

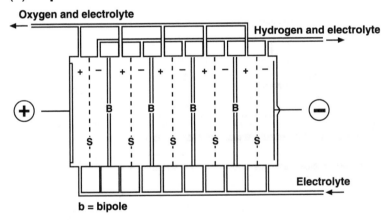

b = bipole

Figure 8.5   *General arrangement of* (a) *monopolar and* (b) *bipolar electrolyzer modules.* (Courtesy of Imperial College Press)

Substantial amounts of oxygen are used in steel making, but this supply is already well established. Other possible major uses of oxygen are in the partial oxidation of hydrocarbons to form syngas (see Section 2.1, Chapter 2), or in the underground gasification of coal. For these applications, oxygen is much preferable to air as the resulting gas is of higher calorific value since it is not diluted by nitrogen.

A concentrated solution of potassium hydroxide at temperatures approaching 100 °C is not an easy material to handle, especially in high-pressure systems. It poses some difficult materials-science problems, which arise from stress-corrosion cracking of steels, degradation of gaskets and sealants, *etc*. In order to avoid these problems, many years ago, in the USA, the General Electric Company introduced a new concept for an electrolyzer based upon the use of a solid polymer electrolyte

(SPE), Nafion®, that was manufactured by the Du Pont Corporation. This is a perfluorsulfonic acid polymer that exhibits good cation-exchange properties and is highly conductive for cations, which include hydrated protons. It is acidic and is available as thin sheets to form a membrane electrolyte. Appropriate electro-catalysts are coated on each side of the polymer membrane and a metal mesh contacts these in order to complete the electrical circuit. Electrolyzers based on this technology are very successful, although there is a cost problem as the solid polymer membranes are expensive. Some applications for these electrolyzers, however, are relatively insensitive to cost compared with other considerations such as reliability and safety. Thus, they are used to generate oxygen for life-support in manned satellites and space probes, and also in nuclear submarines that remain submerged for many days at a time.

A detailed drawing of a design for a bipolar proton-exchange membrane (PEM) electrolyzer stack is shown in Figure 8.6. This type of unit is also known as an SPE electrolyzer. The conducting bipole is made of carbon (graphite) and is corrugated, or ribbed, to form channels through which the gases produced during electrolysis escape from the cell stack. A thin sheet of titanium is used to protect the carbon

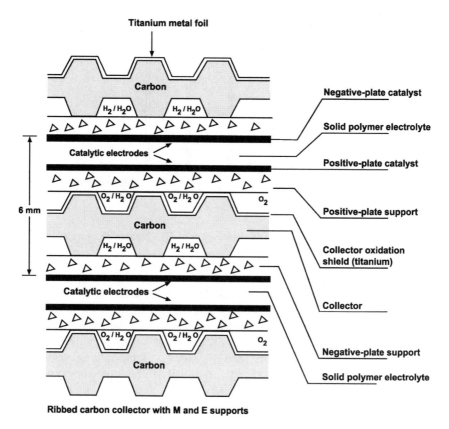

Figure 8.6 *Cell stack configuration of a PEM electrolyzer.*

positive electrode from oxidation. The polymer electrolyte is in the form of a thin, plastic membrane. Sophisticated electrocatalysts have been developed for use in proton-exchange membrane fuel cells (PEMFCs, see Section 8.3) and should be equally applicable to PEM electrolyzers. New techniques for applying the electrocatalysts to the membranes of PEMFCs have also been formulated. These advances have resulted in improved performance at reduced loadings, which is especially important when using expensive electrocatalysts.* In general, PEM electrolyzers are best suited to small plants, especially those powered by renewable energy (*e.g.* solar, wind) where the electricity supply is variable, while alkaline electrolyzers are more practical in larger systems that are connected to the grid.

Drawing also on fuel-cell research, there is the possibility of developing water electrolyzers that are incorporated in furnaces and operate at high temperatures (700–1000 °C) – so-called 'steam electrolyzers'. They use a ceramic ion-conducting electrolyte of the type that has been developed for high-temperature (HT) fuel cells (see Section 8.3). Moreover, as shown in Figure 8.3, HT operation favours electrolysis (*i.e.* improves efficiency) since some of the energy required to split water is supplied by heat. This substantially reduces the electricity demand (at least 25% of the electricity can be saved) but material requirements are more severe. Of course, corresponding amounts of HT heat must be supplied, which makes water electrolysis a potential new application for a dedicated co-generation unit (*i.e.* combined heat and power, CHP) or even, on a huge scale, a high temperature nuclear reactor (see Sections 3.4 and 3.5, Chapter 3).

Commercial water electrolyzers are today manufactured in varying sizes, with power consumptions in the range of kW–MW. The larger sizes are limited to alkaline electrolyzers, and a unit capable of producing 500 m^3 of hydrogen per hour might consume about 2.3 MW of power. Large though this is, it is less than 1/400th of the size that would be suitable for coupling to a typical 1000-MW pressurized water reactor (PWR, see Section 3.5, Chapter 3). By contrast, solar arrays with power outputs in the kW–MW range could readily be coupled to present-day water electrolyzers.

Large-scale water electrolysis plants have long been installed in certain countries (*e.g.* Norway, Canada) to make use of the surplus hydroelectric capacity. The hydrogen generated was produced for the manufacture of ammonia-based fertilizers rather than for electricity storage. With the advent of cheap natural gas as a source of hydrogen, this process has become much less significant. In 2003, a filling station for hydrogen-powered vehicles opened in Iceland. The hydrogen is produced by electrolysis using electricity generated from hydropower and geothermal springs. Over 90% of Iceland's electricity is derived from these two sources.

---

* Consideration has been given to operating fuel cells in reverse as electrolyzers. The dual-function system has been termed a 'regenerative fuel cell'. Such technology would save on weight and costs compared with a system that employs a separate fuel cell and electrolyzer. It would also offer the prospect of using renewable energy (*e.g.* solar, wind, geothermal) to generate hydrogen that would be stored in the same unit for subsequent production of electricity. Although the feasibility of the regenerative concept was demonstrated at the 50-W level in the 1990s, reliability problems with bifunctional electrodes and efficiency limitations of the 'charge–discharge' process (ways have to be found to optimize the catalyst for both processes) have resulted in little further progress.

Looking to the future, there are possible situations in which it might prove economic to generate bulk hydrogen electrolytically, for example:

- in remote regions of the world where the supply of renewable energy exceeds the local demand for electricity
- as a means of load-leveling in the electricity utility network
- using nuclear power plant dedicated to hydrogen production.

Although electrolytic hydrogen would be used first in the chemicals market, this need not continue to be the case. For example, a country that has a high installed capacity of nuclear power, but is short of indigenous hydrocarbon fuels (*e.g.* France), might find it economic to use nuclear electricity to generate electrolytic hydrogen, which could then be combined with carbon monoxide to manufacture synthetic fuels and chemicals (see Equations 2.7 and 2.8, Chapter 2). This raises the interesting concept of a hydrocarbon or methanol molecule that is part fossil and part nuclear in its origins! At the moment, this situation is far from being realized as there is plenty of natural gas remaining worldwide to manufacture inexpensive hydrogen (see Table 1.5, Chapter 1).

In principle, there are several other ways of decomposing water to form hydrogen, without generating electricity first. These avoid the need for an electrolyzer. The first route, which has been extensively investigated, is to use thermal energy to decompose water by means of cyclic chemical reactions, so-called 'thermochemical cycles'. These indirect routes offer the potential for producing hydrogen at lower temperatures than that required for the direct thermolysis of water. Without the intermediate production of electricity, overall efficiencies of around 40% are expected, *i.e.* higher than the overall efficiency of electrolysis. Although several cycles have been demonstrated in the laboratory (*e.g.* the calcium bromide–iron oxide, copper-chloride and sulfur–iodine processes), they are still far from practical realization.

The remaining options are specific to the use of solar radiation and here there are at least three possibilities. The first utilizes a solar–thermal power plant to decompose water directly *via* a 'photothermal process'. In such a plant with a central collector, the temperature can reach over 2000 °C and this is sufficient to breakdown water into hydrogen and oxygen. Research is being conducted on the use of catalysts to reduce the temperature for water dissociation, and on the development of methods for the efficient separation of the two gases at high temperatures to prevent their recombination.

The second solar route is through a 'photo-electrochemical process' whereby a suitable semiconductor, such as titanium oxide (TiO_2) or cadmium selenide (CdSe), in the form of a thin layer with electrodes attached, is immersed in an aqueous solution and irradiated with sunlight. When the external circuit is completed, water starts to decompose, liberating hydrogen at one electrode and oxygen at the other. This process is still at the research stage and, as yet, the efficiency is very low, just a few percent.

An intriguing idea has been advanced to employ solar–thermal radiation to facilitate and improve the photo-electrochemical decomposition of water. Much of solar radiation lies in the infrared spectrum and is of too low energy (below the band-gap energy of most semiconductors) to be utilized in photo-electrochemical reactions and is therefore wasted. The idea proposed involves the photo-electrolysis of water at high temperatures and pressures where, as seen from Figure 8.3, the required voltage is lower and the entropy of reaction can be supplied thermally. A field of heliostats would be used to focus solar radiation on a receiver mounted on a tower. The radiation would then be split into an infrared component to heat the pressurized water to at least 300 °C, and into visible/ultraviolet radiation to effect the photo-electrochemical water-splitting reaction. Thermodynamic calculations suggest that at high temperature and high pressure, the required voltage is substantially reduced and it should be possible to reach overall efficiencies approaching 20% for the conversion of solar energy to hydrogen. This approach to water splitting is termed 'thermally-assisted photo-electrochemical decomposition'.

The final approach to the direct solar decomposition of water is to utilize natural bacteria, or biological organisms such as blue algae, that have the propensity to decompose water under the action of sunlight and release hydrogen freely into the air. In theory, algae can produce hydrogen with an efficiency of up to 25%. The problem is that oxygen is also formed and inhibits the hydrogen-producing enzyme, hydrogenase, so only small amounts of hydrogen are actually generated. Research is in progress to moderate the release of oxygen and to increase both the general feasibility and cost efficiency of such photo-biochemical processes.

All of the above solar-based approaches are too speculative, futuristic and complex to discuss in detail in this book, although conceivably they might become important before 2020.

8.2 Hydrogen Distribution and Storage

Gaseous Hydrogen

In principle, hydrogen is an ideal vector for the transmission and storage of energy. One might imagine, in the post-fossil-fuel age, huge solar collectors and electrolysis plants located over large areas of desert. It would then be necessary to convey the hydrogen to market. The most obvious means would appear to be *via* pipeline. Indeed, as long ago as 1938, conventional mild-steel pipelines were installed in the Ruhr district of Germany to convey hydrogen between refineries and chemical works. The initial system was about 24 km in length and fed two plants. Since then, the grid has been expanded to a 210-km network that links four producers of the gas and thirteen users. It even crosses the Rhine in two places. The pipes are wrapped in bitumen and plastic, have diameters of 15–30 cm, and operate at pressures up to 1.6 MPa (16 atm.). Remarkably, no major accidents have arisen from escaping hydrogen or potentially explosive hydrogen–air mixtures. There is also a 170-km system in Northern France and a total of some 1500 km in Europe as a whole. North America has at least 700 km, and other smaller networks are in use in South Korea and Thailand. Clearly, the technology of transmitting pressurized hydrogen safely over considerable distances has been demonstrated and used commercially to supply gas for chemical processes.

Hydrogen as an energy vector is a rather different proposition since the distances involved are likely to be much greater and the allowable costs much less. Moreover, it is by no means certain that in-place networks for natural gas can be adequately and safely used for the distribution of hydrogen, given the tremendous diversity of materials that are employed, *e.g.* iron, steel and plastic for pipes, brass for valves, natural and synthetic rubber for mechanical joint seals and meter diaphragms, lead and jute for sealing compounds, and cast aluminum for meter housings and regulator parts. In particular, problems may arise from hydrogen leakage through gaskets and valves, and from hydrogen attack of certain metals (blistering, embrittlement, decarburization). Hydrogen is compatible with aluminum, brass, low-carbon steels and polymers, but not with high-carbon steels, high-strength alloys and titanium. Plastic pipes (made from polyvinyl chloride or the newer high-density polyethylene) that are employed in some natural-gas lines are too porous and not suitable for transporting hydrogen. Over a long distance, there will also be a need to re-pressurize regularly, perhaps every 100 km or so, and this will represent a further loss of energy. The distant transmission of energy, whether as electricity, natural gas or hydrogen will involve some losses and it has to be assessed whether the losses are too large and whether they are economically acceptable.

The low volumetric energy density of hydrogen means that, to deliver a given amount of energy, the flow rate through the pipeline must be much faster than that for natural gas, *i.e.* by a factor of 2.8. As a consequence, and despite the lower viscosity of hydrogen, flow resistance is greatly increased so that, for a given delivery rate, more energy is required to move hydrogen relative to natural gas. Nevertheless, at standard and safe pipeline pressures and provided that the above-mentioned problems can be overcome, it is considered that existing networks could deliver as much as 85% of the energy transported nowadays by natural gas to the end user. It should be further noted that hydrogen gives rise to greater problems than natural gas when new lines have to be installed or when old lines have to be repaired. Also, because hydrogen burns in such a wide range of air mixture ratios, new installations have to be purged with an inert gas. Although the basic feasibility of distributing hydrogen *via* extensive pipelines is not in dispute, forecasts of the implementation costs vary widely according to the method used, *i.e.* between 1.5 and 3 times that for natural gas. This uncertainty is not surprising given the profuse number of variables that come into play. These include: choice of materials; pipeline diameters and pressures; type of valves, compressors, sensors and safety devices; spacing and fuel costs of compression stations; embrittlement; the geographical locations of sources of hydrogen.

The next problem is how to store gaseous hydrogen in bulk. To some degree, the pipelines themselves would provide a storage component – just as they do for natural-gas grids in many countries. For many years, natural gas has been seasonally held underground in depleted oil and gas fields on a huge scale (especially in the USA) and this is a low-cost storage option. The ability to store gas underground depends critically on the nature of the rock strata. Porous permeable rock is required, while sealing of the system is accomplished by the capillary action of water in the 'cap-rock' above the reservoir. In addition to using depleted gas fields, it is also possible to contain the gas in porous aquifers simply by displacing water and creating an artificial storage volume, always providing that there is impermeable cap-rock present to prevent gas escape. In France, for example, a country that lacks oil or gas fields, a number of aquifer stores have been utilized for natural gas, which includes one of 200 000 m^3 capacity outside of Paris.

The question then is: could such storage means be used equally well for hydrogen? The principal difference between hydrogen and natural gas lies in the smaller size of the hydrogen molecule and its higher diffusion coefficient, both factors that would tend to facilitate escape from the store. Fortunately, the size of the pores in cap-rock is sufficiently small that the water is displaced only with difficulty and, provided that gas pressures are not excessive, the cap-rock will serve to retain the hydrogen. Indeed, the higher diffusion coefficient of hydrogen should make the reservoir easier to fill and empty than with natural gas. Other options for underground storage are abandoned mines, natural limestone caves, or man-made cavities in underground salt deposits. The salt cavities are made by drilling down into the deposit and injecting water to dissolve out the salt. Since salt is itself impermeable, it forms an excellent container for the gas. Nevertheless, the excavation costs make this option less attractive than the use of existing man-made or natural cavities. In the final analysis, the form of underground storage, if any, that will be employed is determined by the geology. One operational problem with underground storage is that a fairly large amount of 'cushion' gas has to be written-off initially in order to build up and maintain the underground pressure so that the recoverable part can be extracted later. Rock caves allow a 25% turnover in capacity per storage cycle, and wet salt cavities may approach 100% turnover. Despite these considerations, there have been some commercial examples of underground hydrogen storage. For example, ICI has stored 95%-pure hydrogen at 50 MPa in salt caverns at Teeside, UK, for use by industrial customers, and Gaz de France has stored town gas containing 50% hydrogen in an aquifer.

On a much smaller scale, compressed hydrogen may be stored in pressure vessels. This requires the use of specialized compressors. Industrial consumers of hydrogen sometimes use above-ground storage in vertical rows or horizontal stacks of cylinders at pressures in the range of 20 MPa up to a maximum of 80 MPa. Such high pressures necessitate thick-walled and heavy steel containers. In the past, failures have been experienced through stress induced by hydrogen embrittlement and, consequently, cylinder manufacture is now subject to strict standards and codes of practice. Tank storage is modular with little economy of scale. For portable and mobile applications (*e.g.* as a fuel for road vehicles), both weight and volume are vital considerations and it is desirable to have storage densities as high as possible. Conventional steel cylinders are limited in terms of storage energy density. Improvements may be effected by using modern materials in place of steel. This implies high pressures and lightweight vessels. At present, the best lightweight vessels are composed of wound carbon-fiber shells with aluminum liners. The most common design – due to its ease of fabrication and, therefore, lower cost – is the hoop-wound design, in which the composite wrapping is only applied to the liner along the parallel section (vertical walls) of the cylinder. In the fully wound design, the composite covers the entire external surface of the cylinder. The liner acts as a gas-impervious barrier, which for hoop-wound cylinders also provides considerable strengthening. A number of polymeric materials, *e.g.* ultrahigh-molecular-weight polyethylene, have also been examined as possible liners, but permeation/leakage of hydrogen raises safety issues. Both designs of vessel will withstand an internal pressure of 55 MPa and provide a theoretical energy density of 6.9 MJ dm^{-3}, which reduces to a practical value of 3–4 MJ dm^{-3} when account is taken of the volume of the cylinders and their packing. Improvements in technology are required to meet the targets of 5.4 MJ dm^{-3} by 2010 and 9.7 MJ dm^{-3} by 2015 that have been set by the US FreedomCAR Partnership (see Section 10.5, Chapter 10) for hydrogen stores on electric vehicles (EVs) powered by fuel cells. The corresponding US gravimetric targets of 7.2 and 10.8 MJ kg^{-1} are still more difficult to reach.

Storage vessels made from high-tensile steel are not suitable for most portable or transportation applications, but can store hydrogen at up to 80 MPa. Even at this maximum pressure, which necessitates special cylinder manufacture, the theoretical value for the volumetric energy storage (10 MJ dm^{-3}) does not start to compare with that of methane at the same pressure (32 MJ dm^{-3}), or with liquid octane (34 MJ dm^{-3}) or liquid propane (25 MJ dm^{-3}). Clearly, compressed hydrogen gas poses a major storage problem for road vehicle applications in terms of both weight and volume. If the above-mentioned lightweight canisters made from carbon-fiber composite are employed, then the cylinder volume required for a given vehicle range would be about five times greater than that of a petrol tank holding a corresponding amount of energy; an example of hydrogen storage in composite cylinders on-board a car is shown in Figure 8.7(a).

A further problem is the amount of electrical energy needed to operate the compressors. (Note, the compression process could be reduced, or eliminated altogether, if the hydrogen was supplied from pressurized electrolyzers and this is a prospect for the future.) It has been estimated that the electricity required to compress hydrogen to 20 MPa is equivalent to over 7% of the energy content of the hydrogen. This figure rises to 10% for compression to 80 MPa. Added to this, is the fact that cylinders do not pack as well as a single container and it is clear that the proposition of using compressed hydrogen gas as a fuel for most small vehicles will result in restricted driving ranges. For example, the Honda 'FCX' and Toyota 'FCHV' fuel-cell powered cars are equipped with high-pressure tanks and have ranges of only 270 and 288 km, respectively. A more-promising application of this storage technology is in fuel-cell buses, such as the Toyota 'FCHV-Bus 2', as these vehicles operate on fixed routes. Cars and buses powered by fuel cells are discussed in Section 10.5, Chapter 10. It should be noted, however, that the refilling time for compressed hydrogen is similar to that for petrol. A station for the rapid refueling of hydrogen-powered vehicles has been installed at the Powertech Labs in British Columbia, Canada (Figure 8.7(b)). The gas is stored underground at 87.5 MPa and dispensed at 70 MPa.

Liquid Hydrogen

Liquid hydrogen (LH$_2$) is a cryogenic liquid that boils at 20 K. Superficially, this is an attractive way to store hydrogen on account of its compactness compared with

(a)

(b)

Figure 8.7   (a) *Fully wound hydrogen cylinders in the boot of a car.*
(Courtesy of MRS Bulletin)
(b) *fast-fill hydrogen refueling station.*
(Courtesy of the British Oxygen Company)

gaseous hydrogen; the relative density factor is about 850. There are, however, some serious drawbacks. First, the liquefaction process itself requires the expenditure of a considerable amount of energy. The theoretical amount of energy required is around 4 MJ of electricity per kg of hydrogen, although in practice many times this amount is used and depends on the scale of operation. For large liquefaction plants (1000–10 000 kg h^{-1}), the energy input is equivalent to around 30% of the energy value of the hydrogen itself, and rises to equal or exceed this figure in small-scale operations. Next, the cryogenic equipment to liquefy and contain the hydrogen is sophisticated and costly. Finally, even with good insulation, such as multi-layer vacuum super-insulation, the boil-off rate is such that LH$_2$ in kilogram quantities can only be stored for a few days at most. Clearly, liquid hydrogen is not a practical proposition as a fuel for most road vehicles, even before the efficiencies and costs of renewable electricity and electrolyzers are taken into account. Despite this, some automotive manufacturers have persisted with trials of LH_2-fueled vehicles (see Section 2.5, Chapter 2).

Liquid hydrogen has, however, been produced in substantial quantities for use in nuclear bubble chambers (for observing the paths of charged particles), space rockets, *etc.* In the 1970s, NASA required a continuous supply of LH_2 as rocket fuel for the Apollo program. Spherical, vacuum-jacketed containers were built to hold over three million liters of the cryogenic liquid (Figure 8.8(a)). It was also shipped across the USA in cryogenic rail cars. Nowadays, LH_2 is also transported by road in tank trucks (Figure 8.8(b)) and by sea in tank ships. The latter are similar to LNG tankers (Figure 2.2, Chapter 2) apart from the fact that better insulation is required to keep the hydrogen cool over long distances. Hydrogen that is unavoidably evaporated may be used as fuel on board. Although the technology for handling LH_2 in bulk does exist, the limitations outlined above make it unlikely that LH_2 will be employed in situations where other storage forms for hydrogen will suffice. The possible use of liquid hydrogen as a fuel for aircraft has been discussed in Section 2.5, Chapter 2.

Metal Hydrides

An alternative storage procedure, which is much better matched to the likely scale of solar energy installations or hydrogen-fueled road vehicles, is to store hydrogen in the solid state as a metal hydride. A number of metals and alloys absorb hydrogen reversibly to form metal hydrides. The hydrides are classified into five families denoted as A, A_2B, AB, AB_2 and AB_5, where metal A is an early transition metal, (*e.g.* titanium, vanadium), a rare-earth metal or magnesium, and B is aluminum, chromium, cobalt, iron, nickel or manganese. The key to the practical use of metal hydrides is the ability to absorb and release the hydrogen many times without deterioration. The hydrogen is first dissociatively adsorbed on the surface and then hydrogen atoms diffuse into the metal/alloy. These dissolved atoms can take the form of a random solid solution or an ordered hydride structure, both with a high volumetric packing density. The quantity of hydrogen absorbed is expressed in terms of hydride composition, either on a molar or a weight-percent basis. Volumetrically, the hydrogen content may be as high as that in liquid hydrogen. Some of the metal hydrides are of quite variable composition (*i.e.* variable metal-to-hydrogen ratio), whereas others have only a narrow range of composition.

The absorption process is generally exothermic. Thus, in order to absorb hydrogen continuously to the maximum capacity, heat must be removed. The direction of the hydrogen absorption–desorption process is determined by the pressure of the hydrogen gas. If the pressure is above the equilibrium value, then the hydride will be formed. Conversely, below the equilibrium pressure, hydrogen is released and the metal/alloy returns to its original state. The equilibrium pressure, itself, depends on temperature; it increases with increasing temperature and *vice versa*. This relationship is expressed by the van't Hoff equation, *i.e.*

$$\ln P = \Delta H / RT - \Delta S / R \qquad (8.3)$$

(a)

(b)

Figure 8.8   (a) *LH₂ container used in the NASA space program.*
(Courtesy of NASA)
(b) *road transportation of LH₂.*
(Courtesy of MRS Bulletin)

where P is the dissociation pressure; ΔH is the change in enthalpy; R is the gas
constant; T is the absolute temperature; and ΔS is the change in entropy (see Box 8.1).
The value of ΔH can vary widely from metal/alloy to metal/alloy. By contrast,
ΔS does not vary as much. For a given metal hydride, the van't Hoff plot of ln P
versus $1/T$ is a straight line and provides a useful illustration of the hydrogen

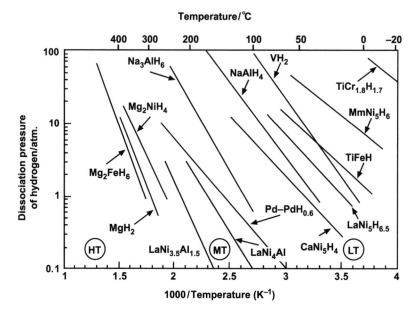

Figure 8.9   *Dissociation pressure of various metal hydrides.*

absorption–desorption characteristics. The dissociation pressure curves for a number of metal hydrides are presented in Figure 8.9. These plots show that hydrides can be broadly classified into three categories, namely: HT, medium-temperature (MT) and low-temperature (LT) hydrides. The LT category is the most useful in that such hydrides offer hydrogen storage closest to ambient temperature and are therefore convenient for supplying fuel cells. In this case, the required enthalpy of hydrogen desorption would be provided by the waste heat from the fuel cell.

A great many alloys have been screened for hydrogen storage with respect to the following criteria: (i) reversible hydrogen capacity; (ii) ease of initial activation of the alloy; (iii) operating pressure–temperature range (van't Hoff plots); (iv) reaction kinetics; (v) stability on repeated cycling of hydrogen; and (vi) cost. One of the first metal hydrides to be studied in detail was $LaNi_5$, which takes up hydrogen reversibly to form $LaNi_5H_{6.5}$. Varying amounts of zirconium or titanium may be substituted on the lanthanum lattice sites and aluminum, chromium, cobalt, manganese, vanadium on the nickel lattice sites. The hydriding reaction is reversible at ambient temperature and the enthalpy of reaction is low (31 kJ per mole H_2), which implies that not too much heat has to be supplied and removed. The alloy may be subject to many hundreds of hydriding–dehydriding cycles without loss of capacity. Cycling is accompanied by a corresponding major expansion and contraction in volume that causes the alloy to disintegrate into a fine powder and, thereby, enhances the subsequent reaction kinetics. Unfortunately, lanthanum is an expensive metal and the hydride contains only 1.5 wt% hydrogen.

Magnesium hydride, MgH_2 (an *A* alloy), contains the highest percentage of hydrogen by weight (7.65 wt%) and is inexpensive. Nevertheless, it has drawbacks.

A temperature above 300 °C is required for the release of hydrogen (Figure 8.9) and the enthalpy of dissociation is high (75 kJ per mole H_2), which necessitates the supply and removal of considerable quantities of heat as the hydride is decomposed and reformed, respectively. Moreover, the kinetics of the initial hydriding–dehydriding reaction are very slow unless an over-pressure many times the equilibrium dissociation pressure is employed. This limitation can be circumvented, at the expense of reduced gravimetric hydrogen content, by using the type A_2B alloy Mg_2Ni to form the hydride Mg_2NiH_4, which contains 3.6 wt% hydrogen. In addition, Mg_2Ni is readily activated, has good reaction kinetics, and cycles reversibly at a rather lower temperature than MgH_2.

Vanadium hydride, VH_2, has a favorable dissociation pressure at near-ambient temperatures, but only one hydrogen atom may be removed reversibly, and this results in a hydrogen storage capacity of just 1.9 wt%. The enthalpy of decomposition is moderate (40 kJ per mole H_2). The titanium–iron alloy, TiFe (type AB), forms a hydride $TiFeH_{1.5}$ that is also usable at ambient temperature and has a low enthalpy of decomposition (28 kJ per mole H_2), but its reversible hydrogen content is only 1.5 wt%.

From the above studies, the following characteristics have been identified as being desirable for alloys to serve as practical hydrogen-storage media.

- the dissociation pressure of the alloy should have a value of 0.1–1 MPa at near-ambient temperature
- the hydride should have a high hydrogen content per unit mass
- the alloy should be of low cost and should be readily prepared
- the system should exhibit favorable and reproducible reaction kinetics
- the enthalpy of hydride formation should be as low as possible
- the bed of reactant should have a high thermal conductivity
- the alloy should not be poisoned by gaseous impurities
- the system should be safe on exposure to air and should not ignite.

No one alloy meets all these specifications and the choice of alloy will be a compromise based upon the intended application. $LaNi_5$ is a good choice for many applications, but it is expensive. Research on alloys for use in nickel–metal-hydride batteries (see Section 9.2, Chapter 9) has led to new hydrides based on mixed rare-earth metals ('misch metal', Mm) that are both superior in properties and cheaper than pure $LaNi_5$, *e.g.* $MmNi_5H_6$ (Figure 8.9).

In order to develop a successful hydride bed on a large scale, it is necessary to investigate not only the basic chemistry of the hydrides themselves, but also the chemical engineering of such beds. During the 1970s, a major program of work was carried out at Brookhaven National Laboratory in the USA to construct and test a large hydride bed that would be suitable to store the hydrogen required to power a hydrogen-fueled road vehicle. The pilot bed contained 400 kg of FeTi alloy and was capable of storing 5.5 kg of hydrogen. A study was made of heat and mass transfer effects in a bed of this size. It was demonstrated that the bed could be charged with hydrogen in 5 h, with an associated temperature rise of 37 °C. A charging pressure of around 3.5 MPa had to be applied. The studies demonstrated the feasibility of scaling-up hydride beds to realistic sizes for use with industrial-scale electrolyzers, although by no means all the engineering problems were solved. In the late 1970s, the Daimler Benz motor company converted a number of Mercedes-Benz vehicles (cars, vans and a minibus) to hydrogen operation and equipped them with hydride beds. In the case of the minibus, a TiFe bed (ambient temperature) was used for start-up and a Mg_2NiH_4 hydride bed (300 °C) for supplying hydrogen when hot. Eventually, this program was abandoned as not being economic in the prevailing circumstances. Further scale-up work must await both outcomes from basic scientific studies on newly discovered hydrides and changing economics of fuel supply.

Chemical and Related Storage

Organic liquids, such as cyclohexane or methanol, can serve as chemical carriers of hydrogen. The gas is subsequently recovered by catalytic decomposition. Methanol (CH_3OH) is usually manufactured from synthesis gas by the catalytic reaction of two molecules of hydrogen with one of carbon monoxide (see Section 2.4, Chapter 2). It is an extremely versatile chemical in its own right and is used to a limited extent as a motor fuel, *e.g.* in certain classes of racing cars. Since established economic processes for the production of methanol exist, it is unlikely that it would be manufactured from electrolytic hydrogen and then decomposed back to hydrogen for use in a fuel cell. The overall energy efficiency of such a cycle would be very poor. Methanol derived from fossil fuel is, however, a prime candidate for fuel cells in road transportation and portable applications (see Section 8.3).

Hydrogen may also be stored chemically in the form of the soluble ionic salts $Na^+[AH_x]^-$, where A represents boron or aluminum; such compounds are generally known as 'complex hydrides'. Sodium and lithium borohydrides are well-known reducing agents that are used in organic chemistry. For hydrogen storage, the aluminum salts $Na[AlH_4]$ and $Na_3[AlH_6]$ (the so-called 'alanates') are the preferred reagents. Thermal decomposition of $Na[AlH_4]$ takes place in two steps, *i.e.*

$$3Na[AlH_4] \longrightarrow Na_3[AlH_6] + 2Al + 3H_2 \tag{8.4a}$$

$$Na_3[AlH_6] \longrightarrow 3NaH + Al + \tfrac{3}{2}H_2 \tag{8.4b}$$

The reactions are reversible only at elevated temperatures and pressures (Figure 8.9). The first step at 50–100 °C, corresponds to the release of 3.7 wt% hydrogen and the second step, at 130–180 °C, to a further 1.9 wt% hydrogen. Research has shown that, in the presence of a titanium catalyst, the temperatures for discharge and recharge of hydrogen may be brought down to acceptable levels. Titanium-catalyzed $Na[AlH_4]$ has thermodynamic properties that are comparable with those of classic LT hydrides (*e.g.* $LaNi_5H_6$ and TiFeH, Figure 8.9). Moreover, even if only the first reaction step can be utilized, the gravimetric hydrogen storage of $Na[AlH_4]$ is still more than that offered by AB, AB_2 or AB_5 hydrides. By contrast, $Na_3[AlH_6]$ requires higher temperatures for hydrogen liberation and might be useful for non-fuel applications such as heat pumping and heat storage. There are also complex hydrides based on transition metals, *e.g.* Mg_2FeH_6 (Figure 8.9). In most cases, they are reversible only with difficulty. The possibility of overcoming this limitation through catalysis awaits further research.

Sodium borohydride, $NaBH_4$, is stable until about 400 °C and is therefore not suitable for providing hydrogen through a thermal activation process. It does release hydrogen, however, on reaction with water:

$$NaBH_4 + 2H_2O \longrightarrow NaBO_2 + 4H_2 \qquad (8.5)$$

This is an irreversible reaction, but has the advantage that 50% of the hydrogen comes from the water – in effect, $NaBH_4$ is a 'water-splitting' agent. Based on the mass of $NaBH_4$, the hydrogen released is 21 wt% – a remarkably high figure. Several of these so-called 'chemical hydrides', *e.g.* CaH_2, $LiAlH_4$, LiH, $LiBH_4$, KH, MgH_2, NaH, are being evaluated for their reactivity with water. One approach to preparing the storage medium is to mix the hydride with light mineral oil and a dispersant to form an 'organic slurry'. The oil coats the hydride particles and protects them from inadvertent contact with water, and also moderates the reaction rate of the hydride with water when desired.

The downside of using chemical hydrides is that the spent solution has to be returned to a processing plant for regeneration of the hydride. From the standpoint of mass, volume and cost, however, the system appears superficially to be attractive as a hydrogen-storage scheme for fuel-cell vehicles (FCVs). For example, DaimlerChrysler has demonstrated that a $NaBH_4$ system, developed by Millenium Cell in the USA, can provide a minivan (the 'Natrium') with a range of 480 km. Much will depend on the difficulty and cost of the reprocessing operation. At the service station, instead of refueling with hydrogen gas, the vehicle would have its tank emptied and refilled with a fresh hydride slurry. This is similar to the procedure proposed for the zinc-air traction battery (see Section 9.3, Chapter 9).

Another possible approach to hydrogen storage stems from recent work on materials whose structural elements have dimensions in the nanoscale range. These so-called 'nanostructured' materials have high specific surface-areas ($m^2 \, g^{-1}$), which can be attained either by fabricating small particles or clusters where the surface-to-volume ratio of each particle is high, or by creating materials where the void surface area (pores) is high compared with the amount of bulk support material. Intense interest in such materials commenced in the early 1990s with the discovery of techniques to produce various types of carbon nanostructures. It was further shown that at the nanoscale, materials can exhibit chemical and physical properties that are characteristic of neither the isolated atoms nor of the bulk material. Moreover, there is the opportunity to engineer the structural architecture to yield desired properties. With respect to hydrogen storage, the small size of nanostructured materials influences the thermodynamics and kinetics of hydrogen adsorption and dissociation by increasing the diffusion rate and decreasing the required diffusion length. The materials can be divided into two categories:

(i) 'dissociative' materials, in which the hydrogen molecules are dissociated into atoms that bond with the lattice of the storage medium (chemisorption), *e.g.* metal hydrides, discussed above

(ii) 'non-dissociative' materials that by virtue of their high microporosity and high surface area store hydrogen in the molecular state *via* weak molecular–surface interactions (physisorption), *e.g.* carbon and boron nitride nanostructures, clathrates, metal–organic frameworks.

Clearly, physisorption is more desirable as it would moderate the pressure and temperature required for the respective uptake and release of hydrogen.

It has been shown that carbon, in the form of nanofibres or nanotubes (fullerenes), is capable of holding reasonable quantities of hydrogen. Various types of graphitic nanofibres have been investigated. These are grown by the decomposition of hydrocarbons or carbon monoxide over metal catalysts. The fibre consists of graphene sheets aligned in a set direction (dictated by the choice of catalyst). Three distinct structures may be produced: platelet, ribbon and herringbone, as shown in Figure 8.10(a). The structures are flexible and can expand to accommodate the hydrogen. Carbon nanotubes are cylindrical or toroidal varieties of fullerene (the generic term used to describe a pure carbon molecule that consists of an empty cage of 60 or more carbon atoms), and have lengths of between 10 and 100 μm. Each end is capped with half a spherical fullerene molecule. 'Single-walled' nanotubes are formed by only one graphite layer and have typical inner diameters of 0.7–3 nm. 'Multi-walled' nanotubes consist of multiple, concentric, graphite layers and show diameters of 30–50 nm. Examples of different types of carbon nanotube structure are shown in Figure 8.10(b). Various pretreatments have been suggested to enhance the storage capacity of such materials, which may amount to several wt% hydrogen. There is still considerable controversy over the findings, however, because of the difficulty in preparing homogeneous, well-defined, pure and reproducible samples. Nevertheless, strenuous efforts are being made to develop methods for producing nanotubes economically on a large scale. For the present, hydrogen storage in carbon- and non-carbon-based nanostructured materials is no more than an outside possibility so far as FCVs are concerned, although it could be of interest in small-scale applications.

Hydrogen Storage on Road Vehicles

As seen above, the four principal options for hydrogen storage in transportation applications are: compressed gas, liquid hydrogen, metal hydride, chemical carrier (at present, it is considered unlikely that nanostructured materials can accommodate the required amount of hydrogen). A comparison is given in Table 8.2 of the probable masses and volumes that would be required to accommodate 11.75 kg of hydrogen, equivalent in energy content (1.4 GJ) to the energy contained in 45 L of petrol. The data may be only approximate, but they give some idea of the magnitude of the technical problems that has to be solved. In terms of mass, the only feasible option would appear to be liquid hydrogen, while in terms of volume

(a)

(b)

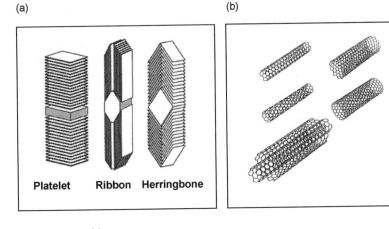

Platelet   Ribbon   Herringbone

(c)

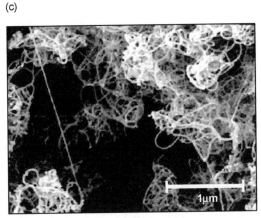

1μm

Figure 8.10   *Schematic representation of* (a) *carbon nanofibres;* (b) *single- and multi-walled carbon nanotubes; and* (c) *electron micrograph of bundles of carbon nanotubes.*

none of the options appear to be practicable, unless a major reduction in vehicle range between refueling stops is acceptable.

It should also be recognized that an efficient method of storage is but one of a list of formidable issues that must be resolved if hydrogen is to be introduced successfully as a fuel for road vehicles. These include:

- inertia to change an existing technology, industry, and way of life
- long time-scale for introducing a new energy industry and the large capital investment required
- difficulties of manufacturing, distributing and storing hydrogen on the megatonne scale in a form suitable for transportation applications
- manufacture of hydrogen at a price that competes with natural or synthetic liquid fuels

Table 8.2 *Approximate comparison of masses and volumes of petrol and hydrogen for equivalent energy content (1.4 GJ)*

	Mass of store and fuel (kg)	Index	Volume of store (L)	Index
Petrol (45 L)	41	1	45	1
Compressed H_2 (16 MPa), conventional steel cylinder	1433	35	1352	30
Compressed H_2 (35 MPa), composite cylinder	360	8.8	370	8.2
LH_2 in cryostat	98	2.4	333	7.4
Ti–Fe hydride bed	1047	25.5	275	6.1
Mg–Ni hydride bed	406	9.9	232	5.2

- technical issues of carrying hydrogen on the vehicle, *e.g.* volume of gas cylinders and their accommodation, mass of metal-hydride beds and their slow refueling
- lingering doubts about the safety aspects of hydrogen, particularly in the context of retail sales to the public at garages
- practical aspects of joining high-pressure gas lines (or, in the case of LH_2, cryogenic lines) to vehicles in a retail environment
- competition from hybrid electric vehicles (HEVs), especially those with a diesel engine.

Given this daunting list of obstacles, it seems unlikely that a supply and distribution network for hydrogen to fuel road vehicles will be widespread in the next 20 years. The situation is even more problematic when considering the production of hydrogen from clean renewable sources. The conversion of renewable electricity to traction effort *via* intermediate electrolyzers, hydrogen distribution-storage and fuel cells is likely to involve low overall efficiency and high costs.

8.3 Fuel Cells to Generate Electricity

If renewable energy is to be stored as hydrogen, it is necessary to have a means of reconverting this hydrogen to electricity. This is the role of the fuel cell. Fuel cells are essentially water electrolyzers working in reverse; a fuel gas (usually hydrogen) is fed to the negative electrode and oxygen (or air) to the positive electrode; for electrode terminology, see Box 8.2 presented above. An electrochemical reaction takes place – fuel is oxidized, oxygen is reduced, water is formed – and a voltage is generated across the cell. A direct current may then be drawn. The electrode at which the water is formed depends upon the nature of the electrolyte and the ion-conducting species. In LT fuel cells, if the hydrogen ion is the conducting species, then water is formed at the oxygen electrode; conversely, if the hydroxide ion (or the oxide ion) is the conducting species, then the water is formed at the hydrogen electrode. In HT fuel cells, water is formed at the hydrogen electrode.

A fuel cell differs from a conventional battery in that the reactants are gaseous and are stored outside the cell. Therefore, the capacity of the device is limited only by the size of the fuel and oxidant supply and not by the cell design. For this reason, fuel cells are rated by their power output (kW) rather than by their capacity (kWh).

The fuel cell was invented in 1839 by Sir William Grove and therefore pre-dated the first internal-combustion engines by about 50 years (see Section 1.1, Chapter 1). Even so, it is only now coming into its own as a practical power source. The fuel cell has been much investigated, particularly in recent times, on account of its perceived advantages for power generation. These advantages are: high thermo-dynamic efficiency (as, unlike a mechanical engine, it does not involve a Carnot cycle), good performance under low-load conditions (in contrast to internal-combustion engines), silent operation, low pollution emission, simplicity of design, and modular factory construction (which facilitates the gradual built-up of large units). Against these advantages must be set the many difficulties that have been encountered in the attempts to bring fuel cells to commercial realization. The primary fuels of natural gas, oil and coal must first be reformed to hydrogen, and the difficulty of developing a fully satisfactory reformer is comparable with the difficulty of developing the fuel cell itself. It is then necessary to integrate the two units and to match their operating kinetics and thermal characteristics so as to obtain good mass and heat transfer balances. There is also the problem of poisoning, both of the catalyst used in the fuel reformer and of the electrocatalyst in the fuel cell, by impurities (especially sulfur) in the primary fuel. Water management constitutes yet another challenge – it is essential that the cell neither saturates nor dries out. Altogether, a fuel cell that uses fossil fuels poses a complex set of engineering problems. Nevertheless, successful compact reformers for methanol and for natural gas have been developed and work is continuing on reformers for petrol and diesel fuels.

There are six principal types of fuel cell, as summarized in Table 8.3. Their mode of operation is illustrated in Figure 8.11. These fuel cells fall into three broad categories as regards their temperature of operation:

(i) LT (50–150 °C): alkaline (AFC), proton-exchange membrane (PEMFC) and direct methanol (DMFC) fuel cells
(ii) MT (around 200 °C): phosphoric acid fuel cell (PAFC)
(iii) HT (600–1000 °C): molten carbonate (MCFC) and solid oxide (SOFC) fuel cells.

The DMFC differs from the other five types in using a liquid fuel. The other fuel cells all burn hydrogen, of varying degrees of purity. The LT cells (AFC and PEMFC) require pure hydrogen, free of sulfur and carbon monoxide because these impurities act as electrocatalyst poisons. The hydrogen supplied to AFCs must also be free of CO_2, which reacts with the alkaline electrolyte to form solid carbonates that reduce the conductivity of the electrolyte and severely decrease the performance of the cell. (Obviously, if air is used at the positive electrode, this too

Table 8.3 *Principal types of fuel cell*

Fuel-cell technology[a]	Electrolyte	Temperature range (°C)	Electrocatalyst		Fuel	Efficiency[b] (% HHV)	Start-up time (h)
			Positive electrode	Negative electrode			
PAFC	H_3PO_4	150–220	Pt supported on C	Pt supported on C	H_2 (low S, low CO, tolerant to CO_2)	35–45	1–4
AFC	KOH	50–150	NiO, Ag, or Au–Pt	Ni, steel, or Pt–Pd	Pure H_2	45–60	<0.1
PEMFC	Polymer[c]	80–90	Pt supported on C	Pt supported on C	Pure H_2	40–60	<0.1
DMFC	H_2SO_4 Polymer[c]	60–90	Pt supported on C	Pt supported on C Pt–Ru	CH_3OH	35–40	<0.1
MCFC	Li_2CO_3	600–700	Lithiated NiO	Sintered Ni–Cr and Ni–Al alloys	H_2, variety of hydrocarbon fuels (no S)	45–60	5–10
SOFC	Oxygen-ion conductor	700–1000	Sr-doped $LaMnO_3$	Ni- or Co-doped YSZ cermet	Impure H_2, variety of hydrocarbon fuels	45–55	1–5

[a] PAFC, phosphoric acid fuel cell; AFC, alkaline fuel cell; PEMFC, proton-exchange membrane fuel cell (also termed a polymer electrolyte fuel cell, and sometimes a solid polymer electrolyte fuel cell, SPEFC); DMFC, direct methanol fuel cell; MCFC, molten carbonate fuel cell; SOFC, solid oxide fuel cell.
[b] Reported values of the efficiency of a given type of fuel cell vary widely and often no information is provided on whether the HHV or the LHV value is used. This issue is discussed in Box 8.3. The values that are taken here from the literature should be treated with caution as to their exact meaning and are simply included to provide an approximate comparison of the performance of the respective systems.
[c] Proton-conducting polymer: perfluoro-sulfonic acid polymer.

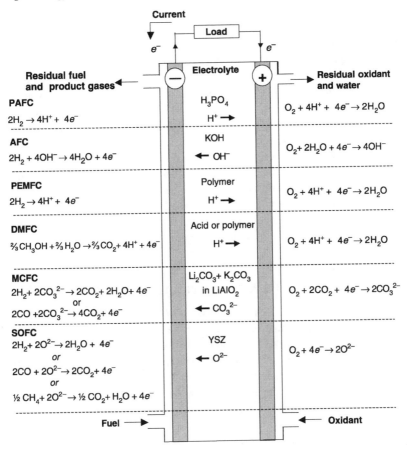

Figure 8.11   *Electrochemical reactions occurring in different types of fuel cell.*

must be purged of carbon dioxide.) The MT fuel cell (PAFC) does not pose quite so rigorous a purity specification, while the HT fuel-cell types (MCFC and SOFC) can accept fairly impure hydrogen derived, for example, from coal by reforming (see Section 2.3, Chapter 2).

In order that the internal resistance of the fuel cell be as low as possible, it is necessary that the electrolyte be as thin as possible. For cells that use liquid electrolytes, *viz.* PAFC (phosphoric acid), AFC (potassium hydroxide), DMFC (sulfuric acid), MCFC (molten lithium and potassium carbonate salts), the electrolytes are all contained within a thin layer of a porous matrix that wicks electrolyte over the entire surface of the electrode. Often a paper or thin felt matrix is employed, although for the HT MCFC the matrix is a porous ceramic, $LiAlO_2$. The PEMFC system, as well as modern designs of DMFC, has a thin non-porous polymer membrane that is highly conductive to hydrogen ions (protons), while the traditional electrolyte in the SOFC is a thin ceramic membrane made from yttria-stabilized zirconia (YSZ), which is also non-porous and is conductive towards oxygen ions at high temperatures.

Since the combustion of hydrogen is exothermic and involves a contraction in the volume of the gases, thermodynamics favor operation at low temperature and high pressure for maximum equilibrium conversion to water, but kinetic factors call for higher temperatures and a compromise must be made. At high temperatures, the overall energy efficiency can be improved by waste heat recovery, (even though the net voltage efficiency is lower), and impurities in the fuel are not so significant. The fuel cell is then operating as a CHP unit. The overall efficiency of fuel cells is often a matter of some controversy and widely differing figures may be found in the literature. In part, this derives from whether the starting point is a fuel that has to be purified and reformed, or whether the starting point is pure hydrogen. Then, there is the question of whether the heat released is utilized (as in a CHP system) or goes to waste. A further complication is that all fuels have a higher heating value (HHV) and a lower heating value (LHV), as described in Box 8.3. For hydrogen, these values differ by 18%. Altogether, the question of defining and predicting the efficiencies of fuel cells is something of a minefield.

Box 8.3 Higher and Lower Heating Values of Fuels

All chemical compounds have a 'heat (enthalpy) of formation' (ΔH_f) that equates to the heat liberated or absorbed when one mole of the compound is formed from its constituent elements. An element in its standard state is defined as having zero heat of formation. The standard molar heat of formation, ΔH_f^0, of a compound is then the change in enthalpy, positive or negative, when 1 mol of the compound is formed at standard conditions (see Box 8.1) from the elements in their most stable physical forms (gas, liquid, or solid).

For hydrogen, the heat of combustion equates to the heat of formation of the product water, *i.e.*

$$H_2(gas) + \tfrac{1}{2}O_2(gas) \longrightarrow H_2O(liquid)$$

ΔH_f^0 for liquid water $= -285.83$ kJ per mole of water

The negative sign indicates that heat is evolved in the process.

When a fossil fuel is burnt completely to CO_2 and water, the heat of reaction (or heat of combustion) per mole equates to the difference in enthalpy of the products and the reactants. Thus, for the combustion of methane:

$$CH_4 \; + \; 2O_2 \longrightarrow \; CO_2 \; + \; 2H_2O$$
$$\Delta H_f^0 : -74.85 \qquad 0 \qquad -393.5 \qquad 2 \times -285.83 \qquad (8.7)$$

$$\text{Heat of combustion} = -393.5 - 571.66 + 74.85$$
$$= -890.3 \text{ kJ per mole methane}$$
$$= -55.6 \text{ MJ per kg methane}$$

Box 8.3 (cont'd)

These values for the heats of combustion of hydrogen and methane are fundamental physical quantities and represent the maximum quantities of heat that can be evolved in the combustion process. They are termed the HHVs for the fuels.

In many practical appliances, ranging from steam engines to internal-combustion engines, the CO_2 and H_2O products are released to the atmosphere at comparatively high temperature and the heat that these gases hold above the reference temperature of 298 K is lost. This heat consists of both the 'sensible heat' (*i.e.* the heat carried by the substance, its 'heat capacity') and the latent heat of condensation of the steam. For practical engineering purposes, it has become customary, therefore, to define a LHV that corresponds, arbitrarily, to the maximum heat recoverable when the gases are emitted at 150 °C. When comparing the efficiencies of various appliances that are using the same fuel, it is convenient to take the LHV since this is the maximum possible heat that can be recovered in the appliance itself.

When using different fuels, which have different sensible or latent heats, this comparison is no longer valid. Nor does it hold when considering appliances with waste-heat recovery systems (*e.g.* condensing gas boilers). In such cases, one should always use the fundamental HHVs rather than the practical LHVs. The difference between the two figures varies with the fuel. Usually, the sensible heat is small and it is the heat of condensation of steam that predominates. It follows that the richer the fossil fuel is in hydrogen, the greater is the deviation between the LHV and the HHV. The ratio of LHV to HHV is almost 1.0 for carbon monoxide (no hydrogen), 0.98 for coal (a little hydrogen), 0.91 for petrol, 0.90 for methane, down to 0.85 for hydrogen.

In absolute numbers, the HHV and LHV for hydrogen are 142 and 120 MJ kg^{-1}, respectively, *i.e.* the HHV is 18% greater than the LHV. This is important when considering the efficiency of fuel cells, which generally burn hydrogen. As the heat of the exhaust gases can be recovered (*e.g.* in a CHP scheme), the HHV should be used when calculating the efficiency. Some authors have erroneously employed the LHV, which flatters the fuel-cell efficiency unduly.

Fuel cells are of interest for stationary, portable and mobile applications. Historically, the first successful engineered fuel cells were employed in the NASA space program in the early 1960s. The Gemini series of earth-orbiting missions used PEMFCs to provide power for communications and control systems. Later, for the manned Apollo missions to the moon and for the subsequent operations of the Shuttle fleet, AFC technology was selected to provide mission power, while the water generated in the fuel cell was used for life support (Figure 8.12). The decision to change the technology rested mainly on the greater power-generating efficiency of the AFC. With pure hydrogen produced by water electrolysis, there is no need

Figure 8.12   *Alkaline fuel cell used in NASA Space Shuttle Orbiter. The power plant
produces 12* kW, *weighs 98* kg, *and occupies 154* L.
(Courtesy of NASA)

for a reformer and no concern over impurities. The problem of developing a
satisfactory fuel cell is therefore simpler. For manned interplanetary missions that
would last several years, where it is not possible to carry sufficient fuel or water, it
would be necessary to rely entirely on solar energy and have a closed water cycle.
Thus, both electrolyzers and fuel cells would be required, preferably combined in
a single unit, *i.e.* a regenerative fuel cell, see Footnote 3 in Section 8.1. The
electrolyzer would also act as an alternative (or support) to batteries for storing
solar PV-generated electricity in the form of hydrogen and oxygen, for subsequent
combustion in the fuel cell. Through such activities, space technology is providing
a lead for future terrestrial applications.

Stationary Power

The PAFC is the most advanced for moderately large stationary power units. This
fuel cell typically has electrodes made from Teflon-bonded platinum and carbon,
and uses hydrogen fuel that is reformed externally from natural gas. The
concentrated phosphoric acid electrolyte allows the cell to operate well above the
boiling point of water, a limitation on other acid electrolytes that require water for
conductivity. Moreover, the high operating temperature of around 200 °C enables
the platinum electrocatalyst to tolerate up to 1 wt% (100 ppm) of carbon monoxide
and this broadens the choice of fuel. The use of phosphoric acid requires, however,
that other components resist corrosion. Turnkey systems are available commercially and have been installed at locations in Asia (principally Japan), Europe, and
the USA. These systems supply CHP to major building complexes such as airport
terminals, hospitals, hotels, military facilities, office buildings, and schools. The
PC25^{TM} unit developed by United Technologies Corporation in the USA, is shown
in Figure 8.13(a). Larger plants can be built up from basic modules. For example,

(a) 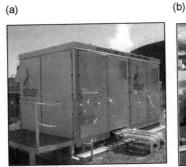 (b)

Figure 8.13 (a) *200-kW PAFC module (the PC25^{®}); (b) 1-MW PC25^{®} power plant in Anchorage, Alaska.*
(Courtesy of United Technologies Corporation)

a 1-MW plant of five PC25^{®} units has been established in Anchorage, Alaska to supply power to the local US Postal Service Airport Mail Facility (Figure 8.13(b)). Experience has shown that the PAFC is extremely reliable and produces power of high quality. Some systems have operated continuously with no shutdown for maintenance. These properties make them ideally suited to situations where security of power supply is critical, for instance at computer centers. The PAFC is also a promising technology for distributed power generation and for integration into the electricity grid.

PEMFCs are also being produced in a variety of sizes for stationary power applications – they are very responsive to changes in electrical load and the start-time is faster than that of PAFCs (Table 8.3). The central component of a PEMFC is the so-called 'membrane electrode assembly' (MEA), in which a sulfonated polymer membrane (usually, Nafion^{®} from Dupont Corporation) is sandwiched between the negative and the positive plates. Both types of plate are made from carbon cloth (or carbon paper) with very finely divided platinum as the electrocatalyst. As well as providing the basic mechanical structure for the electrode, the carbon substrate diffuses gas onto the catalyst and therefore is often called the 'gas diffusion layer'.

The configuration of a single cell, together with its electrochemistry, is illustrated schematically in Figure 8.14. Outside the MEA are flow-field plates. These contain channels to ensure that each gas is in contact with the whole surface of the respective electrode. They also serve to remove unused gas and the water that is produced at the positive electrode. Sulfonic acid groups ($-SO_3H$) are attached to the carbon chain of the polymer and allow protons to pass through the material. To give maximum conductivity, the membrane must be humidified during operation, but must remain sufficiently dry around the electrodes so that the transport of gas is not limited. Clearly, for the PEMFC to function properly, flooding of the electrodes by exhaust water must be avoided through careful management. The platinum catalyst is deposited on the surface of the carbon electrodes that are then hot-pressed to the polymer membrane, so that the membrane extends partially into the porous electrodes. A gas–electrocatalyst–electrolyte interface (the so-called 'three-phase

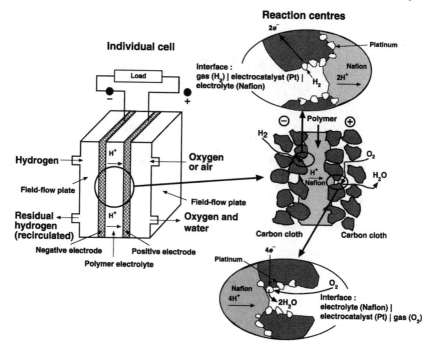

Figure 8.14 *Configuration and electrochemical processes taking place in a PEMFC.* (Courtesy of heliocentris Energiesysteme GmBH)

boundary') is formed, such that the electrocatalyst has simultaneous contact with the hydrogen or oxygen/air, the proton conductor (the electrolyte membrane), and the electron conductor (carbon electrode). The electrochemical reactions occur at these points of simultaneous contact.

The use of the platinum is a major factor in the cost of PEMFC cells and research is being directed at reducing the size of the particles and spreading them more evenly. Modern cells can operate with a total electrocatalyst loading (both electrodes) of 0.8 mg cm^{-2} (2 g per kW peak). The targets set by the US FreedomCAR Partnership are 0.4 and 0.1 mg cm^{-2} by 2005 and 2010, respectively. It should also be noted that, because of the lower operating temperature of the PEMFC, the platinum electrocatalyst is more susceptible to poisoning by carbon monoxide than that used in a PAFC. Accordingly, the carbon monoxide level must be kept below 10 ppm. There is also work in progress to develop new membranes that will enable fuel-cell operation at higher temperatures. This would yield significant energy benefits. For example, heat rejection would be easier and would allow use of smaller heat exchangers in fuel-cell systems. In addition, with reformate systems, tolerance of the stack to carbon monoxide is less problematic at high temperatures.

As with other fuel designs, single PEM cells can be stacked together, in series electrically, to form a module with a higher voltage. A schematic of a typical configuration for a two-cell module is given in Figure 8.15(a). The individual MEAs are separated by bipolar plates, which are usually 3 mm or less in thickness

(a)

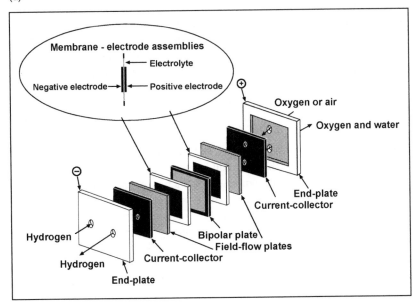

(b)

Figure 8.15 (a) *Structure of two-cell PEMFC module.*
(Courtesy of heliocentris Energiesysteme GmBH) *and*
(b) *portable Nexa™ power module.*
(Courtesy of Ballard Power Systems Inc.)

and are made from graphite foil or graphite polymer composites. Each bipolar plate is in contact with the negative electrode of one MEA and the positive electrode of the next, *i.e.* in a manner similar to that for a bipolar electrolyzer (Figure 8.5(b)). The plates are complex structures because they have to supply the cells with hydrogen and oxygen/air (on their opposite sides), conduct current to the next cells, enable water to escape, prevent leakage of gases, and collect and remove waste heat produced by the cell. Fuel cell stacks using such designs of bipolar plate require only two flow-field plates, one at each end. It should be noted, however, that there are minor variations in stack design between different developers, particularly with respect to the sub-components.

The Nexa™ power module, manufactured in Canada by Ballard Power Systems Inc., is shown in Figure 8.15(b). This unit has a power output of 1.2 kW and weighs 13 kg. Introduced in 2001, it became the world's first volume-produced PEMFC for stationary and portable power-generation applications. A subsequent Ballard product, the AirGen™, has been designed for indoor operation. This 1-kW PEMFC can act as a standby uninterruptible power supply and has a built-in suppression system that protects sensitive electronic equipment from high-voltage surges. The fuel-cell runs on hydrogen that is supplied through either industrial cylinders or canisters according to the required operation time. Field trials have also been conducted on 250-kW PEMFC plants to evaluate their suitability and performance as distributed power generators for commercial and residential consumers.

There are two HT fuel cells, namely, the MCFC and the SOFC. The latter is an all-solid-state device with no liquid components. Both present difficult materials science and technology problems. Molten alkali carbonate at 600–700 °C is the most aggressive medium and corrosion problems are severe in this fuel cell. A further issue is the introduction of carbon dioxide and its control in the air stream since this gas is consumed at the positive electrode and transferred to the negative (Figure 8.11). The SOFC operates at even higher temperatures, in the range 700–1000 °C as dictated by the composition of the solid oxide electrolyte employed, but at least has no liquid components to cause corrosion.

The HT operation of the MCFC and the SOFC systems does, however, offer the following advantages:

- removes the need for a precious-metal electrocatalyst, which reduces cost
- allows the reforming of fuels internally, which enables the use of a variety of fuels, simplifies the engineering (especially heat balancing) and reduces the capital cost; it should be noted, however, that whereas SOFCs can operate on gases made from coal, MCFCs are less resistant to impurities from coal such as sulfur and particulates
- provides high tolerance to carbon monoxide poisoning.

On the other hand, there are disadvantages, such as:

- slow start-up (Table 8.3)
- slow response to changing power demands
- the requirement for significant thermal shielding

- accommodation of the differential expansion coefficients of the electrodes, the solid electrolyte and the inter-connection between cells in a SOFC stack
- stringent durability requirements for materials.

To improve durability, efforts are in progress to develop SOFCs that have thin-film electrolytes and operate at or below 800 °C with fewer material problems, so-called 'intermediate-temperature' cells (IT-SOFCs). It is expected, however, that the lower temperature will result in some loss of electrical power.

Despite the above problems, considerable research progress has been made, and prototypes of both MCFC (300 kW–3 MW) and SOFC (100–250 kW) plants have been built and tested in several countries. These are not usually seen as mobile units, rather as the basis for static CHP installations. The high operating temperatures result in exhaust heat of good quality that may be used to drive steam or gas turbines. When used in this combined cycle fashion, overall efficiencies of at least 70% are achievable.

Portable Power

There are many potential applications for portable fuel cells. In particular, there is a surge of interest in the development of units that generate just a few watts to power a wide range of consumer electronics, as well as in larger cells (up to a few hundred watts) that are suitable for military equipment such as man-pack radios, helmet-mounted image displays, night-vision sights, laser range finders and sophisticated communications systems.

Infantry soldiers require a power pack that is readily portable, has a longer operational life than batteries, and is silent. A small hydrogen fuel cell would be ideal, although the question remains how best to carry and generate the hydrogen. A detailed study of the storage options for hydrogen within a practical weight limit of 3 kg for the total power pack (1.4 kg for the fuel cell and 1.6 kg for the hydrogen store) has shown that the best option by far is the 'one-shot', non-reversible, chemical generation of hydrogen by reacting water with chemical hydrides such as $NaBH_4$, LiH, or $LiAlH_4$. For a fuel cell with 60% efficiency, a power pack based on this concept should provide up to 10 times as much electrical energy (Wh) as a rechargeable lithium battery of comparable mass. The importance of such studies is that they provide a foundation for other, civilian applications for portable power.

Rechargeable batteries are well suited to applications where the energy requirement between recharges is limited. Lithium-ion batteries (see Section 9.5, Chapter 9) have proved their worth in mobile communications (cellular phones) and in laptop computers. With the advent of mobile broadband computing, however, the next generation of portable electronic equipment will demand ever greater amounts of stored energy, most probably at levels that are well beyond the capabilities of batteries. For this reason, attention is turning to so-called 'micro fuel cells' that, again, promise an energy-storage capability several times higher than that of the best batteries, albeit with a lower power output. This disadvantage can be overcome by having a hybrid fuel-cell-battery combination, in which the battery provides the peak power when required and is recharged by the fuel cell.

Although, as discussed above, PEMFC units have been produced in the 1-kW power range, the most likely technology to be used for portable applications is the DMFC. The use of methanol directly as the fuel without first reforming to hydrogen would be ideal, since a liquid fuel is more readily dispensed and carried than hydrogen. Replacement of a spent methanol cartridge in a micro fuel cell will only take a few seconds – rechargeable batteries require periods of hours to be replenished. Furthermore, liquid methanol has a higher stored energy than hydrogen.

To increase the operational temperature of DMFCs and hence improve the kinetics of the negative electrode, a Nafion^{TM}-type polymer electrolyte is now employed in place of sulfuric acid solution (note, it is the protons that migrate from the negative to the positive electrode and not the methanol molecules, see Figure 8.11). Undesirably, however, the membrane is permeable to both methanol and water. The 'crossover' of methanol will result in a loss of cell performance and efficiency due to unproductive fuel consumption by direct reaction with oxygen at the positive electrode. In addition, migration of water across the membrane (by diffusion and dragged by protons and methanol) floods the positive electrode and thus inhibits the access of oxygen. The methanol crossover problem can be reduced to more manageable levels by using dilute aqueous solutions of less than 10% methanol by volume, but this will cause a decrease in power density. New membrane materials and improved MEA designs are being investigated as more effective means to moderate methanol and water crossover. Methanol crossover poisons the electrocatalyst in the positive electrode and this, together with the poor kinetics of methanol oxidation at the negative electrode, requires a DMFC to have at least 10 times more precious-metal electrocatalyst than a PEMFC to achieve a comparable power output. Success in ameliorating these two problems will therefore result in significant cost savings. Finally, it should be noted that in response to concerns over the toxicity of methanol, some companies have embarked on the development of micro fuel cells that operate on the direct oxidation of ethanol.

Despite the above operational concerns, micro-DMFCs are already being produced as complete packages for low-power, portable applications. A typical unit comprises the fuel-cell sub-assembly, a fuel cartridge, a fuel flow-regulator, and a d.c.–d.c. converter. A micro-DMFC under development by Motorola is shown in Figure 8.16. The prototype is 5-cm wide and has four cells that produce 0.35 W. This power is sufficient to recharge a mobile-phone battery. Indeed, manufacturers consider that the first successful application of micro-DMFCs will be in the form of a portable battery recharger rather than as a battery replacement.

Fuel cells are likely to reach commercialization for small portable applications sooner than for FCV. Apart from ready and rapid refueling (for example, *via* capsules purchased at supermarkets, newsagents, *etc.*), the best argument for a micro fuel cell is cost. Rechargeable lithium batteries are exorbitantly expensive on a kWh basis and it is expected that micro fuel cells can be produced more cheaply. As a replacement for rechargeable batteries, micro fuel cells do not have to meet the tough cost criteria that they face as power sources for vehicles.

The ultimate in miniaturization would be a fuel cell on a micro-chip. Self-powered chips would give birth to a new generation of self-sufficient devices such as remote sensors that could telemeter data from the field back to a central station.

 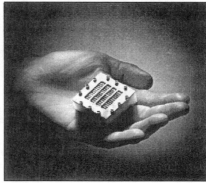

Figure 8.16 *Prototype micro-DMFC developed by Motorola.*
(Courtesy of MTI MicroFuel Cells Inc)

This is still some way from practical realization. Micro fuel cells have no immediate and obvious relevance to the macro-energy scene, but networks of tiny sensors might play an important role in energy conservation and environmental pollution monitoring.

Power for Road Transportation

Fuel cells are promising power sources for almost all types of transportation, *i.e.* from power-assisted bicycles (a few hundred watts), through electric trains (hundreds of kW), to military submarines (several MW). By far the largest potential market, however, is as a power source for road vehicles – both EVs and HEVs. Ongoing developments in this area are described in Section 10.5, Chapter 10, but a few general remarks are timely here.

A particular attraction of fuel cells for transportation applications is that air can be used at the positive electrode and this reactant does not have to be carried around and is free. Unfortunately, this advantage is offset by the difficulty of conveying hydrogen, as discussed earlier. A FCV could, in principle, avoid this difficulty by using liquid methanol as a fuel, with a reformer on-board to decompose it to hydrogen (the so-called 'indirect methanol fuel cell' system). It has to be said, however, that on-board reformers – for any type of fuel – are, at best, a dubious concept. Reformers are essentially micro-refineries and their integration with fuel cells to provide the variable quantities of hydrogen demanded by driving schedules has still to be optimized. There are also concerns over their start-up procedures, reliability, and mass-production cost. It should also be recognized that, in the case of methanol, on-board fuel tanks would store considerably less energy than petrol, and would also be subject to corrosion due to the miscibility of water and methanol. Nevertheless, and despite the fact that some automotive companies are less than enthusiastic about having their vehicles run on methanol on account of its toxicity, some effort continues on the development of DMFCs for road transportation.

Alkaline fuel cells (AFCs) have been developed in multi-kW size for powering FCVs. A problem with using alkaline electrolyte, as mentioned earlier, is that it

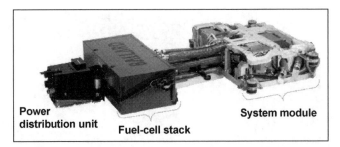

Figure 8.17   *Fuel-cell engine produced by Ballard for Ford Focus FCV-Hybrid.*
(Courtesy of Ballard Power Systems Inc.)

absorbs CO_2 from the air and this will degrade cell performance. It is therefore necessary to pre-scrub the incoming air to remove CO_2. This adds to the complexity, volume and cost of the power source. At present, the PEMFC is the favored system to power EVs and HEVs. Modules suited to various types of vehicle are being produced commercially by Ballard in Canada and also by several companies in Japan. The Xcellsis™ HY-80 fuel-cell system for the Ford 'Focus FCV-Hybrid' (see Section 10.5, Chapter 10) is shown in Figure 8.17.

Aside from the issue of how best to store hydrogen on-board the vehicle, there is also the problem of how to start the fuel cell from cold in sub-zero temperatures. No driver wants to wait the few minutes that are required to heat the unit to its operational temperature, and what will be the source of heat – a propane catalytic burner? Furthermore, there remains the entire infrastructure problem of how to distribute hydrogen and make it available at most service stations. These problems are not insuperable, but they are difficult and costly and should not be under-rated.

The development of fuel cells is a complex subject that involves the collaboration of physical chemists, materials scientists, and chemical engineers. For vehicular use, automotive engineers are also involved as well as strategists, government agencies and many vested interests. Over the past 20 years, development teams have been active in several countries, especially in the USA and Japan, with research back-up in many more countries. Much progress has been made and different sizes and types of fuel cells are becoming commercially available. For further information, the reader is referred to the specialist journals (*e.g. Journal of Power Sources*) and the proceedings of specialist conferences on fuel cells. The applications for different types of fuel cell and some of their technical features are summarized in Table 8.4.

8.4 Prospects for Hydrogen

To summarize this chapter, clear differentiation must be made between:

- hydrogen as a chemical and hydrogen as a fuel
- hydrogen combusted in internal-combustion engines (see Section 2.5, Chapter 2) and hydrogen utilized in fuel cells

Table 8.4 *Overview of fuel-cell systems – applications and technical issues*

PAFC (150–220 °C)	*AFC (50–150 °C)*
Fuelled by hydrogen or reformed natural gas	Fuelled by pure hydrogen
Specifications for hydrogen purity less stringent than for PEMFCs	Used in space missions (high power to weight ratio)
Most commercially advanced technology	Inexpensive (not precious metal) electrocatalyst may be used, although the addition of precious metals improves the kinetics
Medium- to large-scale stationary applications	Removal of carbon dioxide adds to system complexity
Unsuitable for automotive applications	Corrosion problems
Uses precious-metal electrocatalyst	

PEMFC (80–90 °C)	*DMFC (60–90 °C)*
Fuelled by pure hydrogen	Fuelled by methanol
First developed for space missions	No need for external reformer
Portable or stationary applications	Advantages over compressed gas – methanol can integrate with existing distribution systems for liquid fuels
Attracting major investment for automotive applications	Portable electronics applications (micro fuel cells)
Uses precious-metal electrocatalyst	Requires high loading of precious-metal catalyst
Electrolyte membrane is expensive	Methanol crossover and toxicity are issues of concern
Improved membranes required for cell to operate at higher temperatures with greater efficiency	

MCFC (600–700 °C)	*SOFC (700–1000 °C)*
Fuelled by hydrogen, natural gas, petroleum, propane, landfill gas, biogas, diesel, etc. Low S.	Fuelled by hydrogen, or almost any hydrocarbon fuel including coal gas
No need for external reformer	No need for external reformer
Uses less expensive (not precious metal) electrocatalyst	Uses less expensive (not precious metal) electrocatalyst

(Continued)

Table 8.4 *(Continued)*

MCFC (600–700 °C)	SOFC (700–1000 °C)
Medium- to large-scale stationary applications, possibly industrial multi-megawatt plants	Medium- to large-scale stationary applications
Suitable for cogeneration operations	Suitable for cogeneration operations
Slow start-up	Small systems being developed for auxiliary power units in automotive applications
Slow response to changes in load	Slow start-up
Corrosion and durability problems	Slow response to changes in load
Requires carbon dioxide management system	Difficult materials science problems

- hydrogen derived from fossil sources and from renewable energy sources
- hydrogen for stationary applications and for portable or mobile applications.

There is no doubt that hydrogen will continue to be required as a chemical and that it will be derived from fossil sources, notably natural gas, by means of established reforming procedures. Hydrogen as a fuel and energy vector is more problematic, but if fuel cells are to be successful they will require hydrogen. For the foreseeable future, this hydrogen will be derived principally from fossil sources. It seems quite probable that in the next 20 years or so, commercially viable fuel cells will come into widespread use for stationary applications such as CHP units for dispersed electricity generation. Small portable fuel cells for specialist military applications or micro fuel cells for cellular phones and laptop computers may well prove successful. At the other extreme, very large fuel cells for use in submarines – a specialized mobile application – may also come to fruition. The prospects for fuel cells to power EVs are much more questionable, not least because of the technical problems and costs of storing and conveying hydrogen or, alternatively, of integrating reformers (for methanol or petrol) with fuel cells. We do not, however, share the optimism of many of the automotive manufacturers or the US government who are advocating the widespread adoption of FCVs, particularly for private cars, in the next two decades. The reasons for adopting this view are presented in Chapter 11.

As for developing hydrogen as a *general energy vector* to store electricity produced from renewable sources, such as wind or solar energy, and then to use it as a fuel, the energetics are grossly unfavorable and the economics are not commercially viable. At best, the overall electrical efficiency of the electrolyzer-fuel-cell cycle would be about 50% (d.c.–d.c.). Moreover, this performance makes no allowance for energy losses that are incurred in compressing, transporting, pumping and storing the hydrogen, and in inverting the d.c. electricity. This is no way to squander costly solar electricity that is better fed into the grid or stored in batteries. The only exception to this general statement about not coupling renewables to hydrogen might lie in areas of the world (such as Iceland, Norway, or Brazil) where there is an abundance of hydroelectricity and a paucity of fossil fuels. Even then, there might be better things that could be done with the surplus electricity.

CHAPTER 9

Battery Storage

Of the many possible ways of storing electrical energy, the only ones that are truly practical and widely used today are pumped hydro on the medium-to-large scale and battery storage on the small-to-medium scale. Other possibilities (compressed air, flywheels, electromagnetic and electrostatic energy storage, hydrogen) are still in the development stage and, as yet, only have limited application in special situations. It is therefore appropriate that we should devote a Chapter to battery storage.

Batteries are of two basic types: 'primary' batteries that are discharged once and then discarded, and 'secondary' batteries that are recharged many times. Here, we are concerned only with the latter type. Another useful distinction is between 'consumer' batteries as purchased by the individual, and 'industrial' batteries as used in industry and commerce. Most consumer rechargeable batteries are small, single-cell devices (with the notable exception of the automotive battery), whereas industrial batteries tend to be much larger, multi-cell modules.

The manufacture of batteries is a growth industry. The reasons for this are quite clear. There is a huge increase in the sales of portable electronic devices (mobile telephones, laptop and notebook computers, camcorders, *etc.*) that require battery power, and also of mains-connected devices that use batteries for memory back-up. Industry and commerce are creating a strong growth in the demand for stationary batteries to provide instantaneous power in the event of mains failure (see uninterruptible power supplies, Section 6.3, Chapter 6). In the transport sector, electric traction is proving more popular in environmentally sensitive areas and major automobile manufacturers are working towards battery electric and hybrid electric vehicles (see Chapter 10). And finally, the urgent need to harness renewable energy sources (solar, wind, *etc.*) as part of an overall strategy to ensure global energy sustainability will require many more batteries to store the generated electricity and to smooth out fluctuations in power supply and demand.

In a battery, the processes taking place in the so-called 'active materials' of the electrodes are quite complex, although usually they can be represented, to a first approximation, by simple electrochemical equations. During discharge, the equations define a reduction process with take-up of electrons at the positive electrode, and an oxidation process with release of electrons at the negative electrode. The electrons pass from the negative electrode to the positive electrode *via* the external circuit. Recharging a secondary battery is simply a reversal of the processes that occur during discharge. The terminology used to describe the operation of secondary batteries has been described in more detail in Box 8.2, Chapter 8, where it is seen that the discharge of a battery is similar to the operation of a fuel cell.

Rechargeable batteries come in many different sizes – from the small consumer cells used in toys and hand tools, to the huge banks of large cells or modules employed in some standby-power applications (*e.g.* telephone exchanges) and in submarine propulsion. The batteries are also based on different chemistries, some of which are classical and well established while others are of more recent origin and employ novel electrode materials and/or electrolytes. For all types, however, there are on-going development programmes to improve performance and quality, as well as to reduce the cost of manufacture. For example, the classical lead–acid battery, as used in almost all internal-combustion-engined vehicles, is now improved in performance, is lighter in weight, lasts longer, and (in relative terms) is cheaper than it was some 50 years ago.

The minimum cycle-life required from a secondary battery depends on the application, but is often many hundreds of cycles. For instance, a traction battery for an electric vehicle should have a life of at least 1000 cycles, whereas a battery in a low-earth orbit satellite, recharged by solar electricity, is expected to last for more than 20 000 cycles. In chemical terms, this translates into electrode reactions that are fully reversible throughout cycling. Even a 1% irreversibility per cycle (*e.g.* due to a side-reaction) soon compounds into a significant loss of capacity. Restricting irreversibility to a very low level can be difficult, particularly when the electrode reactions involve solid-phase transformations and there is the possibility of other chemical or physical processes that could lead to battery deterioration and failure. For the different battery chemistries, these processes variously include:

- densification of the active materials with loss of porosity and, hence, reduced availability for electrochemical reaction;
- expansion and shedding of active material from the electrode plates;
- progressive formation of inactive phases, which electrically isolate regions of the active material;
- growth of metallic needles at the negative electrode, which give rise to internal short-circuits;
- gassing of electrode plates on overcharge, which can cause disruptive effects;
- corrosion of current-collectors, which results in increased internal resistance;
- separator dry-out through overheating.

Such degradation processes may result in precipitous battery failure, *e.g.* through an internal short-circuit, or in progressive loss of capacity and performance. Generally, the degenerative stages are interactive and accumulative, so that when the performance starts to decline, it soon accelerates and the battery becomes unusable. Despite this catalogue of problems, some remarkable successes have been achieved in designing batteries of several different chemistries to have long cycle-life (>1000 cycles).

Cycle-life is not, however, the only important criterion when selecting a battery for a particular application. Other generally desirable features are:

- a stable voltage plateau over a good depth-of-discharge (DoD);
- high specific energy (Wh kg^{-1}) and high energy density (Wh dm^{-3});
- high peak-power output per unit mass (W kg^{-1}) and volume (W dm^{-3});
- high energy efficiency (Wh output:Wh input);
- a wide temperature range of operation;
- good charge retention on open-circuit stand;
- ability to accept fast recharge;
- ability to withstand overcharge and overdischarge;
- reliable in operation;
- maintenance-free;
- rugged and resistant to abuse;
- safe both in use and under accident conditions;
- made of readily available and inexpensive materials that are environmentally benign;
- efficient reclamation of materials at end of service-life.

Overall, this is a formidable set of specifications, even before going into technical detail, and explains why the development of new rechargeable batteries has proved to be so difficult. The relative importance of the different criteria will depend upon the intended application. For instance, energy efficiency is especially important for solar photovoltaic applications, whereas the ability to accept fast recharge is not a prime consideration. For use in hybrid electric vehicles (Chapter 10), fast recharge is an even more important factor than energy efficiency. One of the complications is that a battery is a multi-variant system with many complex interactions between the variables. Thus, the energy recovered from a battery varies with the discharge rate employed and the ambient temperature; the peak-power output depends on the state-of-charge (SoC) of the battery; and the life is a function of the DoD to which the battery is subjected in each cycle. Batteries with different chemistries vary widely in their performance parameters, but so also do different designs of battery with the same chemistry. Accordingly, it is difficult to make precise comparisons between various battery types. Moreover, the problem is compounded by the fact that the numerical values also depend on the conditions under which the battery is operated. Thus, although comparisons between batteries have been often made, it should be recognised that at best they are only semi-quantitative.

With these general remarks in mind, this Chapter focuses on five generic types of secondary battery that hold most promise for the storage of energy.

9.1 Lead–Acid Batteries

By far the most common type of storage battery is the ubiquitous lead–acid battery, which was invented by Planté in 1859 and greatly improved by Faure in 1881. Over 90% of the world market for medium-to-large rechargeable batteries is still met by lead–acid; most of the rest is satisfied by nickel–cadmium. For small-sized batteries, traditional nickel–cadmium cells are being rapidly replaced by nickel–metal-hydride and lithium-ion cells on account of their superior performance.

Since its invention, the lead–acid battery has undergone many developments, most of which have involved modifications to the materials or design, rather than to the underlying chemistry. The electrode reactions of the cell are unusual in that the electrolyte (sulfuric acid) is also one of the reactants, as seen in the following equations for discharge and charge.

At the positive electrode:

$$PbO_2 + HSO_4^- + 3H^+ + 2e^- \underset{\text{Charge}}{\overset{\text{Discharge}}{\rightleftharpoons}} PbSO_4 + 2H_2O \quad E^\circ = +1.690\,V \quad (9.1)$$

At the negative electrode:

$$Pb + HSO_4^- \underset{\text{Charge}}{\overset{\text{Discharge}}{\rightleftharpoons}} PbSO_4 + H^+ + 2e^- \quad E^\circ = -0.358\,V \quad (9.2)$$

where E° is the standard electrode potential for each cell reaction, *i.e.* the electrode is in a standard state.

The overall reaction is:

$$PbO_2 + Pb + 2H_2SO_4 \underset{\text{Charge}}{\overset{\text{Discharge}}{\rightleftharpoons}} 2PbSO_4 + 2H_2O \quad V^\circ = +2.048\,V \quad (9.3)$$

where V° is the standard cell voltage. On discharge, sulfuric acid is consumed and water is formed, with the converse on charging. The SoC of the battery can, therefore, be determined by measuring the relative density of the electrolyte, which is typically 1.27–1.30 for a fully charged cell. The lead–acid battery is unique in this regard.

The capacity (Ah) exhibited by a lead–acid battery when discharged at a constant rate depends on a number of factors, among which are the design and construction of the cell, the cycling regime (history) to which it has been subjected, its age and maintenance, and the prevailing temperature. Typical discharge curves for lead–acid batteries at varying rates are shown in Figure 9.1. It is immediately apparent how the realisable capacity is strongly dependent on the rate of discharge, *e.g.* the capacity obtained from a 30-min discharge is only a fraction of that from a 10-h discharge. Moreover, the cell voltage is much reduced and, thereby, results in an even greater reduction in available energy (Wh).

The life of a battery is generally defined as the number of charge–discharge cycles that it will sustain before the capacity falls to 80% of its initial value or, alternatively, before the power output at 80% DoD falls below a specified value. For lead–acid batteries, the life is very temperature-dependent. These batteries operate best from 10 to 25 °C – premature failure can occur on continuous operation above 25 °C, while power output falls off sharply at temperatures below 0 °C.

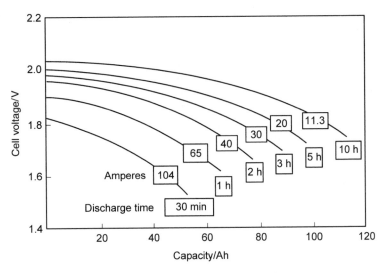

Figure 9.1 *Typical discharge curves for lead–acid traction batteries.*

Battery life is also dependent on the DoD that is employed; this factor is much more critical for some types of lead–acid battery than for others.

There are several generic types of lead–acid battery, as follows.

Traditional Flat-plate Batteries

The first lead–acid batteries were of the 'flooded' design, *i.e.* the sulfuric acid is a free liquid to a level above the top of the plates and above the busbars that connect plates of the same polarity together. Both the positive and negative plates are produced by machine application of a paste of the active materials (lead oxides, additives, sulfuric acid) onto a rectangular, lattice-type grid that is gravity cast from a lead–antimony alloy (see schematic in Figure 9.2(a)). (Note, pure lead is sometimes used for negative grids.) Porous separators, typically made from polyethylene, are placed between each adjoining positive and negative pasted plates to provide electrical insulation and, hence, prevent short-circuits. After assembly, the plates are 'formed' (charged) to convert the active materials to lead dioxide (PbO_2) and spongy lead (Pb) at the positive and negative electrodes, respectively.

Until about 20 years ago, all automobiles used the above design as it is the cheapest form of lead–acid battery. The duty cycle does not normally involve deep discharge and, for most of the time, the battery is in a charged state. The battery is unsealed and liberates gas on charging (oxygen at the positive plate, hydrogen at the negative), and it is therefore necessary periodically to replenish the water ('topping-up') that has been lost from the battery through this electrolysis. Screw caps are provided at the top of each cell for this purpose. The life of the battery, normally several years, is seriously shortened if it is subjected to deep-discharge cycling. The principal reason for this is the molar volume expansion of the positive

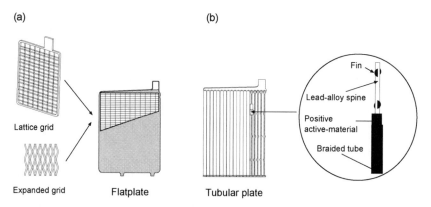

Figure 9.2 *Lead–acid batteries with* (a) *flat plates and* (b) *tubular plates.*

active material on discharge. The repeated mechanical stresses imparted on charge–discharge cycling weakens the active material and cause it to 'soften' and, eventually, to break away from the grids ('paste shedding'). Other debilitating factors are corrosion and cracking of the grids, electrical isolation of active material from the grids, irreversible sulfation of the plates (which can become especially serious if the battery is allowed to stand for excessive periods in a discharged state), and internal short-circuits.

By modifying the design to incorporate glass-fibre mats around the positive plates and by the use of thicker positive grids, it is possible to produce a pasted-plate battery that is suitable for deep-discharge cycling. The glass mat serves to absorb shocks and to prevent shedding of active material from the electrodes. The deep-cycling ability is attained, however, at the expense of increasing considerably the mass and cost of the battery. This design is commonly used for leisure applications (caravans, boats, *etc.*) and for standby-power duties. Heavy-duty versions are manufactured as traction batteries for off-road electric vehicles, as well as for some industrial applications.

Low-maintenance Flat-plate Batteries

In recent years, a new generation of low-maintenance, flooded batteries has been introduced for automotive duties. These lose very little water and require almost no maintenance. The technological advance that made this possible was the replacement of the lead–antimony alloy used for the grids by a lead–calcium or a lead–calcium–tin alloy. These grids are produced either by gravity casting or by slitting and expansion of alloy strip (Figure 9.2(a)). It is the antimony component of the traditional battery that gives rise to excessive gassing and water loss on charge: antimony dissolves (corrodes) from the positive grid, diffuses through the electrolyte, and deposits on the negative plate where it increases the rate of hydrogen evolution on charging. By eliminating antimony, a great improvement is effected. Nevertheless, the batteries are still vented. Their life on deep-discharge cycling is generally shorter than that of the conventional lead–antimony batteries. Not only does the same shedding of paste occur at the positive plate, but also a highly resistive corrosion layer forms at the interface between the grid and the active material. The addition of tin to the positive-grid alloy moderates the influence of this 'barrier' layer. In any event, low-maintenance batteries should not regularly be cycled to greater than 15% DoD, and never beyond 50%.

Tubular-plate Batteries

The tubular battery is a technology of long standing that is designed for deep-discharge cycling. The batteries are used extensively as power sources for many types of low-speed electric vehicles, *e.g.* forklift trucks, golf carts, milk floats, tractors. A lead–antimony alloy casting of parallel rods replaces the positive grid of the automotive battery (Figure 9.2(b)). These rods (or 'spines') are attached to a common header, rather like a coarse comb with well-separated 'teeth'. Each rod is inserted into a vertical tube made of braided glass-fibre that is surrounded by a sheath of perforated polyvinyl chloride. The active material is then packed into the tubes around the rods, which act as the current-collectors. The flexibility in the glass-fibre tubes allows for expansion and contraction of the active material during cycling.

As the positive active material is constrained by the tubes, the batteries can withstand deep-discharge cycling. They are, however, more expensive than flat-plate batteries and still require regular topping up with water. It is possible to minimise maintenance (water replenishment) by lowering the level of antimony in the positive grid, or by using lead–calcium alloys. The latter option restricts operation to shallow cycling. At least one manufacturer does market such a battery specifically for solar photovoltaic applications. This battery would be suited to remote applications where the cost of servicing a conventional tubular battery would be so high that it is more economical to buy a battery pack several times larger than would be required for deep cycling, and then allow only shallow discharge, *e.g.* 20% DoD.

Valve-regulated Batteries

For many battery applications, including a high proportion of solar photovoltaic installations, maintenance-free batteries are an essential requirement. In these cells, the problem of water loss through electrolysis during charging has been side-stepped by arranging for the oxygen released at the positive electrode to 'recombine' within the cell. This is made possible by using a non-flooded ('electrolyte-starved') design in which the sulfuric acid is held in the interstices of an absorbent, fibre-glass mat that also serves as the separator. Only the minimum quantity of acid required for the electrode reactions is used and there is sufficient void in the glass mat to permit oxygen gas to diffuse through the separator and discharge chemically at the negative electrode. This process is known as the 'internal oxygen cycle' and operates as follows.

At the positive electrode:

$$H_2O \longrightarrow 2H^+ + \frac{1}{2}O_2 + 2e^- \tag{9.4}$$

At the negative electrode:

$$Pb + \frac{1}{2}O_2 + H_2SO_4 \longrightarrow PbSO_4 + H_2O \tag{9.5}$$

$$PbSO_4 + 2H^+ + 2e^- \longrightarrow Pb + H_2SO_4 \tag{9.6}$$

Since the negative electrode is simultaneously on charge, the discharge product is immediately reduced electrochemically to lead, *via* Equation 9.6, and the chemical balance of the cell is restored. A corresponding internal cycle for hydrogen is not possible because oxidation of the gas at the positive plate is too slow. This, together with the fact that oxygen recombination is typically 95–99%, requires each cell to be fitted with a one-way valve as a safeguard against excessive pressure build-up. Accordingly, the technology is known generically as the 'valve-regulated lead–acid battery' (VRLA) battery.[1] There are two versions of this battery: in one the sulfuric acid is held in an absorptive glass mat ('AGM' technology), as described above, while in the other the acid is fixed in a silica gel ('gel' VRLA).

Because no water maintenance is required, VRLA batteries are particularly suited to RAPS sites where access is difficult. Although essentially maintenance-free, VRLA batteries are not ideally suited to deep discharge, but they may be used for cycles at medium depths-of-discharge. The latter is gaining favour in the RAPS community and is known as 'partial-SoC' (PSoC) duty. This controls the battery below a full SoC for extended periods between full recharges (Figure 9.3(a)). Such a strategy significantly decreases the overcharge delivered to the battery compared with traditional operating methods, and thus prevents electrolyte dry-out (which increases the internal resistance of the battery), as well as reduces corrosion and other related damage to the positive plate during charging. The gel design of VRLA battery has been found to give particularly good cycle-life under PSoC duty. For example, the lifetime ampere–hour throughput of a gel battery performing PSoC cycles between 40 and 70% SoC has been shown to be three times greater than that obtained under 100% DoD cycling. Given this practical benefit from PSoC operation, the gel battery has become the preferred VRLA technology for many RAPS applications. For example, gel batteries have been chosen for the RAPS facilities that are being installed in the Amazon Region of Peru to provide continuous power to villages. The batteries (Figure 9.3(b)) and their dedicated PSoC management system have been designed and produced by Battery Energy Power Solutions Pty Ltd and CSIRO in Australia.

The first VRLA cells to become commercially successful were made in 1971 by Gates Energy Products, Inc. in the USA, and were marketed under the Cyclon[TM] brandname. This is an AGM design in which a single pair of positive and negative plates is separated by a layer of the separator, and then spirally wound (Figure 9.4(a)). Busbars are welded to the exposed lugs before the assembly is inserted in the

(a)   (b)

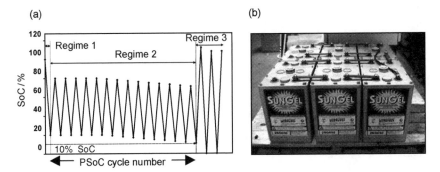

Figure 9.3 (a) *Schematic of PSoC operating procedure;* (b) *gel VRLA batteries awaiting installation in Peru RAPS sites.*

container. The tightly wound plate-group enables the separator to be maintained under high compression and thus helps to retain the positive active material during cycling. In addition, the use of thin electrodes provides an active-mass surface that is significantly larger than in conventional, flat-plate batteries. This reduces the internal electrical resistance and provides exceptionally high power. By virtue of these attributes, interest in spiral geometry has re-awakened. For example, a spiral battery has recently been developed for automotive applications (Figure 9.4(b)).

The possibility of employing spiral-wound lead–acid batteries in hybrid electric vehicles as a more affordable alternative to the present use of expensive nickel–metal-hydride batteries is also being explored. The extraordinary high levels of power demanded by such vehicles are well within the capability of lead–acid, but consideration must be given to optimising the current-collection function of the

(a)   (b)

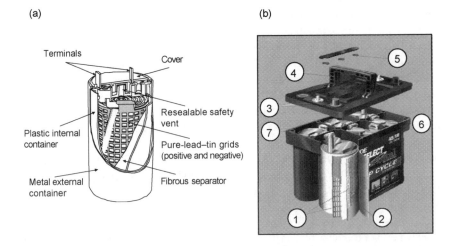

Figure 9.4 (a) *Cyclon™ cell;* (b) *spiral-wound battery for automotive applications:* (1) *lead–tin grids,* (2) *AGM separator,* (3) *terminals,* (4) *fold-away handles,* (5) *pressure-relief valves,* (6) *through-the-wall cell connection,* (7) *busbar.*

grid and reducing heat generated in the cell through overpotential and resistive losses. It is well established that the performance of lead–acid batteries can be adversely affected by a non-uniform distribution of current over the plates. This effect becomes more pronounced the higher the rate of charge and discharge and the larger the plate. At the highest rates, the voltage drop down the grid causes inhomogeneous utilisation of the active material (that near the current take-off 'tab' is worked harder than that further away) and increases the heat produced. Research conducted by CSIRO in Australia has demonstrated that the addition of a second current take-off tab, symmetrically placed opposite the first, results in more uniform utilisation of the active materials and safeguards against the development of high operating temperatures. These benefits, in turn, translate into improved power capability and longer cycle-life. In more recent developments, CSIRO and Hawker Energy Products, Inc. in the UK have applied the 'dual-tab' concept to the spiral-wound Cyclon$^{\circledR}$ cell (Figure 9.5(a,b)). In early 2004, a Honda 'Insight' retrofitted with a pack of these batteries and a purpose-built management system (Figure 9.5(c)) underwent successful road trials in the RHOLAB project, which was funded jointly by the UK government and the Advanced Lead–Acid Battery Consortium (ALABC). For further details of the Honda 'Insight' and other hybrid electric vehicles see Section 10.4, Chapter 10.

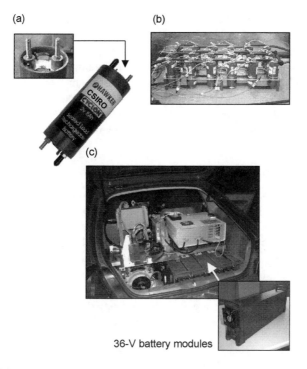

Figure 9.5 (a) *Spiral-wound cell with dual tabs;* (b) *36-V battery of dual-tab cells under test;* (c) *dual-tab battery system installed in Honda 'Insight'.*

9.2 Alkaline Batteries

Alkaline batteries are based on an electrolyte of concentrated ($\sim$30 wt.%) potassium hydroxide (KOH) solution. The positive electrode is normally nickel hydroxide ($Ni(OH)_2$), which is capable of being oxidised electrochemically to Ni^{3+} ('nickel oxide') during charge and reduced back to Ni^{2+} during discharge. Occasionally, for specialised applications, this is replaced by silver oxide (AgO), although this is an expensive option.[*] The negative electrode may be one of three metals – cadmium, iron, zinc – or a complex metal hydride.

By far the best-known rechargeable alkaline battery is cadmium–nickel-oxide, commonly termed 'nickel–cadmium' or 'NiCad'. The Swedish scientist Waldemar Jungner invented this battery at the end of the 19th century. At about the same time, working in the USA, Thomas Edison invented the iron–nickel-oxide (or 'nickel–iron') battery, which has very similar chemistry. The overall cell reaction for each is:

$$M + 2NiOOH + 2H_2O \underset{\text{Charge}}{\overset{\text{Discharge}}{\rightleftharpoons}} M(OH)_2 + 2Ni(OH)_2 \qquad (9.7)$$

where M = Cd or Fe. The standard cell voltage is 1.30 and 1.37 V for nickel–cadmium and nickel–iron, respectively, *i.e.* considerably below that of lead–acid (2.048 V).

Over the course of the past century, the nickel–cadmium cell proved to be more commercially successful than the nickel–iron cell and the latter is now little used. The principal reasons for this are that the iron electrode: (i) gases excessively on charge, which results in low electrical efficiency and high water maintenance; (ii) suffers from a high rate of self-discharge on standing, which is caused by corrosion of the negative plate. On the other hand, the nickel–iron cell is highly robust and is capable of 2000 charge–discharge cycles at 80% DoD. Given this attribute, it was formerly used as a traction battery for industrial trucks and some railway applications where electrical efficiency was not an over-riding consideration.

Nickel–Cadmium Batteries

The practical cell voltage of traditional nickel–cadmium batteries is 1.2 V and their specific energy is 30–40 Wh kg^{-1} (rather similar to that of lead-acid batteries), although 60 Wh kg^{-1} has been achieved in some recent designs. Nickel–cadmium batteries often come as packs of five inter-connected cells that are series-connected to give a 6-V battery. The high-rate and low-temperature performances of nickel–cadmium are both better than those of lead–acid, and other beneficial features are: a flat discharge voltage, long life, continuous overcharge capability, and good reliability. Cells and batteries are available in many different sizes and with pocket-plate, plastic-bonded or sintered electrodes. Sealed consumer cells range in size from

[*] Manganese dioxide, used for the positive-electrode material in primary alkaline cells (with zinc as the negative electrode), is not normally regarded as rechargeable. It is possible, however, to design small (2.5 Ah) rechargeable cells of this type using special components. These cells must not be discharged too deeply and must be recharged with a special charger.

10 mAh up to 15 Ah, while vented cells used as standby power units have capacities up to 1000 Ah. High-power cells, capable of delivering up to 8000 A are available for engine starting and are used in aircraft and some heavy vehicles. The principal disadvantages of the nickel–cadmium system are its high cost and environmental concerns associated with the disposal of batteries that contain toxic cadmium. Nickel–cadmium batteries are gradually being replaced by nickel–metal-hydride alternatives, especially in the smaller sizes, but are holding their own for industrial applications. Particular applications include aircraft batteries, vehicle traction, and standby power. As discussed earlier (Section 6.2, Chapter 6), a very large nickel–cadmium system is being installed in Fairbanks, Alaska to sustain the power supply from the local utility.

Nickel–Metal-hydride Batteries

The nickel–metal-hydride battery also uses a nickel oxide positive electrode, but the negative active material is essentially hydrogen that is stored as a metal hydride. One of the popular metal hydrides is based upon a complex alloy of rare earth metals, nickel, aluminium, and other additives. An alternative alloy is based on titanium and zirconium. The electrode reactions are as follows.

At the positive electrode:

$$NiOOH + H_2O + e^- \underset{\text{Charge}}{\overset{\text{Discharge}}{\rightleftharpoons}} Ni(OH)_2 + OH^- \tag{9.8}$$

At the negative electrode:

$$MH_X + OH^- \underset{\text{Charge}}{\overset{\text{Discharge}}{\rightleftharpoons}} MH_{x-1} + H_2O + e^- \tag{9.9}$$

The operating voltage of a nickel–metal-hydride cell (1.2–1.3 V) is almost the same as that of nickel–cadmium, which allows ready interchangeability, and the discharge curve is quite flat. By contrast, the capacity of a nickel–metal-hydride cell is significantly greater than that of a nickel–cadmium cell of the same mass, with the result that the specific energy (60–80 Wh kg^{-1}) is higher. Moreover, nickel–metal-hydride batteries are capable of producing pulses of very high power. The batteries are resilient to both overcharge and overdischarge, and may be operated from -30 to $+45\,°C$. Another attraction is that there are no toxicity problems with recycling. Disadvantages associated with nickel–metal-hydride are a comparatively high cost, a higher self-discharge rate than nickel–cadmium, and a poor charge-acceptance at elevated temperatures.

In recent years, since their development in the late 1980s, small nickel–metal-hydride cells have become extremely popular for use in portable electronic devices (*e.g.* mobile phones, laptop computers, cassette players, *etc.*), as well as in toys and workshop tools. Coin cells are manufactured for use in calculators. To a significant degree, lithium batteries are now replacing them, but the market is still huge. Large (100 Ah) prismatic cells are made for assembly into 12- and 24-V modules and these have been employed as traction batteries in a number of demonstration electric vehicles. By virtue of their high power, nickel–metal-hydride has been the battery of choice for the new generation of hybrid electric vehicles that are under development in Japan, see Section 10.4, Chapter 10.

Nickel–Zinc Batteries

Zinc is the ideal material for negative electrodes in alkaline electrolyte batteries on account of its high electrode potential. It is the most electropositive of the common metals that can be plated from aqueous solution. The nickel–zinc cell has, therefore, a comparatively high standard voltage, *i.e.*

$$Zn + 2NiOOH + 2H_2O \underset{\text{Charges}}{\overset{\text{Discharges}}{\rightleftharpoons}} Zn(OH)_2 + 2Ni(OH)_2 \quad V^\circ = +1.78\,V \quad (9.10)$$

and a correspondingly high specific energy. A practical nickel–zinc cell discharges at ~ 1.6 V and can attain 90–100 Wh kg^{-1}, while an industrial battery pack would be expected to yield 70 Wh kg^{-1}. This performance is substantially higher than that of lead–acid, nickel–iron or nickel–cadmium, and is attractive for motive-power applications because it virtually doubles the daily range provided by lead–acid batteries.

The nickel–zinc battery has been extensively studied in recent years as a candidate traction battery for electric vehicles. Unfortunately, however, the system suffers from one serious drawback, namely, the greater solubility of zinc in potassium hydroxide, compared with cadmium or iron, leads to a much reduced cycle-life. During charge–discharge cycling, this solubility causes the zinc to migrate and accumulate towards the centre of the negative plate such that the plate densifies, changes shape, and loses capacity. Also, there is a marked tendency for zinc dendrites (needles) to grow from the negative during recharging and these can penetrate the separator and cause internal short-circuits. Much research has been directed towards elucidating and overcoming these two limitations, and some success has been reported. Prototype batteries with stabilised zinc electrodes have sustained up to 500 cycles, but this is still too low for the battery to achieve commercial success as a power source for electric vehicles. If the problem of short life can be resolved, then it would appear that the nickel–zinc battery could find use in many different applications. This is one of the remaining major challenges in the field of alkaline batteries.

Zinc–Air Batteries

The zinc–air battery is novel in two respects: (i) it has a gaseous positive active material, *viz.* oxygen; (ii) although not electrically rechargeable, it may be recharged 'mechanically' by replacing the discharged product, zinc hydroxide, with fresh zinc electrodes. With these features, the battery is akin to a fuel cell, with the 'fuel' being a slurry of finely divided zinc powder suspended in potassium hydroxide solution. During discharge, the zinc is converted to zincate ($Zn(OH)_4^{2-}$)

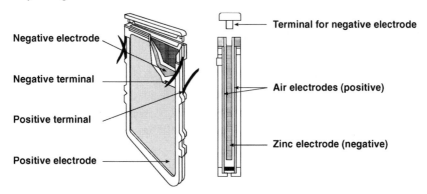

Figure 9.6 *Zinc–air cell and cassette system.*
(Courtesy of Batteries International)

and zinc hydroxide ($Zn(OH)_2$) at the negative electrode. The battery is recharged by removing the spent slurry from the cell and replacing it with fresh electrolyte and zinc powder. The spent slurry is then regenerated in a separate electrolysis unit. An alternative arrangement, for use in electric vehicles, has the zinc powder compacted onto a current-collector frame, which is inserted in a separator envelope that is flanked on both sides by air electrodes (Figure 9.6). The electrode assembly is in the form of a cassette that may be removed from the cell stack and conveyed to a central facility for electrochemical regeneration of particulate zinc. The overall system is composed of three units: a cell stack, a stack dismantling and refuelling machine, and a factory-based plant for fuel regeneration. A traction battery of eight modules (150 kWh) was built and tested successfully in Germany in a Mercedes Benz 410 postal van. The battery had high specific energy (200 Wh kg^{-1}), but only modest specific power (100 W kg^{-1} at 80% DoD). The range of the vehicle was 300 km between recharges.

9.3 Flow Batteries

Flow batteries possess certain characteristics that are typical of rechargeable batteries, but in other respects resemble fuel cells. They may be seen as a hybrid of the two. Like traditional secondary batteries, they are rechargeable and charge–discharge cycling involves the oxidation and reduction of a metal cation (or of an anion). Nevertheless, they are comparable with fuel cells in that the reagents are stored outside of the battery itself, in external reservoirs, and the capacity of the battery is determined only by the size of these reservoirs. Thus, as with a fuel cell, the energy content and the power of the battery are divorced; the latter is determined by the design and size of the cell stack. Moreover, in many designs of flow battery the cell stack is built on the 'plate-and-frame' principle, as shown in Figure 8.15(a), Chapter 8 for a bipolar fuel cell. The separator is either a microporous plastic membrane or an ion-selective polymer membrane of high ionic conductivity; the latter is similar to that used in a proton-exchange membrane fuel cell. The positive and negative electrode systems each have an electrolyte loop for the supply of reagents from the respective external reservoirs.

Zinc–Bromine Batteries

One of the best known of these flow batteries is the zinc–bromine battery (Figure 9.7). In the discharged state, the electrolyte in both loops is a concentrated solution of zinc bromide ($ZnBr_2$). The electrode reactions are as follows.

At the positive electrode:

$$Br_2 + 2e^- \underset{\text{Charge}}{\overset{\text{Discharge}}{\rightleftharpoons}} Br^- \qquad E^o = +1.065 \text{ V} \qquad (9.11)$$

At the negative electrode:

$$Zn \underset{\text{Charge}}{\overset{\text{Discharge}}{\rightleftharpoons}} Zn^{2+} + 2e^- \qquad E^o = -0.763 \text{ V} \qquad (9.12)$$

On charging, Zn^{2+} ions are reduced to zinc metal, which is deposited on the negative electrode, while Br^- ions are oxidised to bromine at the positive electrode. Because bromine is a highly volatile and reactive liquid, it is complexed with an organic reagent to form a poly-bromo compound, which is an oil and is immiscible with the aqueous electrolyte solution. This oil is separated and stored in a special storage compartment in the external reservoir of the positive electrode until needed again for discharge.

The standard cell voltage is 1.83 V, but typically falls to 1.3 V at an operating current density of 100 mA cm^{-2}. An attractive feature of this battery is that it is constructed mostly of lightweight plastic components (frames, tanks, plumbing, *etc.*) and is easily assembled. Research and development has been on-going in

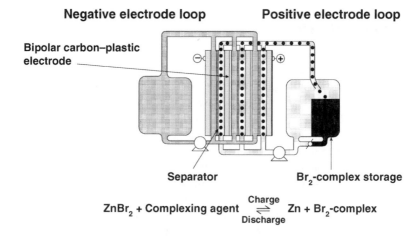

Figure 9.7   *Schematic of zinc–bromine flow battery.*

several countries for a number of years. Traction batteries have been built and tested in electric vehicles, and larger systems have been demonstrated as energy-storage facilities for electric-utility applications. The zinc–bromine battery has, however, yet to prove a commercial success. A major technical problem is the chemical reactivity of bromine towards most plastic components, while there is also concern over safety in the event of leakage of bromine vapour.

Redox Batteries

Other flow batteries are of the redox type in which ions are oxidised and reduced but remain in solution, with no solids, gases or immiscible liquids being formed. One such battery – the vanadium redox system – is being investigated in Australia. This depends for its operation on the fact that vanadium exists in several different valence states. In the charged state, the positive-electrolyte loop contains a solution of V^{5+} and the negative loop contains a solution of V^{2+}. On discharging, the former solution is reduced to V^{4+} and the latter is oxidised to V^{3+}, as follows.

At the positive electrode:

$$VO_2^+ + 2H^+ + e^- \underset{\text{Charge}}{\overset{\text{Discharge}}{\rightleftharpoons}} VO^{2+} + H_2O \qquad E^o = +1.00\,V \qquad (9.13)$$

At the negative electrode:

$$V^{2+} \underset{\text{Charge}}{\overset{\text{Discharge}}{\rightleftharpoons}} V^{3+} + e^- \qquad E^o = -0.26\,V \qquad (9.14)$$

The standard cell voltage is 1.26 V. Under actual operating conditions, however, a concentrated electrolyte is used (2 M vanadium sulfate in 2.5 M sulfuric acid) and the open-circuit cell voltage is 1.6 and 1.4 V at 100 and 50% SoC, respectively. Since both electrolyte loops contain only vanadium, any leakage across the ionically conducting membrane separator will not cause a significant loss in cell performance. There are four storage tanks external to the battery: two for the reagents in the charged state and two for the discharged state. A 12-kWh vanadium redox battery was constructed in 1994 for use in the storage of solar energy. More recently, a 450-kW, 1-MWh facility has been built as a demonstration system for load-levelling service at the Kansai Electric Power Plant in Japan, as well as several other 25-kW units for wind energy storage and other stationary services. The latter units have reported a life of more than 16 000 cycles. The advantage seen for this technology over lead–acid lies in its indefinite cycle-life, which is limited only by the materials of construction and not by the reactants in solution. It should be noted, however, that the vanadium redox battery has a low specific energy (typically, 25 Wh kg^{-1}), and is therefore unlikely to find application in electric road vehicles.

The latest flow battery that has been under intensive development is known as the Regenesys® system. This differs from other flow batteries in that no metal cations are involved in its electrode reactions. Rather, anions are oxidised and reduced at both electrodes. The cell, which is again built on the bipolar plate-and-frame principle, utilises a membrane separator that is permeable to sodium ions but not to sulfide anions. During discharge, the reaction at the positive electrode is the reduction of bromine dissolved in sodium bromide (NaBr) solution to bromide ions, while the reaction at the negative electrode involves the oxidation of sulfide ions to sulfur, which is contained in sodium polysulfide solution. The electrode reactions are as follows.

At the positive electrode:

$$Br_2 + 2e^- \underset{\text{Charge}}{\overset{\text{Discharge}}{\rightleftharpoons}} 2Br^- \qquad E^\circ = +1.065\ V \tag{9.15}$$

At the negative electrode:

$$S^{2-} \underset{\text{Charge}}{\overset{\text{Discharge}}{\rightleftharpoons}} S + 2e^- \qquad E^\circ = -0.508\ V \tag{9.16}$$

Thus, the standard cell voltage is 1.57 V.

A schematic of this battery, which is also considered to be a type of 'regenerative fuel cell' (see Section 8.1, Chapter 8), is shown in Figure 9.8(a). The system has been developed by RWE nPower, a UK utility company, for large-scale energy storage at power stations and within the electricity-supply network itself. The plant is constructed on a modular basis; the individual modules have an output power of 100 kW (Figure 9.8(b)). A 10-MW plant would use 100 of these modules joined together electrically, and have reservoirs to give a stored energy capacity of 100 MWh. Until recently, two prototype plants were under construction – one at a power station in the UK and the other in the USA. These would have been among the largest batteries ever to be constructed but, in early 2004, both projects were postponed for commercial reasons.

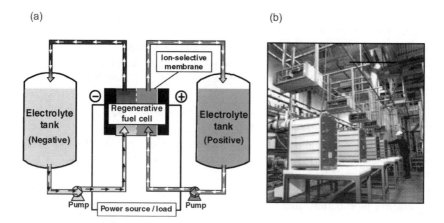

Figure 9.8 *Regenesys® flow battery:* (a) *schematic of operating principle;* (b) *100-*kW *modules.*
(Courtesy of RWE nPower)

9.4 Sodium Batteries

Sodium–Sulfur Batteries

Sodium is a highly reactive metal that can ignite or explode when treated with water; it is therefore not an obvious candidate as an electrode in a battery. On the other hand, it has some attractive features, namely, it: (i) has a high electrochemical reduction potential (-2.71 V, compared with -0.76 V for zinc); (ii) is abundant and low cost; (iii) is non-toxic.

The possible use of sodium in a battery became a practical proposition in 1967 when scientists working at the Ford Motor Company in the USA discovered that a solid ceramic material, sodium β-alumina, is an exceptionally good conductor for sodium ions at temperatures above about 300 °C. This opened up the possibility of a high-temperature battery based on liquid sodium as the negative active material. The same group of scientists also devised the idea of using liquid sulfur as the positive active material – and so was born the sodium–sulfur battery. This battery operates at 300–400 °C and is contained in a heated, insulated enclosure. The cell discharges in two steps as Na^+ ions pass from the sodium negative electrode, through the β-alumina electrolyte, to the sulfur positive electrode, *i.e.*

Step 1:

$$2Na + 5S \underset{\text{Charge}}{\overset{\text{Discharge}}{\rightleftharpoons}} Na_2S_5 \qquad V^o = +2.076 \text{ V} \qquad (9.17)$$

Step 2:

$$2xNa + (5-x)Na_2S_5 \underset{\text{Charge}}{\overset{\text{Discharge}}{\rightleftharpoons}} 5Na_2S_{5-x(0<x<2)} \qquad (9.18)$$

$$V^o = +2.076 \rightarrow +1.78 \text{ V}$$

Thus, the standard cell voltage is constant at 2.076 V for the first part of the discharge, *i.e.* as far as the production of Na_2S_5, after which it declines linearly to 1.78 V. The electrolyte is in the form of a ceramic tube that holds the liquid sodium, while the liquid sulfur is contained within an outer annulus between the electrolyte tube and the cell housing. Since liquid sulfur is a non-conductor, it is held in the interstices of graphite felt that is packed around the outside of the tube. The felt, therefore, provides electrical contact between the electrolyte tube and the outer cell casing. The cell is sealed hermetically so as to insulate the electrodes from each other and to exclude the atmosphere.

While this concept sounds simple enough, in practice it proved to be incredibly complex. Teams of battery researchers in Germany, Japan, the UK and the USA worked on developing the sodium–sulfur battery as an electric-vehicle battery for over 20 years. Finally, in the late 1980s, most of these teams were disbanded as the technical and safety problems were deemed to be intractable. This was despite the fact that the first electric vehicle to be equipped with a sodium–sulfur battery (a 50-kWh unit) was driven in England as early as 1973. The Japanese, however, continued their programme of research and development, but with the emphasis on load-levelling duty. Prototypes of rather large utility batteries have been constructed and tested in Japan by the Tokyo Electric Power Company. It is not clear how the Japanese workers may have overcome the technical problems of durability, reliability and safety that others have found to be insurmountable for electric-vehicle applications.

Sodium–Nickel-chloride Batteries

Drawing upon the research conducted on sodium–sulfur batteries, a new sodium-based battery was invented by workers in South Africa and the UK in the late 1970s. They recognised that many of the technical problems of the sodium–sulfur battery arose from the sulfur electrode and that if an alternative could be found, a high-temperature sodium battery was still a possibility. After some exploratory work, the researchers eventually settled on nickel chloride as the positive active material to give the sodium–nickel-chloride battery, popularly known as the ZEBRA battery in acknowledgement of its South African origins.

The cell reaction is very simple:

$$NiCl_2 + 2Na \underset{\text{Charge}}{\overset{\text{Discharge}}{\rightleftharpoons}} Ni + 2NaCl \qquad V^o = +2.58 \text{ V} \qquad (9.19)$$

The cell retains the β-alumina electrolyte tube and liquid sodium as the negative electrode, but in this case the sodium surrounds the outside of the electrolyte tube and the nickel chloride is on the inside, see schematic in Figure 9.9(a). Further work showed that this cell only works satisfactorily when assembled in the discharged state. There is then no need to purify and handle sodium metal; rather, it is formed in situ during charging as sodium ions pass through the electrolyte from the positive to the negative electrode. The battery operates over a wide temperature

(a)   (b)

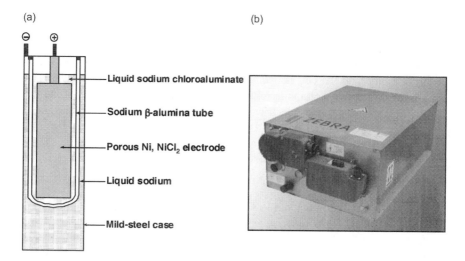

Liquid sodium chloroaluminate

Sodium β-alumina tube

Porous Ni, NiCl$_2$ electrode

Liquid sodium

Mild-steel case

Figure 9.9   (a) *Schematic diagram of a sodium–nickel-chloride cell;* (b) *ZEBRA electric-vehicle battery (278 V, 17.8 kWh).*

range, from 175 to 400 °C. Because the starting materials of the positive electrode (nickel, sodium chloride) are both solid, it is necessary to add a liquid in the positive compartment to make electrical contact with the ceramic electrolyte. The liquid used is molten sodium chloraluminate ($NaAlCl_4$). A traction battery consists of many such cells, wired in series and contained within a double-walled, thermally insulated, battery box.

A 17.8-kWh sodium–nickel-chloride battery that operates at either 278 or 557 V, depending upon the internal wiring configuration, is shown in Figure 9.9(b). Over the past 15 years, many of these batteries have been built in the UK and Germany and used in electric vehicles, which include buses and popular cars (Renault 'Twingo' and 'Clio'; Opel 'Astra'; Mercedes-Benz '190' and 'A' class; BMW '3 series'). These trials have proved highly successful and have demonstrated that ZEBRA batteries are robust, reliable, and safe in operation. Also, they require no maintenance. The vehicles have good ranges of at least 160 km between charges and the driving characteristics are almost indistinguishable from those of the internal-combustion engine versions. Given the encouraging performance obtained from these batteries, a factory is being built in Switzerland for their manufacture.

A limitation to the possible applications for ZEBRA batteries lies in the requirement to keep them hot. The batteries can be cooled and re-heated, but this requires care and would not be a regular practice. They are best suited to vehicles that are used daily on a regular schedule, such as buses and delivery trucks When not in use for more than a few hours, the pack would normally be plugged into the mains for recharging and for maintaining the temperature by means of heaters incorporated in the battery box. The heat loss is small and the battery may be left unattended and disconnected for at least 48 h. A major advantage of high-temperature batteries is that their performance is independent of the ambient temperature and, therefore, they can withstand variations in the latter that conventional batteries would find intolerable.

ZEBRA batteries should also be satisfactory for standby duties, for instance in a telephone exchange or mainframe computer installation. The battery would be floated at 2.58 V per cell until called upon for service. A small input of mains electricity to the in-built heaters would serve to keep the battery at operating temperature. Taken all round, the ZEBRA battery remains a strong candidate for electric vehicles and also for other applications where a medium-to-large battery is required.

9.5 Lithium Batteries

Lithium is the lightest of all the metals. This fact, together with its high electrochemical reduction potential (-3.05 V), makes it particularly attractive as an active material for negative electrodes. The stored energy, expressed as Wh kg^{-1}, is higher than for any other metal. Lithium, like sodium, is chemically reactive and cannot be used in conjunction with an aqueous electrolyte. Possible non-aqueous electrolytes are fused salts, organic liquids, polymers, and ceramics. Unfortunately, the lithium analogue of sodium β-alumina is not a satisfactory ion conductor and no lithium–sulfur or lithium–nickel-chloride battery based on a ceramic electrolyte has been demonstrated. Considerable research has in fact been carried out on lithium fused-salt electrolytes, but progress to date has been limited. Attention has therefore focused on the use of organic liquid and polymer electrolytes.

Small, primary lithium batteries based on organic liquid electrolytes have been manufactured for several decades for use in pocket calculators, and as memory back-up displays for computers, televisions, video cassette recorders, and cameras. These are normally in the form of 3-V coin cells. The negative electrode is lithium metal foil, there are various possible positive electrodes (MnO_2, CF_x, MoO_3, V_2O_5, *etc.*), and the electrolyte is a solution of a salt in an organic solvent. These primary cells are in widespread use on account of their attractive features, which include high voltage, high specific energy, low self-discharge rate, and a wide temperature range of operation.

Initially, attempts to develop rechargeable cells did not meet with much success. The problems were found to lie in the re-plating of lithium metal from organic solution during recharging. Generally, the lithium was laid down as a mossy deposit, which after a few cycles often led to an internal short-circuit or even a fire. Furthermore, the freshly plated lithium was not chemically stable with respect to the electrolyte. Lithium is only stable in a primary cell containing an organic electrolyte by virtue of the formation of a thin film of corrosion product on its surface, which acts as a solid electrolyte and offers protection from further reaction with the organic solvent. Various attempts were made in the 1980s, both in the USA and Japan, to manufacture and market secondary lithium cells that used negative electrodes of lithium metal, but eventually these cells were withdrawn as a result of safety-related concerns.

Lithium-ion Batteries

A significant advance was made in 1979 with the discovery at Oxford University in the UK that the compounds $LiCoO_2$ and $LiNiO_2$ possess a layer structure and that Li^+ ions may be electrochemically withdrawn from the structure and replaced reversibly. This discovery opened up the possibility of using either of the compounds as the active material for a positive electrode in a rechargeable cell. Because these are 4-V electrodes with respect to lithium metal, it is possible to intercalate the Li^+ ions into a suitable negative active material, such as graphite, and still have a 3.5 to 4-V cell. In 1990, the Sony Corporation in Japan announced the 'lithium-ion battery' that was based on this concept. This was the first rechargeable lithium battery to contain no lithium metal, but to depend entirely on the difference in electrochemical potential of lithium ions intercalated in $LiCoO_2$ and in graphite.

The mode of operation of the lithium-ion cell is shown schematically in Figure 9.10. The cell is assembled in the discharged state. During charge, Li^+ ions are withdrawn ('de-intercalated') from the structure of $LiCoO_2$, transported across the electrolyte, and then 'intercalated' into the structure of the carbon. This process is reversed on discharge. In effect, the lithium ions 'rock' back and forth between the electrodes and hence the technology is sometimes referred to as a 'rocking-chair' cell. This 3.8-V cell soon went into mass production in Japan, mostly in small sizes, and is now widely employed in camcorders, laptop

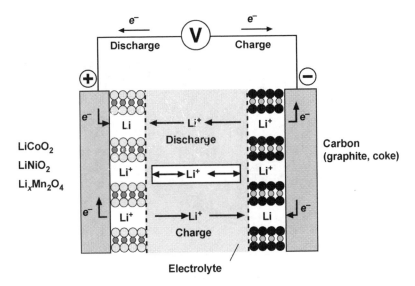

Figure 9.10 *Schematic of mode of operation of a lithium-ion battery.*

computers, and mobile phones. Because of their much higher voltage than nickel–metal-hydride cells (1.2 V), lithium-ion cells store more energy and so provide longer run-times between charges. For this reason, they have usurped some of the markets for nickel–metal-hydride despite being more expensive. Many companies are manufacturing lithium-ion cells with either $LiCoO_2$ or $LiNiO_2$ positive electrodes and the growth in sales, starting in 1991, has been phenomenal. It is estimated that well in excess of 500 million cells are now produced per year worldwide. Ideally, manufacturers would like to replace cobalt (expensive) and nickel (moderately priced) by manganese (cheap and less toxic). Unfortunately, lithium manganese oxide ($LiMnO_2$) is a complex entity that is more difficult to prepare and less stable structurally. Considerable research is presently being devoted to developing a satisfactory intercalation electrode based on manganese and the compound $LiMn_2O_4$ has been employed commercially.

Lithium-ion cells are not, however, without their problems. Aside from relatively high cost, considerable care has to be exercised in controlling the voltage during charge. Overcharging or heating above about 100 °C leads to decomposition of the positive electrode. When cells are coupled in series, or a series–parallel array to form a battery pack, it is necessary to incorporate battery-protection circuits to avoid overcharge and the possibility of fires resulting from the presence of lithium metal. Much larger cells have been constructed, including experimental batteries for electric vehicles and stationary energy storage, but these are not yet commercially available. A 2-kWh battery module produced in Japan and made up of eight 250-Wh cells based on Li-rich $LiMn_2O_4$ positive electrodes is shown in Figure 9.11. For these larger batteries to become viable for motive-power or standby duties, their cost will have to be reduced substantially and they will have to be safe under all conditions of operation.

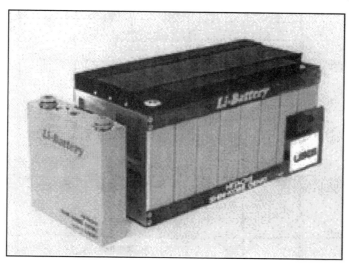

Figure 9.11 *Lithium-ion battery module (8 cells, 2 kWh).*

Lithium–Polymer Batteries

Research on ionic conduction in polymers was undertaken at the University of Grenoble (France) in the 1970s. It was found that polymers with a relatively high dielectric constant, such as polyethylene oxide, could dissolve lithium salts. At ambient temperature, the resulting conductivity was too low for this material to be considered as a possible solid electrolyte, but at about 60 °C there was a transition from a partially crystalline state to a fully amorphous state with an increase of several orders of magnitude in the conductivity. At 80–100 °C, the ionic conductivity was still low compared with conventional battery electrolytes, but this could be compensated if the electrolyte was in the form of a sufficiently thin film. Calculations showed that for an acceptable voltage drop of 10 mV across the electrolyte, the thickness of the polymer would need to lie in the range 10–100 μm. It was well known that polymers could be cast in thin films, as used in packaging, and it was soon found possible to make these new electrolytes as films of appropriate thickness. Small cells were then assembled with negative electrodes made from lithium foil, and with positive electrodes that consisted of an intimate mix of the polymer electrolyte, finely divided carbon and a suitable electroactive material such as manganese dioxide (MnO_2) or vanadium oxide (V_2O_5). Early experiments in the 1970s demonstrated that these small rechargeable cells had some remarkable properties and exciting prospects.

With an ionic conductivity that is low even at 100 °C, it is necessary not only to use thin electrolyte films, but also to restrict the current density of the cell to less than 1 mA cm^{-2}. This poses no great problems, at least in principle, since within a given volume it is possible to pack a large surface-area of cell, typically many square metres. Although the current density is low, the total cell current may be quite high when account is taken of the large surface area. Thus, the manufacturing challenge is to make many square metres of electrolyte film and then apply a coating of the positive electrode mix. Technologies used in the plastics film and the coated paper industries were adapted for this purpose. It was then necessary to add the thin lithium-foil negative and a positive current-collector, and finally to laminate the sandwich together to form an all-solid-state cell. Such operations have to be conducted in a dry-room because of the sensitivity of the components to moisture. Once the laminate has been made, it may be configured in many different ways, as shown in Figure 9.12. Small, flat-plate cells are packaged in composite aluminium–plastic envelopes (*e.g.* of the type used to hold powdered foodstuffs that are moisture sensitive.) Cells of larger area can be rolled into cylinders, or folded, as indicated in Figure 9.12.

The limitations of these cells were found to be the requirement to operate in the temperature range 80–120 °C and the comparatively short charge–discharge cycle-life of early models. Nevertheless, the attractions foreseen for thin, flexible cells based on polymers have encouraged research groups throughout the world to pursue the concept vigorously. One line of approach has been to bring the operating temperature down to ambient by using an organic liquid electrolyte immobilised in a gel polymer, a so-called 'gelionic' electrolyte. When optimised, such electrolytes have most of the mechanical properties of a solid-polymer film and are essentially dry, but provide better conduction of lithium ions. A second approach is to investigate alternative polymers that dissolve lithium salts to give solid solutions that are amorphous (*i.e.* have good conductivity) at ambient temperature. Yet a third is to abandon the lithium-metal negative electrode and aim for the polymer analogue of the lithium-ion cell. This involves a trade-off in accepting some loss of

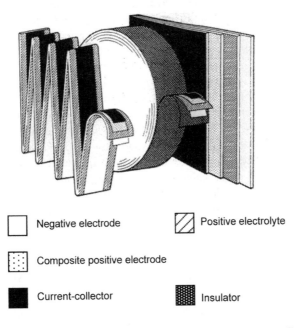

☐ Negative electrode		▨ Positive electrolyte	
▦ Composite positive electrode			
◼ Current-collector		▦ Insulator	

Figure 9.12 *Alternate configurations for large-area, lithium–polymer cells.* (Courtesy of AEA Technology Batteries)

voltage in return for the claimed better rechargability and improved safety of lithium-ion cells.

Although operation at ambient temperature is a pre-requisite for small batteries employed in portable electronics equipment, this is not necessarily the case for large industrial batteries. For example, batteries that operate at 100 °C may be ideal for electric and hybrid electric vehicles. It should be a simple matter to package and maintain a traction battery at this temperature, and so be independent of the prevailing ambient temperature. For operation at extremes of ambient temperature, where aqueous electrolyte batteries are unsuitable, polymer batteries would then be acceptable. A consortium working in the USA and Canada has produced prototype traction modules based on lithium-foil negative electrodes, which have a capacity of 119 Ah, an output of 20 V, and a specific energy as high as 155 Wh kg^{-1}. The modules are said to have a life in excess of 600 cycles and to be safe under all conditions of abuse. If these figures are substantiated, the prospects for lithium–polymer batteries are bright, assuming that the manufacturing costs are reasonable.

9.6 Prospects for Batteries

In the above sections, a brief account has been given of the different types of battery that possibly could be used for the storage of electrical energy for applications that range from portable consumer electronic devices to bulk electricity storage at power stations.* In practice, many of these batteries are at various stages of development and have significant failings, as well as attractive features. Only two, lead–acid and nickel–cadmium, are commercially available today in sizes to match large installations and at prices that are likely to be acceptable. Others, which are commercially available in small or medium sizes (*e.g.* nickel–metal-hydride and lithium-ion), could be scaled-up if the demand existed, although their cost may be a deterrent. Some of the important parameters of the various battery chemistries are summarised in Table 9.1.

Although stored energy and peak power per unit mass are the most commonly cited numerical values for advanced batteries, it should be emphasised that these are not necessarily the most important criteria for particular applications. Even more significant considerations may be initial cost, overall electrical efficiency, reliability and freedom from maintenance, performance under fluctuating ambient temperatures, and effective lifespan under deep-discharge cycling.

What are the prospects of further advances in battery technology? The theoretical limit to the specific energy of a battery is set by the free energy of the electrochemical reaction (which determines the cell voltage), the number of electrons transferred in the reaction, and the mass of the electrodes. Generally, the theoretical cell energy, calculated in this way, is three-to-five times that which is practically achievable. The reason for this huge discrepancy is that the practical value has to

---

* For readers who may have a more detailed interest in batteries, reference is made to two previous publications by the authors: *Understanding Batteries*[2] is an introductory text to all classes of battery; *Batteries for Electric Vehicles*[3] is a more specialised treatment pertaining to traction batteries for electric and hybrid electric vehicles.

Table 9.1 *Summary of battery characteristics*[a]

Battery	*Specific energy* ($Wh\ kg^{-1}$)	*Peak power* ($W\ kg^{-1}$)	*Status*
Lead–acid	35–50	150–400	Widely used, cheapest available
Nickel–iron	50–60	80–150	Low electrical efficiency
Nickel–cadmium	40–60	80–150	Commercially available, but costly
Nickel–metal-hydride	60–80	200–300	Available in small-to-medium sizes; used in hybrid electric vehicles; costly
Nickel–zinc	70–100	170–260	Short cycle-life
Zinc–bromine	70–85	90–110	Under development
Zinc–air	100–200	80–100	Mechanically rechargeable only
Vanadium redox	20–30	110	At demonstration stage for stationary energy
Regenesys®	n.a.	n.a.	Development programme postponed
Sodium–sulfur	150–240	230	Development almost ceased
Sodium–nickel-chloride	90–120	130–160	Battery electric vehicle applications
Lithium-ion	80–180	200–1000	Commercially available in small sizes; careful control of recharging required

[a] As noted in the text, the values for specific energy and peak power are broad-brush only and depend almost as much on the design of the battery as on its chemistry.

take account of the mass of all the other cell components (electrolyte, separators, container, current-collectors, terminal posts, *etc.*), as well as the coulombic inefficiencies arising from side-reactions, such as corrosion and self-discharge, and voltaic inefficiencies associated with overpotential and resistive losses. Most of the inorganic electrochemical couples that are likely to be economically and commercially viable have now been investigated, with the possible exception of new intercalation electrodes and electrolytes for lithium batteries, and it appears unlikely that radically new electrochemical couples will emerge. Nevertheless, there is still enormous scope for improving further the batteries already in existence by means of advanced materials and better methods of construction and quality control. Even the lead–acid battery, after nearly 150 years of history, is still being improved.

9.7 References

1 D.A.J. Rand, P.T. Moseley, J. Garche and C.D. Parker (eds), Valve-Regulated Lead–Acid Batteries, Elsevier, *Amsterdam, The Netherlands*, 2004.

2 R.M. Dell and D.A.J. Rand *Understanding Batteries*, Royal Society of Chemistry, Cambridge, UK, 2001.

3 D.A.J. Rand, R. Woods and R.M. Dell in *Batteries for Electric Vehicles*, Research Studies Press, Taunton, UK, 1998.

CHAPTER 10

Electric Propulsion

Almost 60% of the world's oil production is consumed by transportation, essentially all by petrol and diesel engines. Transportation is therefore a particularly important sector of the economy in which to be looking for the introduction of clean, sustainable energy sources. The most obvious approach is to switch to electric propulsion since electricity is the most direct way of utilizing renewable energy.

In this chapter, we describe electric and hybrid versions of passenger and goods vehicles. These fall into the following five broad categories, as distinguished by the source of the electrical power.

(i) Vehicles supplied directly by mains electricity; this category includes tramcars and trolleybuses, electric trains, and urban metro systems. All of these, with the exception of trolleybuses, run on tracks.

(ii) Diesel–electric traction in which a diesel engine drives a generator to supply electricity to a motor; this form of propulsion is employed mostly in large units such as railway locomotives and ships.

(iii) Battery electric vehicles (BEVs): these are predominantly small, off-road units, although there has been appreciable interest in battery electric cars and vans for urban use. Conventional submarines are also battery-powered when operating submerged.

(iv) Hybrid electric vehicles (HEVs): these have dual power sources, at least one of which is electric.

(v) Fuel-cell vehicles (FCVs): currently, there is great interest among the automotive companies in the use of fuel cells to provide the electricity for motive power.

Electric propulsion has the overwhelming attractions of alleviating the present pollution from internal-combustion-engined vehicles (ICEVs) and of providing silent power for transportation in our cities, in our countryside, and in marine environments.

10.1 Traditional Vehicles

Tramcars and Trolleybuses

Early in the 20th century, electric tramcars ('trams') were a common sight in cities around the world. This was a natural extension of the horse-drawn trams that had existed in the 19th century. It was not until mains-generated electricity became widely available that it was possible to electrify tramways with overhead cables. In many cities, tramlines extended from the business centre into the suburbs and the trams were extensively used by commuters (a term not known at the time) for travel to and from work. A tram that was operated in Bristol, UK in the 1930s is shown in Figure 10.1(a).

There were certain disadvantages associated with trams. Early versions were generally open-topped and passengers on the top deck were therefore exposed to the vagaries of the weather.* Also, the steel wheels on the steel track made a great deal of noise, which was augmented by the frequent sounding of a loud bell to warn other road users of the tram's approach. Perhaps the most significant disadvantage arose from the very fact that they were tracked vehicles. The twin tracks occupied the entire centre of the road and other vehicles had to squeeze by on either side. To reach or leave tram-stops, passengers had to cross the flow of traffic. This was dangerous and held up the rest of the traffic. There was also the difficulty of overtaking other, or broken-down, trams. Finally, the points system in the road was complex and often the driver had to alight in order to switch manually the track to a new direction. Despite all these disadvantages, the tram fulfilled a useful purpose in its day. It should be noted, however, that trams still operate in many cities today, for example, in Vienna (Austria), Melbourne (Australia), St Petersburg (Russia), and Blackpool (UK).

Sometime later, the trolleybus was developed and became well established; one in service in Portsmouth, UK in 1936 is shown in Figure 10.1(b). The trolleybus was a cross between a tram and a bus. There were no tracks and it ran on rubber tyres. Consequently, the motion was smooth and essentially silent. The driver and passengers were enclosed and not exposed to the weather. Not being confined to tracks, this vehicle had much greater freedom of movement on the road and could pull into the kerb for the greater safety of passengers. On the other hand, being mains-operated, there were still the limitations associated with having to remain attached to overhead electric wires. Trolleybuses were very popular and they are still in use in some European cities, and elsewhere. Some cities that introduced conventional buses as replacements in the second-half of the 20th century are even considering reversing this decision in the belief that the advantages of clean and silent operation may offset the disadvantages of a catenary system. One option is to have dedicated trolleybus lanes. Both single- and double-decked trolleybuses are known; the former may tow a second passenger coach, often articulated, and are thus able to seat at least as many passengers as a double-decker. In Germany, the

---

* The open-topped design was a natural progression from the horse-drawn tram in which, obviously, it had been important to minimize weight. Moreover, in the UK, the 1870 Tramways Act gave local authorities the right to enforce the compulsory purchase of any routes that ran through their areas for 21 years after the respective routes were first built, and every seven years thereafter. Thus, many tram companies were reluctant to invest in new, closed-topped fleets.

(a)   (b)

(c)

Figure 10.1 (a) *Early design of UK tram;* (b) *UK trolleybus;* (c) *German 'Duo-Bus',
a mains-battery hybrid.*

'Duo-Bus' was developed (Figure 10.1(c)). This was a mains-battery hybrid bus, in
which the traction battery allowed limited 'off-the-line' travel and thereby
enhanced manoeuvrability. With such a facility, a trolleybus can travel under
mains power from the city centre and then move around a suburb using battery
power when collecting and delivering passengers nearer home.

Railways and Metros

During the course of the 20th century steam trains were largely phased out and
replaced by diesel or electric locomotives. Electric propulsion is now widely
employed for mainline railways and metro systems, and is generally acknowledged
to be a convenient and clean form of transport. Particular operational advantages lie
in the freedom from pollution in the urban environment and in the fast acceleration
of electric trains. The downside of electric traction lies in the capital cost of
installing the electricity-supply network and the lack of flexibility in not being able
to operate on non-electrified track. In principle, it is possible to avoid these
limitations by means of battery electric traction, and small battery-powered trains

have been used in Germany for local journeys. These have not proved popular elsewhere, however, as their range between battery charges is small. A better prospect might lie in an all-electric mains-battery hybrid. The battery would permit limited 'off-the-wire' operation, for example shunting activities on non-electrified track. Sometimes, the electrification of an existing track is difficult as there is insufficient headroom for the catenary in tunnels that were originally built for railways in the 19th century. A hybrid propulsion system would overcome this problem by employing the battery for motive power while in the tunnel. Finally, the addition of a battery to a mains-operated train would permit the recuperation of braking energy when stopping. Despite these superficial attractions, hybrid trains have not yet been widely adopted. This is a possibility for the future.

Yet another option is a diesel-battery hybrid locomotive. By using the battery as an auxiliary power source when accelerating, it would be possible to economise on diesel consumption by employing a smaller engine, which is sized for steady-speed running. Some of the energy lost in braking could also be recuperated. This is the railway analogue of the hybrid electric car (see Section 10.4).

Metro rail systems are invariably mains electric as the fumes from diesel engines in subways would be intolerable. Many capital and major cities throughout the world have their own metro systems that are extensively utilised by commuters, shoppers and tourists. The systems vary considerably in the size of the network, age, cleanliness and comfort, and operational policy. London has one of the largest and oldest systems ('The Underground'), while two modern metros are to be found in Singapore and in Washington, DC, USA. As cities become more and more congested with surface traffic, underground railways become increasingly attractive and it is likely that further cities, where geological conditions are favourable, will consider building metro systems.

Marine Craft

Another application of electric traction is in submarines. Conventional diesel–electric submarines use their diesel engines for propulsion when on or near the surface and also for recharging the traction batteries that power the vessel when submerged. These craft have exceptionally large lead–acid traction batteries that are specially designed for the purpose. Nuclear submarines (Figure 10.2(a)) do not require air for their reactors and so are able to remain submerged and undetected for months at a time. This is a major operational advantage. In the event of reactor shutdown, emergency power is provided by stand-by batteries that are as large, or larger, than those on conventional submarines. Several of the world's navies have considered replacing lead–acid batteries by alternative ('advanced') batteries or by fuel cells. Submarines powered by fuel cells have been constructed for evaluation.

Some modern ships have diesel–electric propulsion. The diesel engine drives a large generator that supplies electricity to the propulsion motors. Many modern cruise ships are of this type. A diesel–electric ship that supplies oil platforms in the North Sea is shown in Figure 10.2(b). Using four, highly controllable, electric motors, it is possible to hold station close to a fixed platform even in heavy seas. This is a necessary feature when transferring staff or supplies.

(a)   (b)

Figure 10.2 (a) *UK nuclear submarine;* (b) *'Big Orange', a North Sea diesel–electric supply ship.*

At the opposite end of the marine spectrum, some small pleasure craft for use on inland waters are electrically propelled. The first recorded attempt to power a boat by batteries was made in St Petersburg, Russia, in 1838. Some 40 years later, an electrically operated outboard motor was developed in France. By 1888, a fleet of 22 electric pleasure boats was operating on the River Thames in London and was supported by an infrastructure of five floating and four land-based charging stations. The largest of these craft was a 75-passenger launch. With the development of the internal-combustion engine, battery-driven boats went out of fashion. Having regard to the noise and pollution caused by diesel engines, particularly in river locks when the engines are idling, it may be timely to consider the re-introduction of electric propulsion.

Off-road Vehicles

Battery electric traction is employed extensively in situations where the pollution and noise associated with internal-combustion engines are unacceptable. Examples are in hospitals, city parks, holiday resorts, retirement villages, factories, warehouses, railway stations, airports, and mines. Most of these vehicles are forklift trucks, platform trucks or tractors for the handling and conveyance of goods. Airports also operate electric vehicles for conveying passengers in terminals, for loading luggage onto planes, and for pushing out planes from the terminal. Other categories of off-road electric vehicle include invalid chairs, golf-carts, and recreational vehicles. Some of these various applications are shown in Figure 10.3.

10.2 Battery Electric Road Vehicles

At the end of the 19th century, with rechargeable batteries becoming mass-produced, BEVs began to replace horse-drawn carriages. By 1912, several hundred thousand electric cars and vans were in service throughout the world in major cities (London, Paris, New York). For a short time, BEVs, ICEVs and steam-propelled cars were in competition with each other. Gradually, the limitations of steam-driven and electric vehicles became apparent and thus ICEVs prospered. Ironically, it was the invention of the battery-operated self-starter that helped to sound the

Figure 10.3 *Various off-road electric vehicles:* (a) *forklift trucks;* (b) *belt loader for aircraft;* (c) *invalid car;* (d) *golf buggy.*

death-knell of the battery electric car. Since that time, there has been continuous experimentation and development of BEVs and countless prototype models have been produced, mostly adaptations of conventional ICEVs. Here, we summarise briefly some of these efforts. Before doing so, it is worth pointing out the two key limitations of BEVs compared with ICEVs.

- the energy content of petrol is around $12\,500$ Wh kg^{-1}, but is only 30–40 Wh kg^{-1} for the lead–acid battery; this disparity results in a large, heavy battery to carry around and a very restricted vehicle range;
- a battery takes many hours to recharge fully, whereas a petrol tank can be filled in minutes.

The consequence is that pure BEVs will never replace the family car or long-distance bus or truck and it is necessary to look for niche applications where these limitations are not too serious and where the advantages of electric traction (quiet operation, no pollution at the point of use) outweigh the disadvantages.

Delivery Vans

There are rather few battery electric road vehicles in general use, although in the UK door-to-door milk delivery has been traditionally by this means, using specially

(a)   (b)

Figure 10.4 (a) *UK milk-float;* (b) *Ford 'Ecostar' electric van.*

designed 'milk-floats' (Figure 10.4(a)). This duty involves repeated start–stop operations that are ill-suited to an internal-combustion engine. The short distance to be covered daily (up to 45 km) is ideal for BEVs. Refuelling with cheap off-peak electricity, and silent delivery early in the morning while customers are still asleep are added attractions. There have been as many as 40 000 milk-floats in use in the UK, although with the advent of supermarkets and universal refrigeration, daily milk delivery is a declining market.

Battery electric propulsion is similarly well-suited to any form of urban drop-off service where the daily range is limited and the stops are frequent. A prime example of this is postal delivery and post offices in several countries (Austria, Germany, UK, USA) have conducted extensive trials with electric postal vans. Other door-to-door services that may be suitable include laundry collection, delivery of newspapers and flowers, domestic-appliance engineers, meter readers, mobile libraries, *etc.* Despite efforts to introduce electric vans, particularly during the mid-1970s, the application of such vehicles has failed to attract much commercial interest. An example of one of these vans, the Ford 'Ecostar', is shown in Figure 10.4(b). A fleet of 105 of these vehicles was launched in 1993 and had logged more than 1.6 million km by the end of the demonstration programme in September 1997.

Buses

Urban buses are an attractive target for electric propulsion since their size is sufficient to carry the number of batteries required to provide a workable range. Also, the public relations aspects of operating silent, non-polluting vehicles in city centres are important, since large diesel engines can cause considerable environmental degradation, which includes damage to the fabric of buildings of historical significance.

Numerous countries have been active in promoting electric buses. One of the earliest was the 'Silent Rider' (Figure 10.5(a)), which operated in the UK in the late 1970s. Germany also was active in the field and, in 1985, had 20 battery electric buses in the Dusseldorf and Munchengladbach urban areas. The USA, too, has shown a keen interest in such transportation and in 1996 there were 81 electric

(a)

(b)

Figure 10.5 (a) *UK electric bus, 1970s;* (b) *shuttle bus, Santa Barbara, CA, USA.*

buses operating in 28 cities. A notable success has been the fleet of eight shuttle buses operated by the Santa Barbara (California) Metropolitan Transit District (Figure 10.5(b)). Electric buses have generally found a favourable response from the public, but less so from the operating companies. As always, the problems lie with the batteries – too heavy (so providing too short a range), too expensive, and too short a life in practical service.

Cars

Although urban delivery vans and buses may well represent the most promising markets for BEVs, another option lies in small cars for commuting or shopping. Such cars might also be used by professional visitors to homes (*e.g.* social workers, district nurses, door-to-door vendors). Many of these 'commuter cars' have been designed and built over the years, either as prototypes or in small production runs, but none has yet succeeded in reaching a mass market – most have finished up in motor museums. There may be several reasons for this, namely: lack of range, especially in winter; poor cabin heating; shortage of recharging points; high cost and relatively short life of traction batteries. Another factor is that, in two-car families, the second car often has to substitute for the first car, in which case versatility and range become important.

A more promising market for small electric cars, especially in the USA, is the so-called 'neighbourhood car' for short journeys up to a few kilometres. These are restricted in speed and are usually one- or two-seaters. Further possible applications for such cars may be found in university campuses, tourist venues, seaside resorts, theme parks, residential communities, and other such locations. Since these small cars are hardly 'traffic-compatible', they are likely to be confined to off-road use or to roads where the speed restriction makes them acceptable.

Since the early 1970s, there has been much research worldwide on developing advanced batteries for use in electric road vehicles (see Chapter 9). Initially, this was in response to the oil crisis in 1973, but later concerns arose over the issue of urban air pollution brought about by cars. In fact, it proved surprisingly difficult to develop an advanced traction battery because of the stringent performance and life specifications that must be met.[1]

In recent years, there has been a renaissance of interest in electric road vehicles – first in EVs and then in HEVs – under the stimulus of legislation that has been enacted in California. Because of the severe atmospheric pollution problems encountered in the Los Angeles basin and elsewhere, the Californian State Legislature mandated in the late 1990s that manufacturers wishing to sell vehicles into the Californian market after 2003 should ensure that 10% of them were zero-emission vehicles (ZEVs), which effectively meant EVs. Failure to meet this goal by a vehicle manufacturer would lead to a prohibition to sell conventional vehicles. All the major automotive manufacturers took this mandate seriously and most started developing electric or HEVs powered by batteries or fuel cells. The major objective was to produce a BEV that would compete in terms of performance and cost with the conventional product. Unfortunately, progress has again been much slower than anticipated, particularly in developing a new traction battery with an acceptable price. Faced with this situation, the Californian Air Resources Board has had little option but to moderate its initial targets, by admitting ICE–battery hybrids ('partial ZEVs') into its programme and also extending the time-scale (for further discussion of HEVs, see Section 10.4). Nevertheless, the requirement to reduce vehicle pollution is still seen as a priority and every effort is being made to meet this goal, within the constraints of what is practical and affordable.

The first modern electric car to go into series-production in the USA, in 1997, was the General Motors 'EV$_1$'. This was a custom-designed, two-seater, sports car and not an adaptation of an ICEV (Figure 10.6(a)). Public charging stations were set up in California and Arizona. The car incorporated numerous novel features, which included inductive charging of the batteries.* Nevertheless, it did not prove to be a commercial success. Using lead–acid batteries, its performance (range and acceleration) was too limited, especially at low ambient temperatures. Only about 375 of the original 1000 EV$_1$ cars built by General Motors were still on the road in August 2003.

---

* This is the same principle as used in the familiar charging of an electric toothbrush, wherein there is no direct contact by wire between the electrical supply and the battery. Elimination of the possibility of 'wet wires' enhances safety.

(a)

(b)

Figure 10.6 (a) *General Motors first-production battery car, EV₁;* (b) *'self-service' electric cars in France at re-charging station.*

In Europe, French and German manufacturers took the lead with electric adaptations of their conventional models. The French manufacturers (Peugeot-Citroën, Renault) tended to prefer nickel–cadmium batteries to lead–acid; it was argued that the significantly higher cost of the former was offset by their longer cycle-life. A fleet of small French cars at an electric recharging station is shown in Figure 10.6(b). The German manufacturers (Volkswagen, Mercedes, BMW) chose lead–acid batteries. In Japan, Toyota and Honda produced electric adaptations of petrol-engined cars. These two companies selected the nickel–metal-hydride traction battery. The various manufacturers produced short series-runs of their electric car models, but these were sold at much less than the manufacturing cost. Although none of these ventures has been a great commercial success, undoubtedly much has been learnt about the broader aspects of designing and operating electric vehicles and advances have been made in motors, control gear, and battery management systems.

The present position with private electric cars is that the development and commercial availability of an advanced performance traction battery would seem to be a necessary, though not sufficient, pre-condition for achieving success. Moreover, only by mass production will the price be brought to a sufficiently low level to compete with conventional models. The advantages and disadvantages associated with the use of electric vehicles are shown in Table 10.1. The striking feature of the data is that all the advantages accrue to society as a whole, but all the disadvantages are borne by the vehicle owner or driver. This is seen as a major stumbling block to the widespread introduction of BEVs. If governments and municipal authorities wish to reap the benefits of electric vehicles, then they will have to take active steps to encourage their introduction and use. This could involve 'stick-and-carrot' strategies. 'Sticks' might include banning ICEVs from city centres or, less contentiously, limiting parking places and making them more expensive, introducing congestion charges, and increasing vehicle licensing charges. 'Carrots' might include exempting BEVs from all congestion and licensing charges, providing exclusive and free parking places for BEVs, subsidising their purchase price and installing frequent recharging points. Some of these measures favouring BEVs already exist in certain countries. Whether or not they will be adopted generally remains to be seen.

Table 10.1 *The impact of electric road vehicles*

Advantages	Disadvantages
Reduce urban air pollution	Limited daily range
Silent in operation	Higher capital cost
Conserve petroleum supplies	Cost of charging equipment
Diversify the energy base of transport	Long recharge times (slow 'refuel')
Help national balance of payments by reducing oil imports	Availability of recharge points
Help to load-level the electricity supply by overnight charging	Battery care and maintenance
	Not suitable for high or low ambient temperatures
	Need for auxiliary heaters and air-conditioning units

10.3. Novel Electric Vehicles

Electric Racing Cars

As long ago as the 1890s, in the very early days of the internal-combustion engine, people were driving electric vehicles and soon an interest in racing them developed. At one stage, a French racing car – 'Jamais Contente' (Figure 10.7(a)) – held the world land speed record of 105 km h^{-1}. Over the years, there has been a continuing interest in racing electric vehicles. As the body design of Formula 1 Grand Prix cars developed, so was it adopted by their electric counterparts.

Another related activity has been EV endurance runs known as 'electrathons'. These events began in 1979 in England and rapidly became popular elsewhere. Unlike conventional competitions, where speed determines the winner, the aim of an electrathon is to drive an EV as far as possible in a set time using a defined mass of batteries. This allows scope for testing advanced batteries of high specific energy.

(a)   (b)

Figure 10.7 (a) *1890s electric racing car, 'Jamais Contente'*; (b) *Japanese high-speed electric vehicle, 'KAZ'.*

A quite different and novel high-speed electric car, the 'KAZ' (Figure 10.7(b)), was developed in Japan at Keio University. This revolutionary vehicle has eight wheels and seats eight passengers. It is 6.7 m long and weighs 3950 kg. Equipped with lithium-ion batteries, it has a maximum speed of 300 km h^{-1} and a range of 300 km at a steady 100 km h^{-1}. While this is unlikely to attract large sales, it is a platform for advanced technology evaluation, which includes the design of frame structures and suspensions, traction motors, intelligent power modules for inverters, and lithium-ion batteries. All the varied attempts to develop electric cars for high speed and good endurance assist advances in EV technology, just as ICE racing cars contribute to advances in the family saloon (sedan).

Solar-powered Cars

A novel form of electrathon has been pioneered in Australia. This is a contest for EVs that are powered solely by solar energy. The 'solar cars' are necessarily ultra-lightweight and can only be operated effectively in a sunny climate, as exists in central Australia. In 1983, two Australians drove a solar-powered EV across the continent from Perth to Sydney (4084 km) in 172 h. This was a remarkable feat that was the first to demonstrate the application of renewable energy to road vehicles. Building on this success, since 1987 a series of solar-powered endurance events – The World Solar Challenge – has been held regularly in Australia, from Darwin in the north to Adelaide in the south, through 3010 km of barren, largely uninhabited, country. The competition is open to all-comers and has proved to be exceptionally popular. Entries have come from major automotive companies across the world and from university engineering departments where students have designed and built their own solar cars. There have also been teams from high schools and even from individuals. The rules of the World Solar Challenge are strictly defined and the competing cars are closely monitored. The principal regulation is both daring and simple: once the race starts, the solar cars must be powered by direct sunlight. It is not even specified how the solar energy should be used to propel a car. This simplicity gives teams the freedom to solve the real technical issues in new and innovative ways.

The World Solar Challenge has witnessed remarkable advances in aerodynamic efficiency, batteries, energy-management systems, lightweight structural materials, motor design, photovoltaics, power electronics, and tyres. Through such progress, the record average speed of 67 km h^{-1} first set by the General Motors *Sunraycer* in 1987 has been raised to 97 km h^{-1} by the Dutch car *Nuna II* in 2003 (Figure 10.8). It is also noteworthy that, in the first five competitions up to 1999, the winning car employed silver–zinc batteries. These batteries have high specific energy ($\sim$125 Wh kg^{-1} compared with $\sim$35 Wh kg^{-1} for lead–acid) but are expensive and short-lived when subjected to deep-discharge cycling. In recent years, lithium-ion batteries have become commercially available in suitable sizes and, with a higher specific energy (over 160 Wh kg^{-1}) and a lower cost than silver–zinc, have become the preferred choice. They were employed in the first-placed cars in 2001 and 2003.

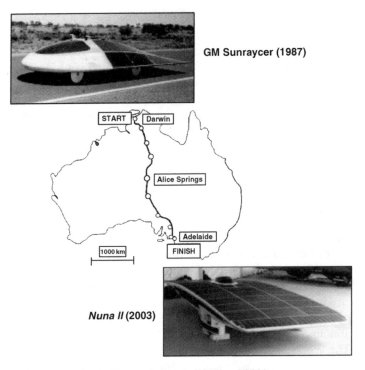

GM Sunraycer (1987)

Nuna II (2003)

Figure 10.8 *World Solar Challenge winners in 1987 and 2003.*

The prime value of the World Solar Challenge is that it provides automobile manufacturers – Ford, General Motors, Honda, Nissan, Mitsubishi and Toyota have all competed – with a valuable test-bed for the development of electric and hybrid electric cars. The use of solar cars is simply a means to this end – nobody has any delusion that, in the future, family cars will be driven directly by solar energy (especially in temperate climates). Nevertheless, it is perfectly conceivable that next-generation BEVs and HEVs will have solar panels incorporated in their roofs and these will help to recharge the batteries during the day. This would be a modest, but direct, contribution of renewable energy to the transport scene.

10.4 Hybrid Electric Vehicles

Given the demonstrated disadvantages of vehicles that are powered by batteries alone, most of the major automotive companies are actively engaged in the development of ICE-battery HEVs as more practical alternatives.* Conventional

---

* Mention has already been made (under trolleybuses and railway locomotives) of all-electric mains-battery configurations, such as the 'Duo-Bus', and of the operational advantages that such hybrids offer. Another all-electric concept, suitable for all types of EV, is the fuel-cell-battery hybrid (see Section 10.5). This combination serves to overcome the limited peak power of the fuel cell and the restricted range of the pure BEV.

ICEVs are grossly overpowered for steady driving, so as to provide the extra performance necessary for acceleration and climbing gradients. The steady-power requirement is often only 10–20% of the peak demand. 'Sports' vehicles, in particular, are highly powered and a prime consideration for many drivers when choosing a car is the time it takes to accelerate to 100 km h^{-1}. This performance is bought at the expense of fuel economy. HEVs provide a solution to this dilemma. In these designs, the ICE provides the steady power for cruising, either directly (mechanical drive) or *via* the battery (electrical drive). The battery also supplies the peak power for acceleration and hill climbing. The electrical drive may be augmented by a capacitor storing electrochemical energy, or perhaps even a flywheel storing kinetic energy. The energy store is recharged by the engine when driving on the level, as well as by the recuperation of some of the energy lost during vehicle braking (so-called 'regenerative braking'). Although batteries are the most commonly used secondary power source, they are not particularly efficient at accepting high rates of charge, as encountered in regenerative braking, and both electrochemical capacitors and flywheels are better in this respect (see Sections 7.5 and 7.3, Chapter 7, respectively). Hybrid power packs of lead–acid batteries and electrochemical capacitors have been demonstrated in two Australian experimental hybrid electric cars (see Section 7.5, Chapter 7).

Classification of Hybrid Electric Vehicles

There are two basic architectures for ICE–battery hybrid vehicles, namely, 'series-hybrids' and 'parallel-hybrids'. In a series-hybrid (Figure 10.9(a)) the drive is all-electric and a small heat engine serves to recharge the battery that powers the electric motor. When used in a purely EV mode (*e.g.* in cities, with the engine switched off), it may, if desired, be re-charged overnight from the mains. An alternative option is to have a larger engine and a smaller traction battery, such that the power is provided mostly by the engine, which is only switched off for limited urban operation.

In a parallel-hybrid configuration (Figure 10.9(b)), there are dual transmission systems: a mechanical one driven directly by the engine and an electrical one driven by the electric motor. The engine is sized for steady highway driving and the battery supplies auxiliary power. The battery also accepts regenerative-braking energy and restarts the engine when stopped in city traffic. A parallel-hybrid corresponds to a conventional automobile with a smaller engine and a larger battery.

To date, most automotive manufacturers have opted for the parallel-hybrid as it makes use of existing vehicle technology for engines, gearboxes, and induction motors. There are several possible versions of the parallel architecture. In one, the mechanical and electrical drive-trains are quite separate, with the engine acting through a gearbox and drive-shaft, as usual, and the battery powering separate electric motors, for example wheel motors. In another version, the electric motor is mounted on the mechanical drive-shaft and serves to augment the power transmitted. This is still a parallel-hybrid, in that each power source may be activated independently, though acting through a common shaft. It is even possible

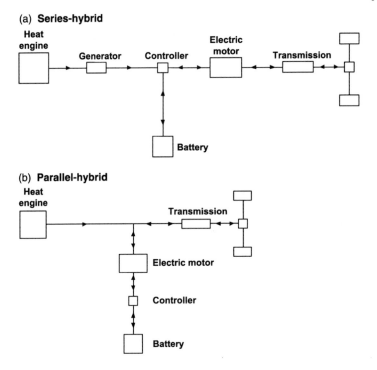

Figure 10.9 *Schematic of* (a) *series-hybrid;* (b) *parallel-hybrid drive-trains.*

to incorporate the rotor of the electric motor in the flywheel of the ICE, and then surround the arrangement with the stator coils.

Hybrid vehicle technology permits various flexible modes of operation. At one extreme, the vehicle operates essentially as an ICEV, using petrol or diesel in a small engine, and the battery acts as an auxiliary power source to boost acceleration or when climbing hills. At the other extreme, when a relatively large traction battery is fitted, the HEV may be operated as a BEV in the urban environment. In this mode, the engine serves merely as a 'range extender' and is only brought into use when the desired trip length exceeds the capability of the battery. Between these two extremes there is scope for flexibility in design and operational practice, and in the proportion of liquid fuel and electricity employed.

The different developments foreseen in the field of vehicle electrics are listed in Table 10.2. At least seven distinct phases have been proposed. These range from dual 12-V battery systems, through a variety of designs in which the vehicle makes use of increasing electrical functionality, to a plug-in hybrid that would provide the vehicle with a substantial range of electric-only drive. The first three options in the list are intended to provide the electrical power that is necessary for an increasing range of electrical ancillaries (*e.g.* drive-by-wire, brake-by-wire). It should be noted that options 2–5 employ a 36-V battery together with a 42-V alternator, and are commonly referred to as 42-V PowerNets. Options 4 and 5, as well as 6 and 7,

Table 10.2 *Proposed changes to electrical systems in automobiles*

Design	Operating characteristics
12-V dual-battery	Lowest cost, lowest-risk approach
	Provides additional power for new vehicle ancillaries
	Improves system reliability
	Increases maximum power and availability
	Already offered in high-end (luxury) cars in Europe
	Can even support stop–start function (designed to shut-off ICE during stop to save fuel and reduce noise level in cabin)
	Can accept regenerative-braking energy
42-V 'upscale' automotive battery[a]	Higher voltage makes it possible to power new comfort and driveability ancillaries without increasing currents
	In some cases, the ancillaries themselves operate more efficiently at higher voltage
42-V with stop–start[a]	Combination of options 1 and 2 for vehicles with large ancillary loads (*e.g.* air-conditioning) during engine-off idle stand
	Supports increased number of engine-start events per drive
	Ensures higher reliability
42-V 'soft' hybrid[a]	Adds soft electrically assisted launch from stop and recuperation of regenerative-braking to option 3
42-V 'mild' hybrid[a]	Electrical motor used for longer periods and more frequently
	Stop–start and regenerative-braking requirements same as for option 4
	Provides power-assist during low-speed acceleration that facilitates down-sizing of ICE and gives fuel, weight and volume savings
High-voltage 'full' hybrid	Low-end requirements overlap those of option 5
	Provides longer and more frequent power assist to high-end cars, *e.g.* sports utility vehicles
'Plug-in' hybrid	Vehicle can be driven in electric-only mode with full power and functionality, but for a limited range

[a] System with 36-V battery and 42-V alternator.

make use of both electric and ICE propulsion. In the final analysis, the choice of hybrid system depends upon the required duty cycle of the vehicle, the degree of engineering complexity, and the capital and running costs. From the viewpoint of *Clean Energy* and sustainability, much depends on the source of the liquid fuel and the electricity used (whether fossil or non-fossil) but, overall, HEVs should consume less energy and produce lower emissions than ICEVs of similar performance.

Practical Hybrid Electric Vehicles

Honda and Toyota were the first automotive manufacturers to commercialize and offer HEVs for sale. The Toyota 'Prius' (Figure 10.10(a)), which was introduced in October 1997, is a conventional five-seater saloon (sedan) with a four-cylinder, 1.5-L, petrol engine that can develop 70 bhp (70 bhp = 52 kW) and with a permanent magnet electric motor that can develop 44 bhp (33 kW). This car, which is a parallel-hybrid, has front-wheel drive and a continuously variable transmission. It may be operated using either or both power sources. There is a mechanical power-split device that acts like a continuously variable transmission. The amount of power provided by each source is electronically monitored, according to speed and load, to ensure that the car always operates at its most efficient. When the demand for power is low, the engine recharges the batteries; when it is high, the motor assists the engine. The car has a battery of 38 nickel–metal-hydride modules (each of 7.2 V, made up of six series-connected cells of 6.5-Ah capacity) that operates at 274 V. The top speed is 158 km h^{-1} and the petrol consumption is only 4.5 L per 100 km on an urban drive cycle. A second-generation Prius was released in October 2003 (Figure 10.10(b)). This has a more powerful motor (50 kW), better fuel efficiency, a novel 'by-wire' (joystick) shift control, navigational aids, and keyless entry and start-up.

The Honda 'Insight', also a parallel-hybrid, is a two-seater coupé (Figure 10.10(c)). It was introduced in December 1999 and has a smaller three-cylinder petrol engine (0.995 L) that can develop 67 bhp and is boosted only to 75 bhp by the battery-electric motor. The car has a top speed of 180 km h^{-1} and a petrol consumption of 4.1 L per 100 km on an urban drive cycle. In December 2001,

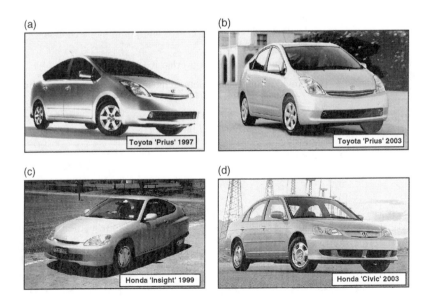

(a) Toyota 'Prius' 1997
(b) Toyota 'Prius' 2003
(c) Honda 'Insight' 1999
(d) Honda 'Civic' 2003

Figure 10.10 (a), (b) *Two generations of Toyota Prius;* (c) *Honda Insight;* (d) *Honda Civic.*

Honda released a parallel-hybrid version of its 'Civic' five-person, family car (Figure 10.10(d)). As with the Toyota 'Prius', both these Honda hybrids use nickel–metal-hydride batteries.

By 2003, over 100 000 'Prius' and more than 13 000 of each of the 'Insight' and 'Civic' hybrids had been manufactured and sold in Japan, the USA, and Europe. Although the cars have excellent fuel economy on the urban cycle, the highway figures are no better than those for a conventional diesel car. Presumably a hybrid with a diesel engine would give an enhanced performance and it is likely that European manufacturers will make these available in due course.

Hybrids like the 'Prius', where the power output of the electric motor makes a major contribution to the overall power, are 'full hybrids'. By contrast, the Honda 'Insight' is a 'mild hybrid', in which the electrical component of the overall output (8 bhp in the case of the 'Insight') is modest compared with that of an ICE. Mild hybrids are much simpler mechanically than full hybrids. They have an in-line electric motor-generator integrated with, or adjacent to, the flywheel. In some models, this motor also serves as both the starter motor and the alternator, which eliminates the need for a belt drive. By comparison, a full parallel-hybrid, such as the 'Prius' has a separate traction motor, epicyclic transmission and complex control systems.

Mild hybrids, although feeble on the face of it, have the potential to make a significant contribution to reducing fuel consumption and exhaust emissions. Not only are they mechanically simple, but they are also efficient and require smaller and lighter motors and battery packs than full hybrids. They are also acceptable to the driver who will hardly perceive any difference from the traditional car, except for improved fuel consumption.

Evidently, the situation is still unclear as to which of the many possible types of hybrid will ultimately prove to be most successful, both technically and commercially. Notwithstanding this uncertainty, there is little doubt that hybrid drive-trains will provide a practical path towards sustainable road transportation.

10.5 Fuel-Cell Vehicles

Fuel cells have been described in outline in Section 8.3, Chapter 8, both in the context of stationary power sources for the distributed generation of electricity and as portable or mobile power sources for applications ranging from advanced cellular phones to electric vehicles. The attractions of fuel cells have been pointed out, as well as some of the difficulties encountered in their development. Here, we consider in more detail the situation as it pertains to electric and HEVs.

As power plant for vehicles, fuel cells face particular problems that are not so critical for stationary applications. The principal difficulties are:

- the requirement for compactness, to fit into a very limited space;
- the availability and supply of a suitable fuel;
- intermittent operation;
- fast start-up from cold;
- high cost.

The severity of these problems varies with the type of vehicle: it is most severe for small cars and least severe for buses and trucks. The first two of the above problems are inter-related. It is generally agreed that hydrogen is the ideal fuel for a fuel cell, but practical methods for its production and distribution and the storage on-board the vehicle in a compact and economic form have still to be devised. Storage schemes under consideration include:

- compressed gas – the containers are bulky and heavy;
- liquid hydrogen – costly technology;
- metal/chemical hydrides – heavy, complex systems; yet to be demonstrated convincingly;
- carbon nanostructures – at a very preliminary research phase.

At present, most of the prototype FCVs being manufactured for fleet trials use compressed hydrogen storage. Recent work on the hydrogen storage tank has permitted the filling pressure to be raised from 25 to 35 MPa, which thereby extends the vehicle range. The Ford 'Focus FCV-hybrid' has demonstrated this technology and is said to have a range of 320 km between recharges (see Figure 10.11(a)). The vehicle also features a high-voltage, nickel–metal-hydride battery pack to help increase performance and efficiency; hence, the designation 'FCV-hybrid'. The FCV version of the Mercedes 'A' class car is shown in Figure 10.11(b).

After hydrogen, methanol is the second best candidate fuel, but this necessitates an on-board reformer that adds to the volume and the cost. There is also the problem of balancing the heat flow between the reformer and the fuel cell. Methanol is toxic, corrosive and water-miscible. It is therefore not favoured by the petroleum companies who would have to distribute it. Finally, there is the option of petrol (or diesel) as a fuel. It is anticipated that an on-board petrol reformer will become available in due course although its development is considerably more difficult than that of a methanol reformer. The choice of a fuel-cell hybrid is a distinct possibility since it would circumvent the problem of achieving fast start-up from cold. In the case of a pure FCV, such start-up may prove to be unacceptably long – certainly for the private car. The fuel cell may be hybridised with either a

(a)  (b)

Figure 10.11 (a) *Ford Focus FCV-Hybrid;* (b) *Mercedes 'A-class' fuel-cell cars; the fuel-system for the Ford vehicle is shown in Figure 8.17, Chapter 8.*

traction battery (FCV–BEV hybrid) or an electrochemical capacitor. The hybrid concept would allow the use of a fuel cell of more modest power output, and therefore would reduce both size and cost.

Most of the major automotive companies that are investigating FCVs have opted for the solid polymer electrolyte fuel cell that uses an acid membrane that is conductive to hydrogen ions, the so-called proton-exchange membrane fuel cell (PEMFC, see Section 8.3, Chapter 8). Many different prototype vehicles have been built and demonstrated. Most US companies have used the Ballard fuel cells, manufactured in Canada. Ford, DaimlerChrysler and Ballard have formed a consortium to exploit the Ballard technology in road vehicles. General Motors, on the other hand, has developed its own PEMFC stacks and has tested them in various car models.

Honda also has developed its own fuel-cell stack; it is of broadly similar design to that of Ballard. In 2003, both Honda and Toyota commenced leasing their first FCVs (Figure 10.12). The Honda 'FCX' seats four people and has a range of up to 270 km. The Toyota 'FCHV' is a 'Highlander' sports utility vehicle in which the internal-combustion engine is replaced with a 90-kW PEMFC and the 'Prius' hybrid electric drive-train is fitted (note, FCHV is fuel cell hybrid vehicle). The system design of the FCHV, together with the arrangement of its components, is shown in Figure 10.13. Nissan is also conducting similar development programmes on FCVs.

In Europe, DaimlerChrysler has been building assorted vehicles equipped with Ballard fuel cells. To date, the emphasis has been on larger vehicles, since these have more space to accommodate compressed hydrogen storage tanks. Mercedes-Benz hydrogen buses are to be built for trials in various cities, which include Hamburg, Paris, Barcelona, Perth (Australia), and Reykjavik. The last-mentioned city was chosen on account of the surplus hydroelectric and geothermal power available in Iceland that may be used to electrolyze water and generate hydrogen. This may be an uncommon situation where it makes sense to convert electricity to hydrogen (as an energy vector) to be re-converted to electricity in a fuel cell. The overall energy efficiency is poor, but when surplus electricity is available at low cost it may make economic sense to adopt this approach rather than to import petroleum. More generally, it is assumed in the USA and elsewhere that, in the early years at least, hydrogen will be produced from natural gas by reforming. The overall energy efficiency of this route from gas to traction effort at the wheels is, however, likely to be little or no better than that of the conventional

Figure 10.12 (a) *Honda FCX;* (b) *Toyota FCHV fuel-cell cars.*

(a)

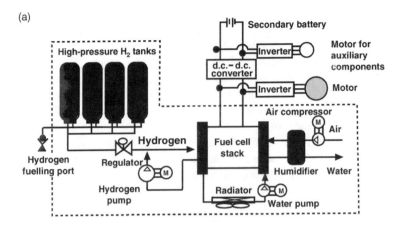

(b)

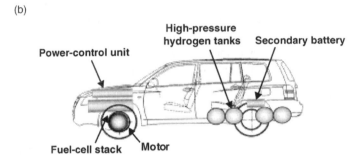

Figure 10.13 (a) *System design;* (b) *components of Toyota FCHV fuel-cell car.*

internal-combustion engine, see Box 10.1. While it is true that overall energy efficiency is not the only relevant consideration, it is an important one that is often overlooked.

In Japan, a fuel-cell bus developed by Toyota, the 'FCHV-Bus 2', went into service as part of Tokyo's metropolitan bus fleet in 2003 (Figure 10.14). It employs two Toyota-designed PEMFC stacks with an output of 90 kW. A nickel–metal-hydride battery stores the energy regenerated during braking and, for highly efficient operation, regulates the electric supply to the motor as determined by the operational status of the vehicle.

Some small specialist companies have opted for different types of fuel cell. In Britain, Zevco produced the 'ZeTek Power' electric vehicle that was based on an alkaline fuel cell. This was demonstrated in taxis and in light commercial vehicles (see Section 8.3, Chapter 8). Zinc–air fuel cells are being developed by Electric Fuel Ltd. in Israel, as well as by the eVionyx Corporation and Metallic Power, Inc. in the USA. This technology is, in fact, a hybrid of a fuel cell and a battery. The 'fuel' is finely divided zinc metal and the oxidant is air. During discharge, the zinc metal is converted to zinc hydroxide. At the end of discharge, the electrolyte

Box 10.1 Efficiency of Fuel-Cell Vehicles

The theoretical voltage for the electrolysis of water at 25 °C is 1.229 V, see Box 8.1 and Figure 8.3, Chapter 8. This value corresponds to the change in Gibbs free energy. If the electrolyzer is adiabatic and the heat associated with the entropy term ($T\Delta S$) is supplied electrically, then the thermodynamic decomposition voltage of water rises to 1.47 V (Figure 8.3, Chapter 8). This corresponds to the energy content of the hydrogen at 25 °C.

The voltage of a PEMFC lies in the range of 0.8–0.6 V, as determined by the current density. If it is assumed (generously) that the average voltage is 0.75 V, then the fuel cell is 50% efficient, *i.e.* half of the energy content of the hydrogen is converted to low-voltage d.c. electricity. In addition, allowances have to be made for parasitic losses in the fuel-cell system (power for pumps, heaters, blowers, controllers, *etc.*), and for energy losses in the vehicle's electrical system (losses in inversion to a.c., the transformer, and the traction motor). In round figures, the collective losses in each system can be taken as 10%. Thus, the overall efficiency of converting hydrogen to traction effort is $0.5 \times 0.9 \times 0.9 = 40\%$. Although this is much higher than for a high-performance automobile (20–25%), as emphasized by advocates of FCVs, the losses incurred in producing hydrogen from primary fuels must also be taken into account.

The steam reforming of natural gas to hydrogen on a large scale is 60–70% efficient (see Section 2.5, Chapter 2). It is next reasonable to assume at least a further 10% energy loss in compressing the hydrogen and 10% in transporting it from the centralized steam reformer to the vehicle-refuelling depot. The overall efficiency from natural gas to traction effort, *via* hydrogen, is then around $0.65 \times 0.9 \times 0.9 \times 0.4 = 21\%$. In other words, FCVs have a performance that is very similar to that of today's petrol engines – they would, however, offer the benefit of zero harmful emissions and therefore would help to reduce urban pollution from road transportation. If the natural gas is steam-reformed regionally or locally, rather than centrally, then there would be some savings in the energy otherwise lost in distribution, but this would be offset by the lower efficiency (not to mention the higher cost) of the smaller reformers. Also, the option for sequestering the carbon dioxide would effectively be lost.

What if the hydrogen is produced by electrolysis rather than directly from natural gas? Here the situation is even more dire. As discussed in Chapter 3, the efficiency of a conventional power station lies in the range of 30% (coal- or nuclear-fired) to 55% (combined-cycle gas turbine). The best electrolyzers are around 80% efficient (see Section 8.1, Chapter 8). The overall efficiency from primary fuel to traction effort is then:

$$\text{coal or nuclear} : 0.3 \times 0.8 \times 0.9 \times 0.9 \times 0.4 = 8\%$$

$$\text{natural gas} : 0.55 \times 0.8 \times 0.9 \times 0.9 \times 0.4 = 14\%$$

Box 10.1 (cont'd)

Note, in these calculations, the 10% energy loss in compressing the hydrogen has been retained and the 10% loss in distributing hydrogen has been replaced with a 10% loss in the electricity-supply system that would result from distribution, voltage reduction, and rectification operations.

The above are only approximate calculations. Nevertheless, from the viewpoint of increasing fuel efficiency or reducing greenhouse gas emissions, the analysis clearly shows that there is no incentive to move from the internal-combustion engine to the fuel-cell vehicle. If and when renewable electricity is available on a large scale, the overall efficiency figures should improve. This is because the conversion of mechanical energy (*e.g.* wind or wave power) to electrical energy does not involve a Carnot cycle and the efficiency should be 80–90% rather than 30–55%. The extent to which such a move to renewable electricity is possible will be determined by cost considerations, by political acceptability and, ultimately, by having a practical means of electricity storage within the grid system.

In the meantime, it should be noted that the efficiency of internal-combustion engines is expected to improve rapidly, as demonstrated by the following projections for various configurations of family-sized cars in 2020[2]

Fuel		*Fuel consumption*	
		MJ km^{-1}	L per 100 km
1996 Reference car	Petrol	2.73	8.46
2020 Technology			
Advanced spark ignition engine	Petrol	1.54	4.79
Advanced compression engine	Diesel	1.36	4.20
Hybrid spark ignition engine	Petrol	1.07	3.32
Hybrid spark ignition engine	CNG	1.03	3.20
Hybrid compression engine	Diesel	0.92	2.86

Thus, by converting to hybrid vehicles – a perfectly feasible proposition by 2020 – it should be possible to reduce fuel consumption by two-thirds compared with the 1996 model family car. This seems a much more realistic option than that of fuel-cell cars, on both energy efficiency and cost grounds.

and the zinc hydroxide are removed and replaced by fresh electrolyte and fuel. The spent fuel is taken to a plant for reclamation of the zinc. With these operating features, the fuel cell is also said to be a 'mechanically' rechargeable battery'. Various vehicles have been demonstrated, *e.g.* a Mercedes Benz 410 postal van in

(a)

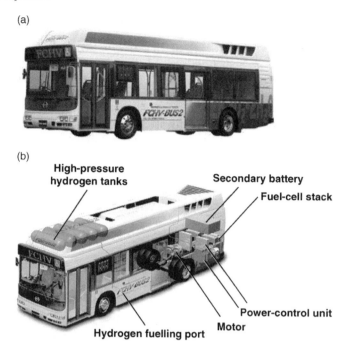

(b)

Figure 10.14 (a) *Toyota FCHV-Bus 2;* (b) *schematic of the bus components.*

Germany (by Electric Fuel Ltd.) and an adapted Honda 'Insight' in the USA (by eVionyx Corp). One of the attractions of this fuel-cell/battery is the long range it offers compared with conventional BEVs; it is claimed that the Honda 'Insight' can run for 1000 km on a single charge. Clearly, entry of these alternative fuel cells into the mass market will require the backing of at least one of the major automotive companies. At present, however, these companies are all focused on the development of PEMFC stacks.

Will FCVs come into common use by 2020? Now that most of the automotive industry is working towards this goal, with large engineering teams devoted to the projects and with encouragement from national governments, it would be precipitous to deny the possibility of success. In the USA, for example, the government and the major automobile manufacturers jointly have set up the FreedomCAR (Cooperative Automotive Research) programme, a national venture that aims to develop FCVs at the earliest opportunity. It is, however, apparent that many technical problems remain with FCVs and that there is a long haul from demonstrating fuel cells in prototype vehicles to mass-producing a vehicle that has all the attributes at a realistic price. Further to that, there is the huge potential cost of setting up a hydrogen production and distribution network across a nation so that fuel is readily available everywhere. There is also the requirement to establish a uniform set of laws and safety regulations that are accepted internationally for the operation and refuelling of hydrogen vehicles. At this stage, the jury is still out on the future of FCVs.

10.6 References

1 D.A.J. Rand, R. Woods, and R.M. Dell, Batteries for Electric Vehicles, Research Studies Press, Taunton, Somerset, England, 1998.
2 M.A. Weiss, J.B. Heywood, E.M. Drake, A. Schafer, and F.F. AuYeung, On the Road in 2020: A Life-Cycle Analysis of New Automobile Technologies, Energy Laboratory Report #MIT EL 00-003, Massachusetts Institute of Technology, Cambridge, MA, USA, October, 2000.

CHAPTER 11

Towards 2020

Attempting to predict the future is notoriously difficult and anyone who does so is providing a hostage to fortune. In 1965, Penguin Books published *The World in 1984*, a paperback that was based on a series of articles that had appeared in *The New Scientist* in 1964. These articles were written by eminent scientists and industrialists of the day, who were asked to predict the likely developments 20 years ahead in their respective fields of specialization. Re-reading this fascinating book today, two general conclusions emerge:

(i) there was much optimism over how quickly new technology might evolve – almost 40 years on, some of the developments are still awaiting realization, for instance: the widespread use of supersonic jets for long-haul flights and electricity generation by magnetohydrodynamics

(ii) some of the really important advances that have subsequently taken place were not foreseen at all, *e.g.* integrated circuits, micro-processors, and the world-wide-web.

Notwithstanding these two generalizations, the various contributors showed good foresight, even if some of the timings were incorrect. History has shown, however, that elsewhere there have been some disastrously wrong forecasts by eminent 'authorities', for example:

"The 'telephone' has too many shortcomings to be seriously considered as a means of communication."
Western Union internal memo, 1876.

'Aeroplanes are interesting toys, but of no military value.'
Marshal Foch, 1911.

'I think there is a world market for maybe five computers.'
Thomas Watson, Chairman of IBM, 1943.

With respect to energy sustainability, the predictions have been uniformly poor. In 1920, geologists forecast that the world's petroleum reserves would be exhausted by 1940. What they failed to anticipate were the major developments in the technology of prospecting that resulted in the discovery of many more oil fields, both on-shore and off-shore. Similarly, in 1971, the 'Club of Rome' commissioned the use of large computer models to map out the future of the world. The resulting report – *The Limits to Growth* – became an international cause célèbre that sold nine million copies in 29 languages. It was concluded that the world would run out of petroleum in 1992. Again this did not happen, thanks largely to further improvements both in prospecting methods and in the technology for exploiting oil fields on the continental shelf. Huge new supplies of natural gas were also discovered, which were not expected and subsequently assumed many of the roles formerly played by petroleum, *e.g.* space heating and electricity generation.

Predicting the future supply and demand for energy and associated advancements in technology is difficult enough in a stable world situation. It is made infinitely more difficult by the intrusion of global political issues such as: the nationalization of Middle East oil fields in the 1970s; the imposition of embargos and sanctions directed against export to (or imports from) specified nations; the Organization of Petroleum Exporting Countries (OPEC) quasi-cartel; the problem of global warming and the Kyoto Protocol; and now, in the 21st century, the fear of global terrorism. When all of this is considered, looking ahead – even for 20 years – lies more in the realm of crystal-ball gazing than of science. Nevertheless, this is no reason not to try. Accordingly, we present here our personal views of the prospects and challenges for world energy to 2020, without claiming any greater insight than many others may have.

One of the incontrovertible facts is the growth in the world's population: two billion in 1939, over six billion today, and heading for 9 or 10 billion before it stabilizes, even if it does then. Associated with this is the ever-growing aspiration for an improved standard-of-living for all, especially for the developing and poor nations. Globalization is slowly bringing this about following the decision of manufacturing industries (and now, given the improvements in telecommunications, also service industries) to move their operations to low-cost countries. Globalization is, however, a controversial issue and whether it will provide true benefits for all remains to be seen. Experience has shown that there is a direct correlation between standard-of-living (or Gross National Product per head) and energy consumption. Unless this link can be broken, the demand for more energy will grow inexorably. If the increased requirement were to be met by the world's 'energy capital' (fossil fuels), the consequences in terms of global emissions of greenhouse gases and resource depletion would probably be catastrophic. Clearly, over the coming decades, it is necessary to implement an entirely new energy infrastructure based on *Clean Energy*. In the near-term, this is likely to entail the clean-up of fossil fuels so as to minimize both pollution and the release of greenhouse gases. The sequestration of carbon dioxide at power plants would be an important medium-term development, if this could be achieved economically. In the longer term, an energy future that depends on sustainable sources will be required. Bearing in mind the enormous difficulties and the long time-scale that would be involved in replacing the entire conventional energy system, not to mention the daunting magnitude and cost of the task, it is none too soon to be addressing the matter seriously. This is one of the major issues facing the world today, but one that tends to be relegated to the 'pending' tray by politicians occupied with more pressing geo-political and social issues, and by industrialists concerned with making a profit and staying in business. It is up to those who are scientifically and environmentally aware, especially a new generation of well-educated young people, to take up the challenge of creating a sustainable energy future for generations yet to come.

11.1 Greenhouse Gases and the Kyoto Protocol

In considering 'global warming', three facts seem incontrovertible:

(i) there are certain gaseous molecules in the atmosphere (including carbon dioxide, methane and nitrous oxide) that absorb and re-radiate infrared radiation – the greenhouse gases
(ii) the concentration of carbon dioxide in the atmosphere has increased steadily since the industrial revolution
(iii) the mean global temperature is rising slowly.

Most authorities link these three facts and conclude that the temperature rise is a consequence of the anthropogenic release of greenhouse gases. While there is compelling reason to make this link, it is by no means proven beyond all doubt. Another greenhouse gas is water vapour, which is more potent than carbon dioxide. The effect of water vapour is well known. One has only to compare the night-time temperature on a cloudless, starry night in winter with that on an overcast night to experience the effect of water vapour in absorbing infrared radiation. Even in the Sahara Desert, when there is no cloud cover, the nights can be very cold. It is at least arguable that the observed global warming is a consequence of a natural long-term trend towards more atmospheric water vapour. This, in turn, would lead to slight warming of the oceans with greater release of carbon dioxide that would augment the anthropogenic gases in the atmosphere. There is still much to be done in elucidating the mechanism of global warming, and the relative contribution of natural phenomena and man-made releases, but the following discussion is based on the premise (on which we have an open mind) that the majority view is correct and that carbon dioxide formed in combustion processes is the principal culprit responsible for global warming.

In 1992, the United Nations Conference on Environment and Development – better known as the 'Earth Summit' – met in Rio de Janeiro and adopted the United Nations Framework Convention on Climate Change (UNFCCC). Article 2 stated an aim 'to achieve stabilization of greenhouse gas concentrations in the atmosphere at a level that would prevent dangerous anthropogenic interference with the climate system'. Unfortunately, nobody was able to define the level of greenhouse gases that would constitute 'dangerous interference'. Consequently, no targets were set. This issue was addressed at a subsequent meeting in Kyoto in 1997 and the industrialized nations of Europe, Japan and the USA were required to reduce their average national emissions of greenhouse gases over the period 2008–2012. The targets were based on the historical emissions that took place in 1990, less an agreed percentage (see Section 1.5, Chapter 1). As part of the Protocol, it was envisaged that a system of 'emission trading' would be set up, whereby nations or companies wishing to exceed their allocation of emissions would be able to purchase permits from other licence holders who had allowances surplus to their requirements. The environmental outcome would not be affected because the amount of permits allocated would be fixed.

In the seven years that have elapsed since Kyoto, there has been an increasing awareness of the difficulty in meeting the targets. To give just one instance, by December 1999, emissions in the USA had risen 12% above 1990 levels, and were on course to increase to 20–25% above 1990 levels by 2008. Add to this the 7% reduction that the USA is required to make, and the total cut required approaches 30% in the next four years. Bearing in mind the long economic life of the major energy consumers (*e.g.* power stations, buildings, road vehicles), the magnitude of the task becomes apparent. Indeed, it is said that 80% of the electricity-generation plant that will be in use in 2010 is already built. Premature replacement of all less-efficient plant would cost huge sums of money and this simply is not going to happen on such a short time-scale. The costs are politically unrealizable.

To solve this dilemma, the USA and other industrialized nations are relying heavily on purchasing emission permits. It has been suggested that Russia and the Ukraine, among others, may hold permits surplus to their requirements. The details of how such an emission-trading system might work are starting to emerge. Even within a single nation, there is the difficult question of how the government should apportion its national permit for emissions among industry, commerce, and individual citizens. An emission permit is, in effect, a property right that may be bought and sold. Companies and individuals will be anxious to maximize the financial benefit to themselves and there is scope for endless argument. Powerful negotiators will benefit at the expense of the less skilled, and a system of arbitration and appeal will be needed. A further complicating factor is that the emissions from a nation vary with economic growth and with technological changes, neither of which can be planned or controlled by governments to meet targets set years in advance.

Trading of emission permits between nations is likely to be even more fraught with difficulty. It has been estimated that, internationally, permits worth a trillion dollars or more would be required, and even then there is no guarantee that the Kyoto targets would be met. Many developing nations were not allocated targets, and it is precisely in these countries that the rate of energy consumption is increasing most rapidly. The Protocol has made no provision for how future targets are to be allocated, or for the monitoring and enforcement of emission trading. Ultimately, it will be necessary to have an international judicial body to enforce compliance, but international law is too weak at present to take on this task effectively. Inviolate targets and time-scales are simply not practicable to enforce in the light of changing circumstances. And meeting the Kyoto targets is just the first step along the road to halting the build-up of carbon dioxide in the atmosphere.

From this brief discussion, it will be clear that there are major political, economic, technical and legal issues to be faced before an effective system for trading emissions will be in place. The situation has all the hallmarks of a bureaucratic nightmare and we doubt very much that it will be completed by 2008, or even by 2012, unless the deterioration in the Earth's climate is so dramatic that minds will be focused away from disputes over property rights. A more straightforward approach would be to impose a heavy tax on all fossil fuels (a 'carbon tax') to pay for the 'externality' of polluting the atmosphere.[*] Given that other wastes have to be recycled, processed or contained, in principle there is no reason why society should feel free to discharge waste carbon dioxide to the atmosphere without payment. This raises the question of what level of taxation should be imposed. As there is great uncertainty over the real external cost to the environment per tonne of carbon dioxide released, it is difficult to approach the question from this angle. A better method might be to allocate some of the tax raised to measures that are designed to promote energy conservation and renewable forms of energy. The remainder could be offset against existing taxes (such as general sales and services levies) so that the overall rate of taxation is not greatly changed. This should go some way to mollify the taxpayers, although obviously there will be winners and losers.

If global warming and the role played by greenhouse gases are as serious as many believe, then international action becomes a matter of priority. A carbon tax has the benefit of being immediate and would provide the incentive for industry to develop forms of energy conservation and renewable energy that are not competitive at today's low prices for fossil fuels. The carbon tax might start at a relatively low level and rise, year by year, according to a pre-published schedule. This would allow time to adjust to evermore costly fossil fuel and to develop new energy technologies. Such a procedure is already in force in the UK for the disposal of municipal waste by landfill, where the tax imposed rises progressively year-by-year from £7 per tonne in 1996 until it reaches £15 per tonne. We consider this to be a better approach than attempting to formulate emission targets years in advance and then setting up monitoring and enforcement agencies. There are, of course, strong political voices against fuel taxes; the road haulage and motorist lobbies are vociferous, while even the poor need to keep their homes warm in winter. It is precisely for the latter reason that domestic fuels in the UK have a reduced rate of Value Added Tax.

11.2 Energy Conservation

As stated in the Preface, this book makes no attempt to treat energy conservation – this merits a volume of its own. Nevertheless, it would be remiss in looking ahead to 2020 not to mention the role that conservation will play in modifying the demand for energy. There are considerable opportunities for energy savings in almost all spheres of human activity.

---

[*] Externality is a term used in economics to describe costs (or benefits) that are not reflected in the price of the goods or service from which the externality arises.

Space Heating of Buildings

In temperate climates, many buildings are poorly insulated and wasteful of energy, both for winter heating and summer cooling. Old houses are usually built of stone (a good thermal conductor) or of brick, without any significant insulation. For example, it is only in the last 20–30 years that double-glazing and cavity-wall insulation have become common in the UK. As is to be expected, the best-insulated buildings are generally found in the coldest countries – Scandinavia, Canada, Russia. If buildings in temperate zones were insulated to the same high standard, the fuel consumption could be reduced substantially. Indeed, experimental homes designed to high standards of insulation have demonstrated that in such regions it is possible to dispense almost entirely with space heating, but rely on the heat generated by the occupants' bodies and by their domestic appliances.

One of the problems of well-insulated buildings is that it is necessary to arrange for several exchanges of air each hour. This is generally effected by means of exhaust fans, particularly in kitchens and bathrooms. Unfortunately, these fans extract warm air and replace it with cold. Thus, there is a clear need for inexpensive heat exchangers so that the exhaust air is cooled and the fresh air is preheated and humidity controlled. This is not a trivial task since the temperature differential is generally small. Also, the present low cost of energy does not provide a financial incentive to adopt such measures.

In many countries, there have been significant improvements in the energy efficiency of domestic appliances. Refrigerators and freezers are now labeled for their energy consumption, as are clothes and dish washers. Modern condensing gas boilers, used for central heating, extract most of the waste heat from the exhaust gases and transfer it to the incoming cold water. They are therefore more efficient in the utilization of energy.

There are also the options, discussed in Chapter 4, of the passive solar heating of buildings, active solar heating of domestic hot water and geothermal heat pumps, all of which serve to reduce the energy demand. Just which of these are adopted in any particular situation is a matter of economics. If the price of fuel rises, as expected within a 20-year time-frame, there will be more financial incentive to reduce energy consumption in the home.

Lighting

Traditional incandescent filament lamps are notoriously inefficient, much of the energy is dissipated as heat. This is important when it is recalled that 25% of the world's electricity is used for lighting. Fortunately, by no means all of this electricity is used inefficiently in filament lamps. Fluorescent tubes play a major role, particularly in industry, commerce and street lighting.

There have been significant improvements in lighting in recent years. The light output of fluorescent tubes has been enhanced by the development of new phosphor coatings. Compact high-energy lamps are becoming commonplace; a 20-W lamp gives the same light output as a 100-W incandescent filament lamp and lasts much longer. If all the traditional filament lamps in the world were replaced by high-efficiency lamps, or by florescent tubes, there would be a massive saving in electricity. In the longer term, there are good prospects for light emitting diodes (LEDs) to replace conventional lighting, but this will necessitate substantial improvements in performance and reductions in cost. Already, red LEDs are used in some traffic lights and in brake lights on cars. With the advent of LEDs based on gallium nitride, which emit in the blue or green, the possibility exists of replacing all the hundreds of millions of traffic lights around the world with LEDs, to give large savings in electricity consumption. New advances in photochemistry and in photoelectrochemistry (see Section 5.6, Chapter 5) hold out the prospect of even more efficient forms of lighting.

Transportation

It has been estimated that between 2000 and 2025 the world population will grow from 6 to 8.5 billion (42% increase). Most of this growth will be in developing countries, where there is huge unfulfilled demand for private cars. On this basis, the Fiat Motor Corporation has projected that over the same period the global fleet will increase from 0.7 to 1.75 billion. If this demand is to be met, it will be necessary to conserve liquid fuels for transportation and there will be a requirement for vehicles that are much more fuel-efficient than at present. Also, there will be a need to exploit non-traditional sources of fossil fuel, as discussed in Section 1.4, Chapter 1. Of course, such predictions of the increase in vehicle numbers neither take account of whether the Earth's atmosphere can absorb the carbon dioxide, nor of the outcome of the Kyoto Protocol negotiations.

For a new technology to succeed in the marketplace, it must not only be sound and appeal to the customers, but must also be backed by major industrial muscle and finance. The Japanese have demonstrated this point well with motorcycles, cameras, and consumer electronics. Given the sizable effort that many automotive companies are now putting into electromechanical and electric drive-trains, there are good prospects that hybrid electric vehicles (HEVs) and perhaps even fuel-cell vehicles (FCVs) (see Sections 10.4 and 10.5, Chapter 10) will become common-place throughout the world by 2020. This should make a significant contribution to energy savings in the transportation sector and will assist in the reduction of emissions of carbon dioxide and other harmful gases.

Aside from technical advances in the design of vehicles to enhance fuel efficiency, the greatest single improvement would be achieved by substituting public transport (mass transit) for private cars, particularly for travel into the city. Already in major conurbations (London, New York, Tokyo) more people travel to work by bus and rail than by private car. As public transport facilities continue to improve around the world, and as cities become even more grid-locked with traffic, it is likely that this trend will increase.

London has short-term plans to purchase 200 more buses and longer-term plans to upgrade its underground system. A congestion charge has already been imposed on motorists coming into the central business district; other UK municipalities are actively contemplating similar action. Many urban communities across the world have introduced 'bus lanes' to give priority to these vehicles. These lanes, in conjunction with 'park and ride' schemes, facilitate access to city centers. Some authorities also give priority of passage to cars with more than one occupant; this encourages 'car pooling', and thereby saves fuel. Finally, with the construction of safe routes, more people are returning to cycling. The Scandinavian countries and the Netherlands have set good examples in this regard. All these are moves in the right direction to remove traffic jams, make cities more accessible, and reduce both fuel consumption and urban pollution. For Europe, this is a move back towards the 1940s and 1950s when the population was only a little less than it is today, when most people went to work by public transport or by cycle, and there were comparatively few cars so that roads were less congested.

The extent to which the motorcar can be replaced by public transport in the short term is generally limited by the distribution of the population. For many, it is impossible to get to work by bus or train, and often this is through conscious choices that they have made, either in where they have bought a house or taken employment. In recent years, people have been willing to drive long distances to work rather than move home or job. Indeed, compared with having a congenial place to live and a desirable occupation, commuting by car has assumed secondary importance for many individuals. At the same time, developments in electronic communications will mean that more people can work from home, at least for part of the time, and not have to travel daily. In large cities, the choice is clear: either private cars are excluded from city centers, or admission fees are imposed at a sufficiently high rate to dissuade drivers from using their cars. Both approaches depend upon the existence of acceptable 'park and ride' schemes. The alternative is gridlock – when, ultimately, many commuters will give up driving in disgust or engage in political protest.

In the field of car design, we foresee a movement away from advertising peak performance to that of fuel economy. This view is based on the assumption that petroleum prices will rise significantly in real terms over the next 20 years. There is also the prospect of an increasing proportion of diesel-engined vehicles on the road, because of their greater fuel economy, and of vehicles fueled by liquefied petroleum gas (LPG), because of their cleaner exhaust (and, at present, the lower tax in some countries). The move towards diesel engines will be given a further stimulus when viable particulate traps have been perfected for cars, as well as for buses and trucks. It will be interesting to see whether the USA follows Europe in marketing diesel-engined cars and light-goods vehicles. This is only likely if petroleum prices in the USA rise nearer to European levels and, thereby, provide the incentive to make the shift. At present, the desire in the USA and elsewhere for sports utility vehicles poses a particular problem of heavy fuel consumption. Hopefully, the popularity of these vehicles will decline as petrol prices climb so that 'gas-guzzling' vehicles will become comparatively few in number.

As discussed above, another probable development in road transportation will be the widespread introduction of the HEV, and maybe also the battery electric vehicle (BEV) for urban use. The HEV will permit smaller and more efficient engines to be fitted without any reduction in vehicle performance. The BEV may not save much primary energy, but will effect a switch from petroleum to electricity, which can be generated from different primary fuels. It is envisaged that by 2020 many private cars will have some form of electromechanical (HEV) or battery (BEV) drive. By that time, most family-sized cars should return at least 60–80 mi per gallon of fuel (3.5–4.7 L per 100 km). These developments will be driven not only by considerations of fuel economy, but also by local authorities choosing to follow the pattern of Southern California (see Section 10.2, Chapter 10) and introducing regulations to reduce urban pollution.

The future for FCVs is still very much open to question. There are difficult technical problems to be solved, but progress has been made by the major automotive companies who are now taking the concept of the hydrogen FCV seriously. The principal challenges remaining are those of reducing cost to an acceptable level, ensuring reliability and lifetime, deciding which fuel is to be used and whether an on-board reformer is required. If a reformer is necessary, then it has to be well integrated with the fuel cell (for good thermal management and for producing hydrogen at the required variable rate), small, and inexpensive. Should hydrogen be employed directly, then there is the question of on-board storage to be resolved, as well as the establishment of a supply infrastructure. There is also the cold-start problem and the need to eliminate impurities from the hydrogen. On the whole, we are inclined to be pessimistic about the rate of developing this technology for private cars, as a competitor to diesels and HEVs, and therefore do not expect to see a major swing to fuel-cell cars by 2020. Even so, one must recognize the dedication of many major automotive companies to the technology, and the power and influence they can bring to bear on the topic. It is possible that they may prove our pessimism to be unwarranted. If FCVs are indeed introduced in significant numbers during this period, it is more likely that they will be buses or trucks, where space to accommodate the power plant and the hydrogen store is not so restricted, and where generally longer journeys are involved.

In principle, the hybrid concept is equally applicable to railway locomotives. By having an electric hybrid locomotive, it would be possible to fit a smaller and more efficient engine that would run at constant speed and release less pollution. The problem here is that locomotives have a much longer service-life than cars so that it will take years to replace the existing stock, even after the concept has been proved in practice.

In the field of air transport, the recent trend towards quieter aircraft with lower fuel consumption will doubtless continue. There are plans to build even larger passenger aircraft than at present, with a view to reducing the operating cost per seat-kilometer. How enthusiastically the public would take to such behemoths of the skies remains to be seen.

11.3 Fossil Fuels

Petroleum

World petroleum prices are generally low in historic terms, thanks to the opening up of new oilfields in Nigeria, around the Caspian Sea, and elsewhere. At present, the supply of oil exceeds the demand, and this situation is expected to persist in the short term. There is, however, always the danger of prices collapsing completely, particularly if Iraqi oil is produced in abundance. The market for oil is extremely inelastic and, as has been seen in the past, small shortfalls in supply can lead to rapidly escalating prices, while over-production results in an equally sharp decline in the spot price. The problem facing oil-producing nations is that each individually wishes to maximize its income by exporting as much oil as possible, but if all do this collectively there is a glut and the price falls sharply. It was for this reason that OPEC was set up in the 1970s, to act as a quasi-cartel and to control the price of crude oil by allocating supply quotas to each of the participating countries. This strategy has been only partly successful. From inception, OPEC has faced the dual problem that not all of the prospective members have joined and that new producers have come on line in recent years. The latter are not constrained and may produce as much oil as they choose, or regard as prudent. The consequence of low oil (and gas) prices is that the bulk energy is cheap and there is little financial incentive to invest in new technology for non-conventional forms of energy. Over the past 30 years, however, the price of crude oil has fluctuated wildly from less than US$ 10 to US$ 40 per barrel. By contrast, OPEC would like to stabilize the price in the range of US$ 20–25 per barrel, which would provide some reassurance for consumers as well as producers. Nevertheless, so long as the possibility of a short-fall exists, whether politically inspired or otherwise, future high prices cannot be ruled out. For this reason, if for no other, it is prudent for oil-importing nations to be developing alternative energy and transportation technologies.*

There is, of course, a close interaction between the technology employed in discovering and producing oil and the size of the reserves available. In the 1920s, a Hungarian physicist, Baron von Eötvös, developed a device for detecting slight changes in gravitational attraction. This led to the discovery of the huge oil reservoirs of Texas and Oklahoma, and to many others since. Then, in the 1940s and 1950s, off-shore exploration and drilling were carried out in the shallow waters of the Gulf of Mexico. As off-shore technology improved, it became possible to look for oil in deeper water and in rougher seas, which led to the development of the North Sea oil and gas fields. By building on this expertise, and using further technical advances in oil prospecting such as 3D seismic analysis and horizontal drilling techniques, oil companies are opening up more off-shore fields around the globe. Greater scientific understanding of the structure of sedimentary basins, and of the interface between oil droplets and the porous rock, has resulted in dramatic improvements in rate of oil recovery, as well as the quantity obtained before a well is no longer economically viable. Much of this technology is now mature, but there is no reason to believe that further research and development will not lead to improved techniques for the exploration, drilling and recovery of oil. Although society should not be complacent about the future availability of oil supplies, especially in the face of political uncertainties and growing demand, neither should it rely upon shortages of petroleum in the period to 2020 to drive the alternative energy scenario.

Another aspect of the developed world's almost total reliance on oil and gas is that individual countries or regions are vulnerable to interruptions in supply caused

---

* It is salutary to note that at US$ 20 the cost of a barrel (159 L) of crude oil is of the same order as the retail price of 1 L of whisky – and the latter did not take geological time to mature!

by factors quite distinct from resource availability. Such factors might include unusually severe weather, war or terrorism, mechanical breakdown or fire at the refinery or power station, and industrial action by operatives or delivery drivers. Disruption through industrial action has already been experienced in the UK – in 1974, when a general strike in the coal mines had a major impact on the electricity generating industry; and again in September 2000, when a strike of petroleum tanker drivers disrupted supplies of fuel to service stations. On such occasions, the public becomes acutely aware of its dependency on fossil fuels for all aspects of modern life. Similarly, there have been occasions in France and the USA when supplies have been disrupted locally and have led to long 'gas lines' at service stations. With these experiences in mind, security of supply is an important consideration; diversity of energy type and source enhances this security.

With regard to oil supplies, it is worth observing that much of the world's crude oil has to pass through two narrow straits on its way to market. In 2000, 15.5 million barrels per day passed through the Strait of Hormuz and 10.5 million barrels per day through the Straits of Malacca. The latter is only 0.5 km wide at its narrowest point and carries 20 000 tankers (oil or liquefied natural gas) annually. Any obstruction of these two seaways, whether as a result of accident, natural disaster or political action, would constitute a major disruption to energy supply.

Since the USA has been obliged to import oil, its consumption pattern has changed radically. Before the 1970s, 20% of the US electricity was generated from petroleum; now it is less than 1%. In 1973, 25% of homes were heated by oil; now it is less than 10%. Today, most of the output from the US oil refineries is used in the transportation and chemical sectors of the economy. This is a trend that is likely to occur worldwide, and by 2020 it is expected that most of the liquid fuel will be consumed in these two sectors.

One of the problems for any new energy technology in competing with fossil fuels is that the users of the latter are not generally required to pay for the cost of disposal of the products of combustion. Carbon dioxide is a greenhouse gas but, as noted above, there are at present few restrictions or cost penalties on releasing it to the atmosphere. Contrast this situation with that of nuclear electricity where the radioactive waste has to be stored in perpetuity by, and at the expense of, the generating company. Clearly, this is unfair competition. A carbon tax, as discussed in Section 11.1, would go some way towards redressing this imbalance, and it seems possible that such a tax will be imposed in the next 20 years, at least on the major fuel users. The level of tax will be determined by political considerations rather than by cost estimates of the externality. As a very small step in this direction, vehicles in the UK are now taxed on the basis of the amount of carbon dioxide they emit. This is an inducement to purchase smaller cars with more efficient engines.

Natural Gas and Liquefied Petroleum Gas

Natural gas is attractive since, among fossil fuels, it liberates the lowest amount of carbon dioxide per unit of heat produced. Several factors are contributing towards its greater use in industry, in commerce and in the home for space heating:

- discoveries of massive amounts of gas have been made in many parts of the world, both on land and off-shore;
- gas pipelines have been laid to bring supplies to centres of population;
- as a medium for heating, it is clean, convenient, and cheaper than liquid fuels;
- the user does not require a storage tank.

Wherever natural gas is available, it will be preferred to liquid fuels for heating. Over the next 20 years, we expect to see this trend accelerate as natural gas is brought to more people around the world. More long-distance pipelines will probably be laid, for instance from Russia and some of the former Soviet republics to Western Europe. A pipeline already exists under the Mediterranean Sea from North Africa to Europe, and in due course this might link up with one from the West African oil and gas fields. Where distances are too great, for instance to supply gas to Japan and Korea, there will be an expansion in the shipment of liquefied natural gas, which is fast becoming a major item of commerce.

We also anticipate a continuation in the move towards using more natural gas to generate electricity, both centrally in large power stations and locally in combined heat and power (CHP) schemes. Centrally, the driving factors are the high efficiency of combined-cycle gas turbines for electricity generation and the tightening restrictions on the liberation of sulfur dioxide from coal-fired power stations.

Finally, on a much more modest scale, we envisage a growth in the market for LPG, which is a clean fuel used traditionally for portable applications in leisure activities and more recently as a vehicle fuel. It appears likely that LPG will be employed more extensively as an automotive fuel, particularly in cities. In Europe, increasing numbers of service stations are installing LPG pumps and this wider availability of the fuel will encourage its greater use.

11.4 Electricity Generation

It seems inevitable that by 2020 coal will still be the basis of much electricity generation, worldwide. This is because some countries have large reserves of coal, but are short of other fossil fuels. Also, many large coal-fired power stations already exist and are expected to be still operating in 15 years time. The challenge faced by technologists and the business community is to reduce emissions of sulfur dioxide and nitrogen oxides from these existing plants within a competitive cost framework. High-sulfur coal will become of very little value unless a low-cost method is found to remove the sulfur before combustion, or for trapping the sulfur dioxide when released. At present, flue-gas desulfurization units (where fitted) impose a significant cost penalty on coal-fired power stations. Looking further ahead to a time when supplies of natural gas start to dwindle, the vast coal stocks will have to be used in an environmentally friendly fashion and therefore will require effective 'clean coal' technologies (see Section 2.3, Chapter 2).

It is equally important that methods for the sequestration of carbon dioxide be found. Underground storage or disposal in the sea, for example, would require new utilities to be built near suitable reservoirs or the coast, otherwise a transport system would have to be established for the conveyance of carbon dioxide, either by pipeline for the gas or by tanker for the liquefied form.

Gas-fired, combined-cycle, power stations are now preferred to those fueled by coal, on the grounds of both higher efficiency and lower emissions. The extent to which such plants can be introduced depends on many factors such as: the availability of gas supplies; fiscal considerations of the cost of importing gas (where necessary) rather than using indigenous coal; political issues where coalminers' jobs are at stake; the matter of diversifying the fuel base of electricity to ensure security of supply. These factors will vary from nation to nation.

The growing dependence of Western Europe on natural gas imported from countries of the former USSR and from North Africa is a potential cause of concern. These sources involve very long transmission pipelines that carry massive quantities of gas and are open to disruption as a result of accident or sabotage. In the event of restricted gas supplies, the electricity industry would be the first to be rationed and priority would be accorded to domestic and commercial users. This would be decided on safety grounds. When gas supply is interrupted and taps are left open, air can back-diffuse into the line and lead to the possibility of an explosion. With millions of households this is a real danger, whereas professional users, such as electricity utilities, have safe shut-down procedures. The more dependent a nation is on imported gas for its electricity, obviously the more serious would be the consequences following the cessation of power due to a gas shortage. Constructing gas-storage facilities might mitigate short-term disruptions in supply. One example would be to re-inject Russian gas into depleted North Sea gas fields as a large-scale store. With the benefit of hindsight, a better option might have been not to deplete gaseous resources so quickly in the first place! The swing to gas-fired electricity plant has undoubted advantages in the short term, both economic and environmental, but may be storing up problems for the longer term.

Distributed generation should make a growing contribution to overall electricity supply during the next 20 years. Nevertheless, we suggest that distributed generation and electricity derived from renewables (excluding hydroelectricity, which is already well established) will still constitute only a minor component of the worldwide production of electricity.

Another growth area in electricity generation will be that of CHP. Obviously, it makes sense to utilize, where practical, the waste heat associated with electricity generation. The rate of growth of this sector will be determined by cost considerations and by the availability of a suitable market for the heat. Whereas the quantity of heat that is potentially available from a 1 to 2 GW power station is huge, the distance over which it can be conveyed is limited. Thus, district heating is only a practical proposition in situations where the station is adjacent to a city. Moreover, installing district heating in a city is both capital intensive and highly disruptive. Although the overall efficiency of a CHP plant (electricity+heat) is high, the requirement to operate with exhaust gases at a higher temperature results in a reduced efficiency for electricity generation. For all these reasons, it is likely that CHP installations will be confined to relatively small distributed systems and not to large central power stations. Similarly, stationary fuel cells, if they come to pass, will almost certainly be relatively small units.

By far the largest uncertainty lies in the future of the nuclear industry. Whether or not more nuclear stations will be approved and built is essentially a political question, that is unlikely to be resolved until there is a consensus on the reprocessing of nuclear fuel and how best to store radioactive waste indefinitely. With so much public opposition to nuclear power, despite its record of reliable and safe operation in many countries, it may be difficult for governments to approve the construction of further nuclear stations. This situation will certainly vary from country to country and will be determined by a given nation's energy needs and resources, as well as the strength of public opinion. There is also the separate question of the large up-front capital cost and the long lead-times in constructing nuclear stations. Now that responsibility for electricity generation is moving from the public to the private sector in many countries, this may be a deterrent to further major investment. At present, then, the future of nuclear power is very uncertain, but by 2020 the issues should be resolved one way or the other and the industry will either be in terminal decline or in a growth phase where ageing plant is being replaced. The success (or otherwise) of the pebble-bed modular reactor (see Section 3.5, Chapter 3) may also be a pertinent factor. Countries that derive a high proportion of their electricity from nuclear sources (*e.g.* France) will have a particular problem when reactors reach the end of the their life and have to be replaced.

On the horizon, there is the prospect of generating electricity by nuclear fusion. Steady research progress is being made in major laboratories and the next significant step will most probably be a single world demonstration project. Not even the most optimistic of proponents, however, see this technology contributing to world electricity supplies by 2020.

11.5 Renewable Energy

It is anticipated that the harnessing of renewable energy will expand rapidly, in the light of widespread concern over global warming and the remedial actions that governments are taking. Despite such good intentions, however, there is every indication that, overall, renewables will still make only a modest contribution in 2020. It is vital, therefore, that society continues to develop the various technologies and gains experience in their operation as a step towards growth later in the 21st century.

Combustion technology – agricultural and forestry waste, municipal solid waste, energy crops – may make a useful contribution to energy supplies in many countries. Common factors that will limit its take-up will be the low cost of competing fossil fuels (unless carbon taxes are introduced), resource availability, the capital cost of constructing facilities and, in some instances, public opposition to the siting of these facilities. In the case of agricultural and forestry waste, the scope for expanding operations is strictly limited by the resource available and by the cost and the amount of energy consumed in collection. A further factor to be considered is that much of this biomass when left *in situ* decays and helps to enrich the soil. All new landfill sites will be expected to have gas collection and combustion facilities. It should be noted, however, that the recycling of consumer products* is an important and growing trend that will reduce the amount of waste going into landfill and consequently will decrease the amount of combustible gas that is generated. Schemes for growing energy crops will face competition from those wishing to use the land for agricultural purposes or buildings. Moreover, there is a public perception that turning over agricultural land to energy crops is not a good idea. Similar conversion of traditional forest or wild land is also likely to meet some opposition, due to the concern over loss of biodiversity and damage to the eco-system through soil degradation and depletion of essential minerals. In addition, there is a strong case for planting more forests to sequester carbon dioxide, rather than cutting them down in infancy to burn. In Europe, the situation is made more complex by the controversial Common Agricultural Policy that effectively determines agricultural land use, but which will probably be modified as further countries join the European Union.

Energy crops may also be grown to produce bio-fuels (methanol, ethanol, bio-diesel), as well as for direct combustion to generate electricity. The economic incentive for alcohols as a petrol extender depends on the competing cost of petroleum. If fuel cells assume a significant role in road transportation, then methanol is a candidate fuel. In the foreseeable future, however, it would appear that methanol will be manufactured from natural gas rather than from energy crops.

A final word of caution regarding energy derived from biomass. Renewable bio-energy is not necessarily the same as sustainable energy. Careful account must be made of the input of fossil fuel in the form of fertilizers, and also of petroleum for the machinery to harvest and convey the biomass to the processing plant. Furthermore, there may be environmental and social impacts in the growing and harvesting of crops that outweigh the renewable benefits.

Recalling that it takes hundreds of large wind turbines to replace one major power station, it is doubtful that wind energy, despite showing rapid growth in percentage terms, will become more than just a minor contributor to overall electricity generation. A particular problem with on-shore wind farms is that of gaining planning permission for construction. Experience has shown that nearby residents often form pressure groups to oppose the erection of large wind turbines, power lines and pylons in their 'backyard'. Off-shore wind farms are not so open to objection and significant numbers of turbines are being installed off the North Sea coast. Nevertheless, enthusiasts in the UK who advocate the building of many thousands of such turbines over the next decade have been taken to task in a report from the Royal Academy of Engineering. This emphasizes the severe engineering problems to be faced, as well as the impracticality and high costs, in harnessing wind energy to meet most of the UK target of 10% electricity from renewable energy by 2010 (note, existing hydroelectric supplies are not counted). By contrast, wind power is ideal for many isolated communities provided that there is a grid to provide back-up. Otherwise, it is necessary to install battery storage and this adds significantly to the cost. An alternative is to have a hybrid system that comprises a wind turbine and a petrol-driven generator.

---

* A companion volume in this series written by J. Aguado and D. Serrano is entitled *Feedstock Recycling of Plastic Wastes.*

Marine-based technologies (tidal flows, wave energy, ocean thermal energy) will have less of an impact. For example, it is unlikely that the capital investment required to build major tidal barrage schemes will be forthcoming in the next 20 years and, in any event, there are very few suitable sites. Tidal barrages are not necessarily confined to rivers with large tidal ranges, they can also be set up in shoreline lagoons that flood. Many potential sites are available around the coast of Britain. Tidal marine currents present an opportunity and there may be a few of these constructed to generate electricity, probably under government stimulus to help encourage renewables rather than as a result of direct commercial competition with fossil- or nuclear-generated electricity. Wave energy also requires substantial capital investment if it is to be implemented on a large scale. No doubt some small wave-energy machines will be built and demonstrated, but they are very unlikely to provide a significant source of global electricity in the near future.

Finally, there are the solar technologies. Solar heating of buildings and domestic hot water offers many opportunities, as discussed above. Solar photovoltaic (PV) generation of electricity also has made important strides in recent years as the efficiency of silicon PV cells has improved and their cost has fallen. Building-integrated PV panels appear to be the most economic way forward. By 2020, this technology may well be widespread in sunny climates, particularly in countries where fossil fuels have to be imported. Polymer PV materials and dye-sensitized photoelectrochemical cells are at an early stage of development, but success in these ventures could lead to a dramatic fall in the cost of such electricity. This is an exciting area of research that should be pursued vigorously. Moreover, solar panels based on these new materials could be brought to market rapidly, building on the skills and experience of the existing PV industry.

11.6 Energy Storage

The various physical and chemical techniques for energy storage will all continue to be investigated and developed. Of the physical techniques, pumped hydro and compressed air energy storage are the most promising for peak-shaving and load-leveling within the electricity supply network, provided the terrain and other conditions are suitable. For smaller-scale storage, further research will be conducted on flywheels and on electromagnetic and electrostatic devices. Of these, electromagnetic storage is too expensive for general use. Flywheels may prove suitable for some specialized uses, but we doubt that they will find substantial widespread application. Electrostatic devices (electrochemical capacitors) complement batteries in being high-power, low-energy devices and show considerable promise for use in hybrid systems.

Hydrogen energy, the so-called 'ultimate' form of energy, is the Holy Grail for environmentalists – clean, abundant, non-polluting. This dream has been around for over 30 years. The principle of producing hydrogen in an electrolyzer (using a renewable source of electricity), storing it as a chemical hydride, and regenerating the electricity in a fuel cell when needed, sounds attractive at first acquaintance. The practice and the economics are quite a different matter. In the early days of the dream, cheap abundant nuclear power was to have been the most practical means of generating the hydrogen. As this no longer seems likely, it will be necessary to fall back on solar- or wind-generated electricity. The requirement for three separate devices (electrolyzer, hydride-store, fuel cell) merely to store and utilize small quantities of electricity is not at all efficient from an energy viewpoint. Such an approach would therefore be a gross misuse of renewable energy. Moreover, the activity would be capital intensive and there would be the added cost of the power-conditioning equipment.

We do not see hydrogen being produced from renewables on a significant scale in the next 20 years. Rather, hydrogen for fuel cells will be produced, as now, from fossil fuels. Meanwhile, electrolyzers will continue to be used mostly for the production and processing of chemicals and metals, and for life-support oxygen in submarines and manned spacecraft. Recently, the largest hydrogen production plant in the UK, based on natural gas and producing 32 000 t of hydrogen per year, has come on-stream at a chemical manufacturing site in North Teeside. From the point of view of greenhouse gas emissions, however, the use of fossil fuels to generate hydrogen for chemicals manufacture or for use in fuel cells is useful only if the carbon dioxide that is inevitably produced can be sequestered. Practical technology for this does not yet exist and its development is an area for immediate attention.

The realization of a 'Hydrogen Economy' is linked irrevocably with that of the fuel cell. There is no doubt that fuel cells work best on hydrogen and this requires any other fuel to be converted to hydrogen, at least for use in low-temperature cells. Unless the fuel reformer is tied directly to the fuel cell and produces hydrogen at exactly the rate that the fuel cell demands, as is proposed in some of the electric vehicle concepts, it is necessary to have a buffer store for hydrogen. This may be a metal hydride or a chemical carrier. The alternative concept is to have a much larger, industrial-scale reformer, divorced from the fuel cell, and to establish a distribution system for the hydrogen. Some proponents of FCVs favour this approach and are considering setting up a chain of service stations where hydrogen is supplied on tap. There is then the problem of storing the hydrogen on-board the vehicle. The two options are high-pressure storage in gas cylinders – bulky and heavy, through rapidly improving – or in a hydride storage bed. The latter is feasible in theory, but there are some complex heat and mass-transfer problems to solve. As mentioned earlier, we are pessimistic about fuel cells for cars, less so for buses and trucks. It should also be noted that automobiles (particularly diesels) are becoming increasingly efficient. Clearly, the fuel cell is aiming at a moving target.

Stationary fuel cells are quite another matter and it is possible that within 20 years these will be installed widely, with hydrogen piped in from a centrally sited reformer. From an environmental standpoint, however, such an arrangement would not be ideal. All the carbon atoms in the fuel used by a reformer finish up as carbon dioxide so that there is no saving in greenhouse gas emissions, except in so far as the fuel cell is more efficient than an engine.

Finally, we turn to the role of batteries for energy storage. Here, some real progress is being made. In the past 10–20 years, there have been major improvements in the lead–acid battery, the nickel–metal-hydride battery has been invented and commercialized, and the lithium-ion battery has made its début. In small sizes, the last-mentioned battery is sweeping the electronics market, and much work is in progress worldwide to scale it up to larger units, to ensure its safety, and to reduce its cost. Provided these goals are achieved, the lithium-ion battery and its off-shoot, the lithium–polymer battery have a promising future. Some larger storage batteries also appears encouraging. For example, sodium–nickel-chloride is targeted at the market for battery electric-vehicles while sodium–sulfur batteries (still being developed in Japan) would be appropriate for the storage of distributed electricity. Technically, the battery scene is looking promising for small-scale electricity storage, although there is still the issue of cost to be faced.

11.7 Concluding Remarks

It is perhaps trite to observe that the laws of science that determine technical feasibility are immutable, whereas engineering design and manufacturing costs vary from place to place and from time to time. Thus, many goods that formerly were made in developed countries are now produced in lower cost, developing nations. Technical feasibility is, however, the *sine qua non* of any proposed new technology. Although outline costings are needed to determine whether a project promises to be economic, detailed manufacturing costs (with extrapolation to high-volume production) can only be ascertained after technical feasibility has been established and a prototype built. A technology that is too expensive for a particular application in a given nation at one point in time may be acceptable for a different application, or in a different place, or at a different time. This is one of the key problems facing anyone who attempts to predict the future prospects for technology on a global basis, and the task is complicated further by the need to factor in sociological and other considerations. Inevitably, therefore, some of the above-generalized predictions will not apply to specific applications in certain countries. A good example is Iceland or Brazil, where an abundance of cheap hydroelectric power might favor the production of hydrogen by electrolysis. Other forecasts will point in the right direction, but the timing will be wrong, as mentioned at the start of this chapter. Even so, we hope to have provided the reader with food for thought and stimulated some debate and discussion concerning energy futures.

Looking further ahead to 2050, the crystal ball becomes even more cloudy. Nevertheless, the lead-times for new energy technologies are such that it is necessary to take a long-term strategic view. The UK government has accepted this and stated that, by 2050, the country will need to have a clean and secure supply of energy that does not rely too heavily on fossil fuels. In 2001, it commissioned a comprehensive energy review to which 400 different organizations and individuals submitted evidence. Each of these had different, and often conflicting, viewpoints as to the future. Some favored an expansion of nuclear power as the only realistic alternative to fossil fuels and emphasized the importance of following this course soon, before all the expertise and trained staff are lost. Others took a diametrically opposed view and advocated the phasing out nuclear power permanently while vigorously developing all forms of renewable energy. Almost all parties accepted the need for a review of energy policy and agreed that tackling concerns over the security of the nation's energy supply and the environmental impact of greenhouse gases would necessitate changes to the energy-supply infrastructure. The area of disagreement was over the exact form that these changes should take. As might be expected, the general conclusion of the review was that options should be kept open and that the UK should spend more on energy research and development programs. As a working strategy, a good target model for electricity generation in the UK might be 30% coal-based, 30% gas-based, 30% nuclear, and 10% renewables. This would ensure that the nation had a diverse and secure base for its electricity supply.

In other countries, the situation may be quite different. France, for example, has little fossil fuel and is firmly committed to its nuclear program. Germany has plenty of coal, but little gas of its own. The Netherlands and Denmark have good wind resources and are therefore enthusiastic about renewables. Outside of Europe, the USA has both coal and gas in copious amounts, so that there is little incentive to reinvigorate its nuclear programme. Japan has almost no indigenous fuel and is orientated towards nuclear technology and the importing of liquefied natural gas. Australia, like USA, is rich in both coal and natural gas. Each country has to look to its own position to optimize its electricity generation. This makes for difficulties in global forecasting, especially in meeting emission targets for greenhouse gases. In those countries where electricity generation is state-controlled, it is at least possible for the government to exert some influence over the fuels used. On the other hand, in countries such as the UK and the USA where a free and competitive market exists in electricity generation, the State has comparatively little control except through legislation or subsidies. Liberalized electricity markets are hardly compatible with government energy planning.

One general conclusion can be drawn from this discussion of the developing energy scene to 2020 – there will be no overall shortage of fossil fuels. The world has ample reserves of oil and gas for the present and these are widely distributed, although still with a preponderance in the Middle East. Other fossil fuels (coals, tar sands, asphalts, oil shales) are even more widely distributed, but their extraction and utilization impose technical and environmental problems. Moreover, barring political upsets or the imposition of a high carbon tax, fuels should remain comparatively cheap with modest increases in price above inflation. This will define a cost base against which renewable energy has to compete for business in most situations. The comparatively low cost of fossil fuels does nothing to address the greenhouse gas issue and there appears to be no easy answer to this problem. High carbon taxes, such as might make an impact, would be disruptive to the world economy and would be politically unacceptable. Unless and until the causative relationship between greenhouse gas emissions and global warming is established unequivocally and accepted by all, it is unlikely that the world will change dramatically its dependence on fossil fuels. Nevertheless, the time-scale for developing and implementing renewable energy technologies – decades – is such that efforts directed towards this goal should be continued and, moreover, enhanced.

In summary, some of the major energy problems facing the world today are:

- how to reduce greenhouse gas emissions to acceptable levels while there is cheap fossil fuel still available to compete with renewables;
- how to persuade reluctant politicians and the general public of the need for a carbon tax;

- how to develop a practical and economic route for the sequestration of carbon dioxide without its release to the atmosphere;
- how to increase public awareness of the seriousness of the future energy situation and the need to start investing and planning now for a complete break from the present near-total dependence on fossil fuels; this is primarily a socio-political matter but does involve technological developments and choices;
- how to raise the huge amounts of capital investment that will be required to bring new sources of natural gas to market, to burn coal more cleanly, to sequester carbon dioxide, to build new nuclear facilities (if that route is chosen) and, in the longer term, to establish an entirely new, sustainable industry based on renewable sources of energy.

Whereas it is encouraging that schools, universities, interest groups and the media are enabling a new generation of young people, worldwide, to gain a greater understanding of environmental and energy issues, the practical difficulties of moving from fossil fuels to renewables remain enormous. The world's scientists and engineers are striving to develop the required new energy technologies, but in the final analysis politicians, financiers, bankers, industrialists and the general public must act together to establish an economic climate in which sustainable forms of *Clean Energy* can flourish in competition with traditional fuels.

Appendix

Conversion Factors

1. Energy conversion factors

To: From:	TJ multiply by	Gcal	Mtoe	GWh
TJ	1	238.8	2.388×10^{-5}	0.2778
Gcal	4.1868×10^{-3}	1	10^{-7}	1.163×10^{-3}
Mtoe	4.1868×10^{4}	10^{7}	1	11 630
GWh	3.6	860	8.6×10^{-5}	1

Note: 1 kWh = 3.6 MJ.

2. Volume conversion factors

To: From:	Gallon US multiply by	Gallon UK	Barrels	Litres	m^3
US gallon	1	0.8327	0.02381	3.785	0.0038
UK gallon	1.201	1	0.02859	4.546	0.0045
Barrel	42	34.97	1	159.0	0.159
Litre	0.2642	0.220	0.0063	1	0.001
Cubic metre	264.2	220.0	6.289	1000	1

Note: There are approximately 7.4 barrels of crude oil to a tonne; the conversion factor depends upon the density of the crude oi!.

3. Pressure conversion factors

10 standard atmospheres pressure = 10.13 bar = 146.96 psi = 1.013 MPa

4. Hydrogen data

Higher heating value (HHV) = 142 MJ kg^{-1} = 39.4 kWh kg^{-1}
Lower heating value (LHV) = 120 MJ kg^{-1} = 33.3 kWh kg^{-1}

Subject Index